Colin Burgess and Rex Hall

The First Soviet Cosmonaut Team

Their Lives, Legacy, and Historical Impact

 Springer

Published in association with
Praxis Publishing
Chichester, UK

Mr Colin Burgess
Spaceflight Historian
Bonnet Bay
New South Wales
Australia

Mr Rex Hall M.B.E.
Education Consultant
Council Member of the BIS
London
UK

SPRINGER–PRAXIS BOOKS IN SPACE EXPLORATION
SUBJECT *ADVISORY EDITOR*: John Mason, B.Sc., M.Sc., Ph.D.

ISBN 978-0-387-84823-5 Springer Berlin Heidelberg New York

Springer is part of Springer-Science + Business Media (springer.com)

Library of Congress Control Number: 2008935694

Cover design: Jim Wilkie
Project management: Originator Publishing Services, Gt Yarmouth, Norfolk, UK

Printed on acid-free paper

Contents

APPENDICES

Dedication

While this book relates the stories and experiences of all 20 Soviet Air Force pilots selected in the first cosmonaut detachment in 1960, sadly not all of them would realise their dreams. With profound respect, this book is therefore dedicated to the eight spaceflight candidates from that first group who are forever a part of spaceflight history, but who, for a number of reasons, were once discarded as unsuitable and condemned to anonymity. Unlike their more illustrious peers, they were never able to ride the rockets and share in the glory of viewing our blue planet from orbit.

Ivan Nikolayevich Anikeyev
Valentin Vasilyevich Bondarenko
Valentin Ignatyevich Filatyev
Anatoly Yakovlevich Kartashov
Grigori Grigoryevich Nelyubov
Mars Zakirovich Rafikov
Valentin Stepanovich Varlamov
Dmitri Alexeyevich Zaikin

Foreword/Предисбовие

В 1959 г. я был в группе летчиков, проходивших медицинский отбор для включения в секретную программу, которая в будущем создаст тип летчика, именуемого космонавтом. В марте 1960 г. вместе с другими 19-тью летчиками я был отобран в первый отряд космонавтов, которым предстояло быть командирами и управлять первыми космическими кораблями, выведенными на космическую орбиту. Подготовка к полету была трудной и сложной, так как подобной программы вообще не существовало. Не было опыта подготовки людей для такой цели. Руководил нами полковник Евгений Карпов, который, являясь первым начальником Центра подготовки, отвечал за создание первого отряда космонавтов, обучение и подготовку к полетам на новом типе летательного аппарата в жестких условиях космического пространства.

Имена некоторых космонавтов этого отряда впоследствии стали легендами, известными личностями во всем мире: Гагарин, Титов, Николаев, Попович, Быковский, Комаров, Беляев, Волынов, Хрунов, Шонин, Горбатко и Леонов. Мы совершили 21 космический полет, многие из нас во многом были первыми с целью превратить научные фантазии в действительность.

Эта книга лишь небольшая часть повествования о 8 товарищах, которые так и не слетали в космос. Они были нашими коллегами и входят в историю космонавтики. Поэтому их имена и дело, которым они занимались, описаны в этой книге. Они заслуживают чести, чтобы о них знали и помнили. Это - Карташов, Филатьев, Варламов, Нелюбов, Аникеев, Рафиков, Бондаренко, Заикин. Особая признательность за предоставленную возможность вспомнить своих товарищей.

Генерал-майор авиации
Дважды Герой Советского Союза
Летчик-космонавт СССР
Космонавт с 1960 г. по 1975 г.
Космонавт набора 1960 г.

Алексей А. Леонов

Major-General Alexei Arkhipovich Leonov. (Photo courtesy David Meerman Scott)

In 1959 I joined a group of pilots who were being medically tested for what was a secret programme which would create a type of pilot called a cosmonaut. In March 1960, together with 19 other pilots, I was selected to join the first cosmonaut selection that would fly, command and control the first spacecraft launched into space.

Training was hard as no programme existed which would prepare people for this task. We were under the command of Colonel E.Y. Karpov who, as first commander of the training centre, was charged with creating the first cosmonaut team, teaching them the skills to fly a new type of vehicle in the harsh environment in space.

The names of some of the men who were part of this group have become legends and are known to everyone: Gagarin, Titov, Nikolayev, Popovich, Bykovsky, Komarov, Belyayev, Volynov, Khrunov, Shonin, Gorbatko and Leonov. In total we made 21 spaceflights, many creating space firsts and turning science fiction into fact.

This is only part of the story, as eight colleagues did not make that launch. They were part of us and the history of cosmonautics. Their names and the part they played are told in this book. They deserve to be honoured, and the names of Kartashov, Filatyev, Varlamov, Nelyubov, Anikiyev, Rafikov, Bondarenko and lastly Zaikin should be known. This book covers their contribution to our flights and their roles in the space programme of the 1960s.

It is an honour being asked to remember my colleagues.

Major-General of Aviation Alexei A. Leonov
Twice Hero of the Soviet Union
Pilot-Cosmonaut of the Soviet Union
Cosmonaut 1960 to 1975.
A member of the 1960 selection

Other works

Other space exploration books by Colin Burgess in this series:

With David J. Shayler:
NASA's Scientist-Astronauts (2006)
ISBN 0-387-21897-1

With Chris Dubbs:
Animals in Space: From Research Rockets to the Space Shuttle (2007)
ISBN 0-387-36053-0

Other space exploration books by Rex Hall in this series:

With David J. Shayler:
The Rocket Men: Vostok & Voskhod, The First Soviet Manned Spaceflights (2001)
ISBN 1-85233-391-X
Soyuz: A Universal Spacecraft (2003)
ISBN 1-85233-657-9

With David J. Shayler and Bert Vis:
Russia's Cosmonauts: Inside the Yuri Gagarin Training Center (2005)
ISBN 0-38721-894-7

Authors' preface

Located on the Dnieper River between Kiev and Odessa, the south-central Ukrainian city of Zaporozhe (or Zaporizhia) was once home to the Cossacks, a race of colourful and tenacious horsemen who ruled the country during the 17th and 18th centuries. A sprawling industrial city, Zaporozhe is also home to a massive hydro-electric dam which dominates the riverscape from most vantage points in the city. Once heavily polluted by the insidious residues of several motor car and aircraft engine manufacturers, Zaporozhe—like so many other former Soviet cities—has recently undergone an extensive transformation, and even though typically ugly, ubiquitous concrete tower blocks of apartments still dominate the skyline, they are now partly hidden behind tree-lined avenues. In a quiet corner of Zaporozhe is the small, mostly neglected Kapustyany graveyard where, until recently, a weather-beaten gravestone closely surrounded by ragged trees had risen above the bare dirt and weeds, marking what was said to be the final resting place of a man who came tantalizingly close to his dream of one day becoming the first person to fly into space.

Quite remarkably, the fact that Grigori Nelyubov ever bore the title of cosmonaut was nowhere to be found on the gravestone, which has now been replaced. In fact, it would actually be, with the exception of close family members, virtually nobody who knew anything at all about the significance of the life of the uniformed person whose photograph was embedded in the obelisk. Nelyubov died in virtual anonymity, and this was reflected in according him such a meagre resting place. By way of contrast, the remains of many of his one-time colleagues and friends who have passed away now occupy honoured burial places within the Kremlin Wall, or in Moscow's renowned Novodevichy cemetery, where they are clearly identified and lionized as pioneering cosmonauts.

Why the sad anomaly, one might ask? Nelyubov, a jet pilot of substantial skill, determination and courage, enjoyed the same Soviet Air Force background as his colleagues in the first cosmonaut group. Then, selected as one of an elite group of 6 from that initial cadre of 20 candidates for specialized spaceflight training, he was

Grigori Nelyubov's former grave in Zaporozhe.

firmly in line to fly an early Vostok mission, one which would have not only have immediately endeared him to the Soviet public, but immortalised him in the history books. Given the right circumstances, he could even have been selected as the first person ever to fly into space. Had one of America's famed Mercury astronauts died before flying, there is very little doubt a national outpouring of grief and commemoration would have ensued across the United States, yet Grigori Nelyubov died a lonely death and a broken, forgotten man in a once nondescript grave.

Curiously, however, the gravesite in Zaporozhe no longer contains Nelyubov's remains. Instead, a few of his belongings are buried there, as well as some earth from his final resting place located nearly 4,500 miles away. After he died, hit by a train northwest of Vladivostok in February 1966, Nelyubov's body was initially buried in that simple grave in Zaporozhe. Some time later his remains were disinterred and reburied in yet another nondescript grave in Kremovo, a village in the Mikhailovski District, near Vladivostok. Then, after being formally identified as a cosmonaut in 1986, his grave at Kremovo underwent a more befitting transformation. A high, black polished granite stone in which his helmeted image has been engraved now rises above the fenced-off grave, and the inscription below this reads:

Cosmonaut VVS [Soviet Air Force]
2nd Backup Y. A. Gagarin
Pilot First Class

Captain Nelyubov
Grigori Grigoryevich
8.4.1934–18.2.1966

The fact of his second grave on the other side of the country remains something of a mystery, as does the birth date on the Kremovo memorial, which differs from his official birth date of 31 March 1934. On 24 October 2007 the obelisk above the cleaned-up gravesite in the Kapustyany cemetery was also replaced with a more befitting stone. Nelyubov's brother Volodimir was there to place flowers by the new gravestone, as was one of the cosmonaut's school friends, Aris Pecheritsa. On this stone the birth date is given as 31 March 1934. Such are the many imponderables associated with determining the facts behind this first cosmonaut group.

On 12 April 1961, a 27-year-old Russian military pilot whose identity to that time was only known to a few of his friends, family and Air Force colleagues became the first person to fly into space. On that glorious spring day, Senior Lieutenant Yuri Gagarin (promoted to the rank of major during his mission) also became the first member of the first cosmonaut group to achieve spaceflight. Others from the detachment, known as Chief Designer Sergei Korolev's "Little Eagles", would one day follow in his path: Gherman Titov, Andrian Nikolayev, Pavel Popovich, Valery Bykovsky, Vladimir Komarov, Pavel Belyayev, Alexei Leonov, Boris Volynov, Yevgeny Khrunov and Georgi Shonin. Each would achieve lasting fame as one of their nation's pioneering cosmonauts.

It was actually some eight years after Gagarin's historic accomplishment that the final member of that first cosmonaut group would fly into space, when 34-year-old research engineer Viktor Gorbatko was launched into orbit as a third crewmember aboard the Soyuz 7 spacecraft, on 12 October 1969. Much had happened during those interceding years: four more groups of cosmonauts had been selected; Yuri Gagarin was tragically killed in a mysterious aircraft accident; Vladimir Komarov had perished on his second spaceflight when the Soyuz 1 spacecraft he was commanding slammed into the ground at high speed after a troubled spaceflight; and two Americans named Neil Armstrong and Edwin "Buzz" Aldrin had walked on the surface of the Moon. Gorbatko was the 12th member of the first cosmonaut team to fly into space, but that group had initially numbered 20. Who were the missing cosmonauts, and why did they remain unknown candidates on the ground as their colleagues achieved glory and lasting fame?

With many clues to tantalize them, Western researchers would eventually realize that other cosmonaut candidates had been selected in that first group. But penetrating the shroud of secrecy that the Russian space chiefs maintained around their programme was an incredibly difficult proposition, filled with false leads and misinformation. Names were hinted at in some memoirs and publications on Soviet space activities, while the faces of unknown participants were found to have been crudely airbrushed out of photographs showing members of the first cosmonaut team. One of these mystery men would prove to be Senior Lieutenant Grigori Nelyubov, sometimes present but unidentified in photographs accompanying Gagarin and his back-up Gherman Titov on the bus ride out to the Baikonur launch pad in 1961. Officially

At top, Nelyubov's newer grave at
Kremovo. Below is the refurbished
gravesite and replacement marker at the
Kapustyany cemetery in Zaporozhe. (Both
photos courtesy Ivan Ivanov from *http://
astronaut.ru*).

Gagarin (foreground) travelling on the transfer bus out to the launch pad, 12 April 1961. Seated behind him is his backup pilot, Gherman Titov. Behind Gagarin is his second backup Grigori Nelyubov, while Andrian Nikolayev is behind Titov.

recognized as a cosmonaut many years later, Nelyubov would finally be named as a member of the elite group selected from within that first cosmonaut cadre for advanced training, a group that space historian and investigator James Oberg came to call the Sochi Six.

Nelyubov is actually mentioned as a potential cosmonaut candidate in Evgeny Riabkchikov's 1971 book, *Russians in Space*, together with Ivan Anikeyev, another pilot who would join the first group of cosmonauts. One of the photographs in Riabchikov's book shows chief Soviet rocket designer Sergei Korolev relaxing with what is described as "a group of cosmonauts" in May 1961. Apart from several known cosmonauts, the photograph also features unflown cosmonauts who would later be identified as Nelyubov, Anikeyev and Rafikov.

Speculation that there had been a number of unknown cosmonauts in the first group gathered considerable strength with the publication of a book by Georgi Shonin, a flown member of that group. His autobiographical book, the title of which translates to *The Very First Ones*, was published in 1976, and in it he revealed that the first cosmonaut group numbered 20 men. He even gave the first names of his unflown colleagues: Anatoli, Ivan, Dmitri, Grigori, Mars and three others, all named Valentin. In order to distinguish the latter three he referred to them as "Number One", "Junior" and "Gramps". The pieces of the puzzle were slowly starting to come together.

In 1984, an émigré orthopaedic surgeon named Vladimir Golyakhovsky released his memoirs in a book published in New York called *Russian Doctor*. In it he told of a harrowing night in a Moscow clinic when a young man identified to him as a cosmonaut trainee was brought to the hospital, literally burned from head to foot and in a critical condition. He would die soon after. Golyakhovsky's account of that evening's dramas added considerable credence to lingering rumours of a young cosmonaut dying after being extensively burned in a training exercise. This cosmonaut would finally be identified some years later as Valentin Bondarenko, the "Junior" mentioned in Shonin's book.

Two years on, in 1986, a series of officially sanctioned articles appeared in the Soviet newspaper *Izvestia* in which the full names and some biographical information was finally revealed on all eight of the "missing" cosmonauts. By then, a full quarter century had passed since Yuri Gagarin had flown in space.

By way of contrast, NASA's seven Mercury astronauts had been openly introduced to the American public amid much fanfare at a Washington, D.C. press conference on 9 April 1959. It was all part of the open information policy the civilian space agency had adopted and maintained in consultation with the White House, in which the nation could participate by knowing what space feats were being planned, when they were scheduled to launch, and who would fly them. It was an agreeable and workable policy, but one that would continually frustrate NASA officials as it gave considerable propaganda advantage and leverage to their Cold War adversary, the Soviet Union. In the early days of the adversarial Space Race to the Moon, as soon as NASA announced an ambitious manned space mission, Soviet leaders from Premier Nikita Khrushchev down would emphatically demand that this feat not only be carried out ahead of the Americans, but surpassed. And while America and the entire world were familiar with the names of the American astronauts well in advance of their flights, no one apart from a privileged few would know the identities of cosmonauts assigned to these increasingly hazardous flights.

In qualifications, expertise and experience, the Mercury astronauts (and later astronaut groups) far outshone their Russian counterparts. America's astronauts were highly qualified military test pilots with solid engineering backgrounds and hundreds of hours' experience in supersonic aircraft. As they were expected to participate in the design and actual operation of their craft in space, only the best possible candidates would be chosen, and the selection process was a long and arduous process. Conversely, the first Soviet manned spacecraft, Vostok, would be automatically controlled from the ground, and the first cosmonauts were expected, in the main, to be little more than sightseeing passengers taking part in some relatively innocuous onboard experiments and monitoring systems. Therefore, the search for suitable candidates was a relatively easy one, carried out and finalized within the ranks of jet pilots in the Soviet Air Force and Navy. No real engineering skill was required, and while test pilot experience was useful, it was by no means mandatory.

Overall, the Soviet candidates were much younger than their Mercury counterparts and also shorter in stature, in order to comfortably fit into the confines of the Vostok craft—also reflecting Soviet aircraft design. They would need to have some experience in parachute jumping as (unknown to Western observers at the time) they would not be landing in their spacecraft. The Soviet spacecraft designers were still working on a satisfactory rocket landing system, and harboured concerns that the impact with the ground on landing would be far too hard on the cosmonauts. As a consequence they would be automatically ejected from their craft as they neared the ground. In Gagarin's case this would later provoke controversy in regard to Soviet claims that he had orbited the Earth. Under Fédération Aéronautique Internationale

(FAI) regulations, a record—and orbit—could only be claimed if a participant actually landed in the same craft in which they had been launched. Pragmatists therefore continue to argue that while Gagarin may have been the first person to fly into space, he cannot be considered the first person to have actually completed an orbit of the Earth. That honour, they still say, should be bestowed under FAI rules on U.S. Astronaut John Glenn.

Less than a year after NASA officially named the space agency's seven Mercury astronauts the first Soviet cosmonauts were reporting for spaceflight training, initially in temporary facilities located in Moscow. Later, when completed to occupancy stage, they would continue their training in a special complex outside Moscow that came to be known as Zvezdny Gorodok, or Star City. And less than a month before Alan Shepard became the first American to travel into space, a member of the cosmonaut team—one of Korolev's "Little Eagles"—would enter the history books as the first person to be launched into space.

Apart from a burning desire to achieve spaceflight and weightlessness, Russia's cosmonauts shared one thing in abundance with the Mercury astronauts: an extreme competitiveness to not only be the best, but to fly first. Many, however, would miss out. While America's astronauts were for the most part highly disciplined individuals, a reckless arrogance often fuelled by alcohol and high spirits soon crept into the much larger company of the first cosmonaut team, resulting in four of their number being summarily dismissed without ever flying into space. A physical problem during centrifuge training cost another man his place in the team, yet another suffered a serious, disqualifying neck injury in a simple accident, and a third candidate would be diagnosed with a stomach ulcer at the wrong time in his career. Finally, the youngest of all 20 candidates, 24-year-old Valentin Bondarenko, would suffer a horrible death just a few weeks before Gagarin's flight when his body was engulfed in flames during a training exercise in the oxygen-enriched environment of a pressure chamber.

Told in narrative form for the first time, the story of the original Soviet cosmonaut team is a truly compelling study of human endeavour. It is not only one of extra-ordinary courage and high ambition, but of over-confidence and sometimes foolish behaviour, extreme disappointment and the abject bitterness of failure. The lives, inspirations and aspirations of these 20 men before and after their selection can now be fully related for the first time in factual detail. But while some of their number would come to know greatness and global adoration, others would sadly and anon-ymously fall by the wayside through accidents, illness, arrogance, disciplinary action and even death.

Many of the inherent mysteries of the Soviet space programme still abound. Frustratingly, investigations into these are often hindered by the determination of many early participants to adhere to time-worn fallacies surrounding their training and flights. However, one of the greatest fallacies still causing conjecture is that of the so-called "phantom cosmonauts". This 20th century legend thrives on persistent but unfounded rumours that several men and women died in Soviet space disasters before and after Gagarin's orbital flight. These rumours, and the names of those allegedly involved, will be examined in this book and due homage paid to many of the very real technicians, designers and test engineers who actually participated in successfully creating the hardware, equipment and procedures that later exemplified the Soviet manned spaceflight programme.

This is a book that had to be written and needs to be read in order to gain a more complete understanding of the incredible technological era known as the Space Race and those humans who were the first to train to leave our planet. One could not imagine visiting a library without finding a profusion of books on the Mercury programme and its astronauts, yet, apart from some excellent Who's Who–style books, principally those of Michael Cassutt, Doug Hawthorne and Gordon Hooper, no publication has ever specifically told of the first group of Soviet cosmonauts. With this book, the authors have hopefully redressed that anomaly. It has been researched and written with the utmost respect for those who flew atop the Soviet rockets, and for those once-forgotten few who were left in their fiery wake, never to realise their dreams or potential.

During the period in which this book was being researched and written the world celebrated the 50th anniversary of the Soviet satellite *Sputnik*, a feat which truly ignited the incredible era known as the Space Race. Less than four years later a man would follow that basic satellite into orbit. At the time of writing this book, only 6 of the original 20 Soviet cosmonauts who might have been the world's first spacemen are still with us as living ambassadors of one of the most exclusive fraternities of explorers ever assembled: Valery Bykovsky, Viktor Gorbatko, Alexei Leonov, Pavel Popovich, Boris Volynov and the unflown Dmitri Zaikin. All of them have now been inter-viewed, and their words are an intrinsic part of this book. There are many revelations contained in what they told the authors, giving even more personal and historical impact to the remarkable story of the first Soviet space team.

This collaborative effort had its genesis in the lifelong interest of two people on either side of the world in the wondrous history of spaceflight. Both became entranced by space exploration at a time when human activity in this new arena was in its infancy, and for both it has remained an enduring interest. Some years back a mutual Dutch friend by the name of Bert Vis provided the catalyst for introducing the two authors of this book to each other. Bert, a fireman from The Hague, is a long-time and devoted researcher into Soviet/CIS space activities, and on many occasions he would travel on self-funded trips to Moscow and other world capitals with the specific aim of conducting in-depth interviews with dozens of cosmonauts and other leading figures involved in the origins, and continuance, of this remarkable era in human history. Many of those personalities are no longer with us, which lends these interviews an even greater historical significance. On occasion, Vis would be accompanied by other space historians such as Gordon Hooper, Chris van den Berg, Neil Da Costa and Rex Hall, together with Rex's wonderfully supportive life partner Lynn.

Many of the details within this book represent the extraordinarily incisive and sometimes difficult work carried out by this small band of self-funded enthusiasts, which is gratefully recognized and readily acknowledged by the authors.

About the authors

COLIN BURGESS I owe an incalculable debt for much of my fascination with human endeavours in space to my late and beloved grandmother, Beatrice Morgan. In my early teens I used to treasure any time spent with her as precious days filled with wonder and excitement. We would play old records and discuss episodes of human triumph and tragedy, and together look through a modest collection of newspapers she had collected over the years pertaining to these events. To me, they were a goldmine of information.

Sometime during an Australian summer school break (I believe in January 1962) my grandmother and I fell into a discussion on the much-delayed Mercury flight of Marine Lt. Colonel John Glenn. She said I should follow the progress of his flight, suggesting that this would be a truly pivotal event in history, and that I might begin my own collection of historic newspapers with his safe return from space. From that time on I found myself propelled into the interest and fascination of a lifetime. I not only began clipping out newspaper and magazine articles on Glenn and his mission, but started tracing back and reading up on earlier manned spaceflights; those of Yuri Gagarin, Alan Shepard, Gherman Titov and Virgil "Gus" Grissom.

The following year I entered the workforce with a job near the notorious streets of Kings Cross in Sydney, and one of my guilty pleasures each pay day was to visit a small Red Star bookshop near where I worked. This tiny, ill-lit shop sold all manner of magazines and books about life in the Soviet Union. After a while the elderly proprietor came to know me well, and on each visit he would happily point out magazines containing stories on the cosmonauts which I would purchase and take home to add to my growing collection. I know my mother feared for my mortal soul, and often told me that the FBI would have me on a list of suspected Soviet sympathizers. In a cultural sense that was true, because in reading these magazines I would come to know a great deal about the lives of the people of the Soviet Union, and even though this was at the height of such worrying episodes as the Cuban missile crisis, I never really feared our Cold War adversaries. I knew a lot about the way they lived, but mostly I thrilled to the exploits of their cosmonauts, and regarded them—along with America's astronauts—as heroes.

The interesting thing about growing up in the first few years of human spaceflight was the absolute competitiveness of the Space Race. NASA would openly announce its plans for each successive mission, but there were only ever very broad hints leading up to each Soviet space spectacular. While I empathized with the Americas, there was always a certain thrill in walking by a news stand and seeing a banner headline about a Russian walking in space, three cosmonauts aboard a single spacecraft, or a manned link-up in space.

Yet there were always persistent, dark rumours about a number of cosmonauts who had either gone to glory in training accidents, or who had perished in spaceflight catastrophes before Gagarin's successful mission. Indifferent Soviet officials never really bothered to deny these rumours, or if they did it was to simply dismiss them as complete fantasy, so one never really had any idea whether there was an unexpected truth lurking behind these stories. However, I kept all of these articles, plus a number of magazine photos of men purported to be the missing cosmonauts, hoping that one day the truth about the Soviet space team would finally emerge.

James Oberg's revealing book *Red Star in Orbit* was released in 1981, and it became a source of fascination for spaceflight enthusiasts the world over. Oberg not only discussed (among many topics) the life and premature death of the mysterious so-called Chief Designer of the Soviet space programme, Sergei Korolev, but for me he provided the most intriguing narrative when he described how some men had been deliberately but clumsily airbrushed from some photographs of the first cosmonaut group, effectively going "down the memory hole" of Soviet history. Oberg even gave these mystery men names: there were three known as Valentin, and others were named Anatoli, Ivan, Dmitri, Grigori and Mars. In this he proved to be totally correct, although it would be several more years before their full names and what befell them was officially documented and released. That would occur in 1986, on the 25th anniversary of the history-making flight of Yuri Gagarin.

We even learned the fate of young Valentin Bondarenko, who, at the tender age of 23, died in a horrifying fire in a soundproof pressure chamber just three weeks before his cosmonaut colleague made mankind's first-ever flight into space. To this day, he is still the youngest male candidate ever selected to any nation's space team.

As the horizons of my interest in human space exploration widened, so I came into contact with many fine people who shared my enthusiasm and passion for the subject, and friendships of lasting tenure evolved. Two such chums are Simon Vaughan from Canada and Bert Vis from The Netherlands, with whom I shared a wonderfully productive and enjoyable week at the 1993 Association of Space Explorers' Congress in Vienna. Both were friends with British space historian Rex Hall, whose name I already knew well, and they encouraged me to get to know him. Thus, over the years, another great friendship ensued. As my airline job meant I was in London several times a year, Rex and Lynn would always throw their home open to me for a visit, an animated talk-fest in their living room, and a local takeaway Greek dinner washed down by a splendid bottle of Australian red wine. Truly an international evening!

Rex has always proved to be of great assistance to me in almost everything I've written to this time on the Soviet/Russian space programme, and I am therefore delighted that he so readily came onboard when I first broached the concept of this book with him. Like me, he feels that this story needs and deserves to be written—not only to recognize the many accomplishments of those members of the first cosmonaut group who were able to fulfil their ambitions of flying in space, but also the eight men whose names and achievements were held in limbo for so many years, and who have never been properly accorded their place in spaceflight history.

This, then, is our respectful salute to them.

REX HALL It was in the summer of 1961 that a Soviet touring exhibition came to London, and being swept up in all the excitement of the early days of the so-called Space Race I decided to attend. One of the exhibition's centrepieces was a full-scale representation of a spacecraft, duly marked as a Vostok vehicle. It was in fact Sputnik 3, but the organizers were giving away a small Novosti booklet on the Soviet space programme which I eagerly accepted. I was hooked.

Having had my curiosity aroused, I decided to seek out more information on the men who were flying these craft, so I wrote to NASA and to my joy received a large package of photos and biographical material on the astronauts. It was easy and much appreciated, but where, I thought, would I get the same material relating to the Soviets? I had no idea. I tried writing to the Soviet embassy in London but did not receive a reply. However, I discovered a book shop that had a set of cards on the subject which I purchased. My interest in the cosmonauts was reignited.

In the mid-1970s I discovered American space researcher Jim Oberg through his great article on missing cosmonauts in the British Interplanetary Society's *Spaceflight* magazine, which showed through his investigations that some cosmonauts had been selected in 1960 along with the known group members, but had not flown. Some had even had their images removed from group photographs. He attempted to identify men from the first selection in part from photographs which are reproduced in this book. I was in fact compiling a similar list with backups missing from early missions. It was the start of the "sleuths" who tried to make sense of a Soviet programme set against a background of secrecy. Then, in the mid-1990s, I discovered for myself the intrigue of these matters while sitting in the kitchen of a 1965 military cosmonaut who disclosed that for 20-plus years, both his involvement and his identity had been kept a State secret.

These sleuths, all of whom became firm friends, included some mentioned earlier by Colin such as Bert Vis and Dave Shayler, but this eclectic group also included noted researchers Michael Cassutt, Gordon Hooper, Neville Kidger, Phillip Clark, Antony Kenden, Bart Hendrickx and Geoff Perry. There were others who in many ways have also contributed to understanding the Soviet programme. It is thanks mainly to those amazing people with their enthusiasm, talents and persistence that a new openness came about, which has not only shown how much we did know but sometimes did not understand, as well as uncovering some of the secrets which still exist and hopefully one day will be in the public domain.

My interest, for me, came full circle in 1996 when I visited the cosmonauts' training centre known as Star City for the very first time, and I have subsequently returned a number of times. On these visits I have had the honour and privilege of interviewing a large number of cosmonauts covering the entire spectrum of the Soviet/CIS manned spaceflight programme, asking them how they became a cosmonaut, as well as their trials and tribulations, successes and failures. I now call a number of them friends. Little did I realize that momentous day back in 1961 how my sparked interest in spaceflight history would one day set me on a road that eventually led to the gates of Star City, and the chance to meet, interview and even befriend many of the men and women whose names and exploits had meant so much to me all those years ago.

I would have to say that my proudest moment came when I was laying flowers on the graves of the cosmonauts in the Kremlin Wall and Novodevichy cemetery along with their families and friends on Cosmonautics Day in 2001; it was 40 years to the day since Yuri Gagarin became the world's first human space traveller.

We trust that this book does bring to a wider world the story of those men who were, or could have been, pioneers of our new frontier of space.

Acknowledgements

With more decades of interest in space exploration between us than both authors would care to acknowledge, much of the extensive research carried out for this book required little more effort than simply reaching out for a particular book or file in our respective studies. But as always there are a multitude of unanswered questions, and it is wonderfully reassuring to know that there is always a host of good people out there ready and eager to assist where they can: perhaps in a large way, perhaps even in a very small way, but on each and every occasion very much appreciated. Worthy of particular appreciation are those unsolicited messages that usually began with a salutation, and something like, "I came across this and wondered if it might be useful to you for your book." Almost invariably, yes, it was.

Our individual helpers need to be acknowledged with gratitude. They are, alphabetically, Michael Cassutt, John B. Charles, Kyra Collins, Francis and Erin French, Dr. Vladimir Golyakhovsky, Bart Hendrickx, Ivan Ivanov, Anne Lenehan, Tom Neal, James Oberg, Alex Panchenko, Tony Quine, David M. Scott (not the astronaut), David Shayler, and last, but certainly not least, to our Dutch chum Bert Vis. A vast amount of information and quotes in this book are the result of Bert's many investigative forays into Star City, where he is now widely recognized, admired and treated as a trusted friend for his work in interviewing cosmonauts, engineers, designers and other folks over many years in order to transcribe and retain for posterity a social history of the entire Soviet/CIS space programme. Specifically for this book, Bert interviewed Marina Popovich in Star City, Moscow, and Alexei Leonov at a space conference in Edinburgh, Scotland. Our thanks also go to those two people for graciously consenting to be interviewed.

We would also like to offer profuse thanks to Elena Esina, curator of the museum in the House of Cosmonautics in Star City, who always comes up trumps as a friendly liaison person with the residents and workers at that remarkable place. We also acknowledge the staff of the Yuri Gagarin Training Centre and the cosmonauts of the Air Force detachment, without whom this book would definitely not have been possible.

Kudos also to the staff and work done at *Novosti Kosmonavtiki*, Russia's leading magazine on space exploration, and to the amazing Spacefacts website (*www.spacefacts.de*) so capably administered by Joachim Becker. Many thanks as always to the Council and Staff of the British Interplanetary Society in London for once again allowing us access to their extensive library and photo archive.

And finally, our love and thanks go to our respective First Ladies, Pat and Lynn, for their ongoing but sometimes strained patience and understanding of our shared passion for spaceflight history.

Figures

While the vast majority of photographs came from the personal photo files of the authors, and ownership or permission of others has been acknowledged, it has proved impossible to correctly identify the source of a number of photographs taken during the Soviet regime. The authors wish to extend their apologies to those whose work cannot be properly identified and formally acknowledged in this work.

Abbreviations and acronyms

ARS	American Rocket Society
ASTP	Apollo–Soyuz Test Programme
ATC	Assembly Testing Complex
CapCom	Capsule Communicator
COSPAR	Committee on Space Research
CPSU	Communist Party of the Soviet Union
DM	Descent Module
DOSAAF	Freewill Society for the Army, Aviation and Navy Support
Elint	Electronic intelligence
EN	Everything normal
EVA	Extra-vehicular activity
FAI	*Fédération Aéronautique Internationale* (International Aeronautical Federation)
GCTC	Gagarin Cosmonaut Training Centre
GMK	Chief Medical Commission
GMVK	State Interdepartmental Commission
HAFP	Higher Air Force Pilots aviation school
IAD	Air division
IAP	Interceptor fighter regiment; training fighter regiment
IBMP	Institute of Biomedical Problems
ICBM	Intercontinental Ballistic Missile
IGY	International Geophysical Year
IMBP	Institute for Medical and Biological Problems
IP-1	*Izmeritelny Punkt-1* (Tracking Point 1)
KGB	*Komityet Gosudarstvennoy Bezopasnosty* (Committee for State Security)
KS-4	*Korabl-Sputnik 4* (Spaceship Satellite 4)
LII	M.M. Gromov Flight Research Institute

LK	*Lunniy Korabl* (Lunar Spaceship)
MIAN	V.A. Steklov Institute of Mechanics, U.S.S.R. Academy of Sciences
MOUSE	Minimum Orbital Unmanned Satellite of the Earth
NKVD	*Narodny Komissariat Vnutrennikh Del* (Soviet Secret Police)
OKB	*Opytnoe Konstructorskoi Byuro* (Development Design Bureau)
OM	Orbital Module
RNII	*Raketnynauchno-Issledovatelski Institut* (Reactive Scientific Research Institute)
RRS	Retrorocket system
RSC	Rocket and Space Corporation (Energia)
SAF	Soviet Air Force (see VVS)
SAS	Space adaptation sickness
TsAGI	*Tsentralny Aerogidrodinamicheskiy Institut* (Cental Institude of Aerohydrodynamics)
TsIAM	*Tsentralny Institut Aviatsionnogo Motorosroenya* (Central Institute of Aviation Motors)
TsPK	*Tsentr Podgotovka Kosmonavtov* (Cosmonaut Training Centre)
TsVLK	Central Medical Aviation Commission
TsVNIAG	Central Aviation Institute of Medicine
VMF	*Voyenno Morskoy Flot* (Soviet Navy)
VNA	Vietnamese News Agency
VPB	Ventricular premature beat
VVS	Soviet Air Force

1

Sparking the Space Age

Historically speaking, Sputnik was an entirely appropriate name for the spacecraft. Over countless eons leading up to 4 October 1957, our Earth had been accompanied on its celestial journey by the Moon, nature's own satellite. But on that momentous October day the world would unexpectedly have a second satellite; an instrument package hermetically sealed and filled with gaseous nitrogen within a polished, 22.8-inch-diameter steel sphere that had been given a Cyrillic name translating to "fellow traveller". That day the history of our planet would change dramatically and forever. The Space Age had truly begun.

THE SPUTNIK SURPRISE

Up to that time, no object created by humans had travelled in excess of around 7,000 miles an hour, yet here was a beach ball–sized Soviet satellite impudently orbiting the Earth at more than twice that speed, constantly emitting a simple but distinctive *beep-beep* signal to an awestruck world below. A triumphant Soviet Union had become the first nation to launch an artificial satellite into orbit, and the success of that venture would not only usher in an entirely new and exciting era of space and planetary exploration, but give notice that humans now stood on the verge of moving out into the cosmos on pioneering space missions.

Losing the high-ground advantage

At the Third Symposium of Space Travel held in the city of New York three years earlier on 4 May 1954, Dr. S. Fred Singer had put forward a radical proposal for placing into orbit a 64-pound American satellite. Singer, a professor of physics at the University of Maryland, had called his pet project Minimum Orbital Unmanned Satellite of the Earth, known less awkwardly by the acronym MOUSE.

4 October 1957, and the first Sputnik satellite is launched into orbit.

In the hope of getting a *bona fide* space programme under way, Dr. Singer had estimated the cost of launching such a satellite at a mere one million dollars, or less than the cost at that time of a jet bomber aircraft. He envisaged using a three-stage rocket to place the MOUSE into orbit, but his proposed satellite was found to be far too heavy for any of the booster rockets then available. Consequently, the MOUSE never got any further than the drawing boards. Even so, several of the features embodied in his satellite were later incorporated into the design for the U.S. Navy's Vanguard satellite.

By the following year an easing of tensions between America and the Soviet Union saw the latter nation participating in scientific conferences, while American scientists soon came to realize that their counterparts were pursuing a very vigorous space programme.

On the afternoon of 29 July 1955, a group of reporters was led into a conference room at the White House, intrigued by the prospect of a news story that the president's press secretary James C. Hagerty had promised would be of "some importance". As they filed in at 1:30 PM they found Hagerty already seated, together with a group of men he introduced as scientists from the National Science Foundation and the National Academy of Sciences. Hagerty would then read out an announcement from President Dwight D. Eisenhower, which said that plans had been approved "for the construction of a small, unmanned, Earth-circling satellite vehicle to be used for basic scientific observations" [1]. This had followed a recommendation to the White House by the U.S. National Committee on the International Geophysical Year (IGY) that the United States should launch such a satellite during the 18 months of the IGY, which would last from 1 July 1957 to 31 December 1958. The Department of Defense, Hagerty announced before calling for questions, would provide the logistical support needed to ensure the U.S. Navy programme, known as Project Vanguard, would become a reality. The White House announcement stated that the first Vanguard satellite in a series of launches would take place within three-and-a-half years, and certainly by the end of 1958.

Just four days later, at the sixth International Astronautical Congress in Copenhagen, a leading Soviet science spokesman and head of the Soviet delegation, Leonid Ivanovitch Sedov, smugly revealed that his nation also had plans to launch a satellite during the IGY period [2]. Sedov, an astrophysicist, was the newly appointed chairman of a Soviet Academy of Sciences Commission on Interplanetary Communications, and he spoke with what seemed at the time to be compelling authority when he declared that the Soviet Union had the technical know-how to create a large Earth-orbiting satellite, adding that "the realisation of the Soviet project can be expected in the comparatively near future." He would not elaborate on a timetable.

There had been two serious contenders to launch America's first satellite, and their proposals went before a specially convened selection committee. First, there was the United States Army, who had a spaceflight research team largely comprised of German rocket scientists and technicians captured towards the end of the Second World War, now operating under the erudite leadership of Dr. Wernher von Braun. That team had proposed a project named Orbiter, which would have utilized a Redstone rocket surmounted by an upper-stage cluster of small Loki rockets to launch a simple, five-pound payload into orbit. It would be quick, clean and relatively inexpensive. The other contender was the Navy's Project Vanguard.

The concept of space exploration in the United States was at first kept non-military, in order to prevent any dilution of the nation's ballistic missile development programme with less imperative ventures. Top military officials believed that a space programme based on ballistic missiles might drastically impinge on their primary job, which was to produce efficient deterrent weapons. Furthermore, the United States had agreed along with other nations to make their initial space exploration efforts part of the IGY programme. This in effect would obligate them to publish details of their ballistic rockets—a scenario that made the military brass squirm with uneasiness.

Subsequently, a decision was made to develop a research programme called Vanguard, under the auspices of the Naval Research Laboratory. Project Vanguard would employ an advanced version of the Viking research rocket which had been developed at the White Sands Proving Ground in New Mexico, with an Aerobee Hi sounding rocket used for the second stage, and a solid-fuel third stage.

To the acute disappointment of the von Braun team the Navy won the day with their Vanguard proposal, but as later events would prove it was the wrong option for America in relation to spaceflight history and national prestige. The committee's decision to let the Navy proceed with its more sophisticated Vanguard, rather than the Army's tested and reliable Redstone, meant the honour of being the first nation to orbit an artificial satellite was basically handed to the Soviet Union on a platter.

As America's initially muted anticipation of a satellite launch slowly grew with the development of the necessary hardware, Soviet scientists also followed Vanguard's progress with interest. Later, Western analysts would complain that there had been no indication of Soviet plans to launch their own satellite into orbit, yet the signs were there for anyone who cared to read Soviet publications, which gave precise details of their proposed satellite, down to the point of specifying what radio transmission frequency it would use.

In the 8 June 1957 issue of *Pravda* newspaper the person who had been president of the U.S.S.R. Academy of Sciences for the previous six years, academician Dr. Aleksandr Nikolayevich Nesmeyanov, confidently declared that the Soviet Union had "created the rockets and all the instruments and equipment necessary to solve the problem of the artificial Earth satellite. Soon, literally within the next few months, our planet will acquire another satellite" [3]. In his position he was the ultimate public authority on Soviet space plans and yet, with very few exceptions, no one seemed to be listening.

One of those exceptions was a team of experts working for the RAND Corporation, a U.S. Air Force contractor. They prepared a report based on a number of confident predictions, announcements and strong rumours emanating from Soviet sources. Promulgated in the summer of 1957, the report iterated their concerns about an imminent space spectacular. "The red-letter date on the Soviet astronautical calendar is September 17, 1957," it stated. "This is the 100th anniversary of the birth of K.E. Tsiolkovsky, the founder of the science of astronautics. Though it comes rather early in the current International Geophysical Year, there could be no more fitting way of celebrating this occasion—from the Soviet point of view—than to establish the first artificial Earth satellite in honour of Konstantin Eduardovich Tsiolkovsky. The prestige and propaganda value to be gained from a premier launching of an Earth satellite, whether instrumented or not, undoubtedly present a circumstance too attractive for the opportunists in the Kremlin to ignore" [1].

The commemorative date came and went, but the Russians had been actively preparing to make history. That same month a group of designers, researchers and technicians gathered in a large building at the Tyuratam test base launch site to finalize preparations, and witness the mating of the protective nose cone bearing the Sputnik satellite with the SS-6 Sapwood booster rocket (which came to be designated the R-7), which had already been loaded onto a rail flatcar for the final journey to the launch pad. The four antennas were carefully folded flat for insertion into the nose cone; they would fully extend once the spacecraft was in orbit. Meanwhile, the Soviet space programme's enigmatic Chief Designer, a man named Sergei Korolev, was on hand to oversee proceedings.

Then came the day to test Sputnik's radio transmitter *in situ* for the first time; it was a momentous occasion later described by engineer Aleksey Ivanovitch. "Silence reigned in the big room. The members of the State Commission, Korolev, the chief designers of the engines [Valentin Glushko], the guidance systems [Nikolai

Pilyugin], and other basic systems were standing silent beside the rocket. The command was given, and in that huge room we could hear the clear, distinct signals: *beep, beep, beep*. Later, the whole world would hear them. But when they came through the loudspeaker of the testing system, those signals were really exciting" [3].

The multiple-stage SS-6 rocket, now with a total mass of around 285 tons, was slowly trundled out of the building. On the morning of 3 October 1957 the assembly was raised at the launch pad in preparation for its historic flight.

"One small ball in the air"

It was the evening of 4 October when Radio Moscow suddenly interrupted its normal programmes with a special bulletin. Bold music played, and then came the electrifying announcement that *Iskustvennyl Sputnik Zemli* (artificial space traveller around the world) had been successfully inserted into orbit. Lift-off had occurred at 10:28:04 PM Moscow time.

Sergei Korolev, the Soviet Union's enigmatic but anonymous Chief Designer.

Sputnik 1 (originally known simply as Sputnik) had separated as planned from the carrier rocket after orbital insertion, and a mechanism had been actuated which released its antennae. The third stage of the SS-6 rocket, now drained of all fuel, also went into orbit, as did the satellite's protective nose cone. The world could now tune in to the satellite's beeping transmission, and watch in awe and admiration as it passed overhead, although it was mostly sighting the larger third-stage booster that led people to assume they could see the actual satellite. Western observers tracked Sputnik's initial elliptical orbit at 65.1°, ranging from 145–583 miles, with an orbital duration of 96.1 minutes. The satellite was travelling at an average speed of 17,896 miles an hour; its perigee velocity was around 18,000 mph, and around 16,200 mph at its apogee.

The launch of the world's first artificial satellite predictably produced a diverse crop of reactions. An effusive Sir Bernard Lovell, first director of the Jodrell Bank radio telescope south of Manchester, U.K., referred to Sputnik as "absolutely stupendous ... the biggest thing in scientific history ... [and] the highest scientific achievement of the human intellect." Far less graciously, the U.S. Chief of Naval

Sputnik: The world's first artificial satellite. (Photo: NASA)

Research, Rear Admiral Rawson Bennett, petulantly called it "a hunk of iron almost anyone could have launched" [4].

President Eisenhower, never a great proponent of space science, merely humbugged the news. "One small ball in the air," he grouched when questioned about Sputnik. "Something that does not raise my apprehensions; not one iota." He would soon change his mind. Lieutenant James M. Gavin, the US Army's Chief of Research and Development, and a leading figure in American military rocketry programmes, described the launch of Sputnik into orbit as America's "technological Pearl Harbor."

The Eisenhower administration immediately went on the defensive as it sought to quell a tidal wave of criticism, although the President could personally see little value in engaging the U.S.S.R. in some sort of competition for the high ground of space. "We never thought of our satellite programme as one which was a race with the Soviets," his White House Press Secretary James Hagerty stressed to reporters. "Ours is geared to the IGY and is proceeding satisfactorily in accordance with scientific objectives."

On 15 October, Vice President Richard M. Nixon gave a major policy speech in San Francisco, in which he declared that "militarily the Soviet Union is not one bit

The future president and vice-president, Dwight D. Eisenhower and Richard M. Nixon, photographed in happier times, July 1952. News of the launch of Sputnik would cause a crisis of concern for them and the American people. (Photo: Associated Press)

stronger today than it was before the satellite was launched." However, he stressed that it would be a mistake "to brush off this event as a scientific stunt", but rather to perceive it as "a grim and timely reminder" that Russia "has developed a scientific and industrial capacity of great magnitude".

Eleven days earlier, on 4 October 1957, several members of a cadet unit attached to the 1st Chkalov (Orenburg) Air Force Flying School were touching down at the military airfield after completing what for many would be their last solo training flights in MiG-15 fighter jets. After two years of intense aerial and classroom tuition they were due to graduate the following month, provided they passed their final set of eight practical flying examinations four days later.

Moments after one particular 23-year-old cadet had taxied his MiG to its parking position and shut down the engine, his good friend and fellow cadet Yuri Dergunov ran up to the cockpit with a huge smile on his face and excitement in his eyes. Once the canopy had been opened and the pilot had begun to unbuckle his seat belts, Dergunov yelled out to him, "We've done it! We've sent a satellite into space!" That evening the two men sat around a radio with other starry-eyed cadets as they listened attentively to further details of the great event.

The young cadet was tremendously excited by the news of this space triumph, and as the evening wore on he and the other cadets eagerly speculated on how long it would be before a Soviet man was sent into space. Each of them said they would like to be that person. Little did they know the person who would one day realize that particular dream was in their midst, and in just three-and-a-half years they would be astounded to hear his name broadcast around the world. For his part, the young flying cadet couldn't wait to discuss this amazing news of the Sputnik satellite with his fiancée Valentina Goryacheva. He knew she would be delighted, but he also knew there would be very little time to dwell on the subject. Apart from his completing his final exams they were due to be married just three weeks later. Later that night, when he had finally overcome his excitement, Yuri Gagarin finally fell asleep knowing October of 1957 would prove to be a very momentous month. He could not have known just how much impact it would one day have on his life.

America's dismay at being beaten into space by the Russians was compounded by reports of an article that appeared in *Pravda* five days after the launch of Sputnik. It calmly stated that: "In order to make the transition to manned spaceflights it is necessary to study the effect of spaceflight conditions on living organisms. To begin with, animals will be used for studies. As was done with the high-altitude rockets, the Soviet Union will launch a Sputnik carrying animals as passengers. Detailed observations will be made of their behaviour and their physiological processes" [2].

They were as good as their word. On 3 November, Sputnik 2 was successfully launched into orbit. The massive spacecraft not only contained over a ton of payload, but this time, as forecast, there was a passenger onboard—a feisty little street dog named Laika. While there was jubilation in the streets all over the Soviet Union, it was a sobering time for the Americans. They saw themselves as coming second to their Cold War adversaries in an important arena of science and technology, and rapidly being left behind.

A spectacular failure

Some salvation seemed at hand on 6 December 1957, as America's focus centred on a launch pad at Cape Canaveral. A successful lift-off and orbital insertion of the Vanguard satellite would finally propel the United States into the Space Age, and in a bold move the launch was to be televised live across the nation. As the countdown wound down the Navy rocket team and all of America collectively crossed fingers and prayed for a trouble-free mission.

As the countdown reached zero the pencil-like missile roared into life, and America cheered. Then, as it rose from the launch pad, the rocket suddenly seemed to hesitate and lean over. Moments later it began to collapse downwards onto the pad, engulfed in a massive orange fireball of exploding propellants. While the world watched and listened, the much-publicised effort to place Vanguard into orbit had ended in an embarrassing, morale-shattering disaster for the United States space programme. "It seemed as if all the gates of Hell had opened up," the team's propulsion engineer Kurt Stehling would later write. "Brilliant stiletto flames shot out from the side of the rocket near the engine. The vehicle agonisingly hesitated a

The catastrophic loss of
Vanguard, December 1957.
(Photo: NASA)

moment, quivered again, and in front of our unbelieving, shocked eyes, began to topple. It also sank like a great flaming sword into its scabbard down into the blast pit. It toppled slowly, breaking apart, hitting part of the test stand and ground with a tremendous roar that could be felt and heard even behind the two-foot concrete walls of the blockhouse" [5].

While the world's attention was being distracted by Vanguard and its demoralising aftermath, the first Sputnik was rapidly approaching the time when it would irretrievably fall under the influence of Earth's gravity and burn up during re-entry. From the outset, Sputnik and its carrier rocket had orbited the Earth every 96.2 minutes, but this began to slowly decrease under the influence of air resistance and other factors, particularly so for the much larger rocket. Fifty-eight days after launch the rocket plunged into the atmosphere and was incinerated, while Sputnik's elliptical orbit gradually brought it closer to Earth with each revolution, eventually losing half a mile a day in altitude.

On 4 January 1958 the history-making satellite's fate was sealed when it was dragged into the fringes of the atmosphere and re-entered. Sputnik ended its life as a spectacular, white-hot fireball. It had lasted in orbit for three months, travelling around 37.5 million miles through space—roughly equivalent to the distance from Earth to Mars.

A model of Explorer 1 is held aloft in triumph by (from left) Director of NASA's Jet Propulsion Laboratory, Dr. William Pickering; Dr. James Van Allen from the State University of Iowa; and rocket designer Dr. Wernher von Braun. (Photo: NASA)

Twenty-seven days later, Wernher von Braun's Huntsville, Alabama rocket team, under the auspices of the Army Ballistic Missile Agency, successfully launched America's first satellite Explorer I (officially known as Satellite 1958 Alpha) into orbit atop a Jupiter-C rocket, a special modification of the Redstone ballistic missile.

Within just a few weeks—a brief moment in time—the Space Age had evolved into the Space Race.

REFERENCES

[1] Ralph E. Lapp, *Man and Space: The Next Decade*, Secker & Warburg, London, 1961.
[2] William Shelton, *Soviet Space Exploration: The First Decade*, Arthur Barker, 1968.
[3] Evgeny Riabchikov, *Russians in Space*, Novosti Press Agency, Moscow, 1971.
[4] A.P. Herbert, *Watch This Space: An Anthology of Space Fact*, Methuen & Co., London, 1964.
[5] Arthur C. Clarke, *Man and Space*, Time-Life Books, New York, 1970.

2

A few good Soviet men

One evening late in 1956, Sergei Korolev and his OKB-1 Deputy Chief Designer Konstantin Bushuyev were enjoying a rare night of relaxation, having a pleasant but thoughtful discussion with their chief engineering theoretician Mikhail Tikhonravov on the possibilities of one day sending a man into space. As Vladimir Yazdovsky, founder of Soviet space biomedicine would later recall, "Their mental indulgence was contagious, and soon we were all dreaming about putting a human being on top of a sounding rocket" [1].

AN IDEA GROWS AND EXPANDS

It seems that this conversation might have touched off considerable debate around the various design bureaux, for not long after Yazdovsky received written requests from several leading scientists, including Abram Ghenin and Aleksandr Seryapin, for his opinion on the feasibility of selecting and utilising human subjects on ballistic sounding rocket flights. But this flurry of interest would soon diminish in light of projects already consuming the time and talents of the designers and engineers.

"At that time," Yazdovsky related, "Korolev's design bureau was fully occupied with the development and manufacturing of a multi-stage intercontinental ballistic missile, and therefore any plans to launch a human being aboard a single-stage sounding rocket had to be suspended." Yazdovsky further pointed out that while the cost of conducting a manned flight aboard a sounding rocket would not have been appreciably higher than using a multi-stage rocket, Sergei Korolev displayed little interest in the proposal, which he saw as a wasteful expenditure of time, resources and talents, and considered it a retrograde exercise. He was already looking to future uses of his multi-stage rocket, and felt that this was where any serious proposals for biomedical and manned space missions should instead be diverted. "On balance," Yazdovsky concluded in discussing ballistic manned flights, "we considered the

In 1947, Mikhail
Tikhonravov (left) and
Sergei Korolev
photographed at a
function.

technical, medical and scientific problems too onerous at that juncture." Seeds of
further thought, however, had been sown.

Following the success and astounding propaganda benefits of the first orbiting
Sputnik satellites, Korolev and Tikhonravov held a meeting in May 1958, at which
they formally discussed the use of a modified spacecraft atop a multi-stage rocket to
place a man in Earth orbit. Within weeks they had put together a well-constructed
memorandum intended for government perusal and response on possible future
directions of rocket research and cosmonautics, with an emphasis on the need to
conduct a broad range of scientific studies leading up to eventual manned missions.
That November, the Council of Chief Designers and Scientific Leaders empowered

Korolev with the task of designing and developing an appropriate spacecraft, and this task was handed over to Tikhonravov and Konstantin Feoktistov, who had been studying and developing optimum uses for Korolev's heavy rockets. In early 1959, as Yazdovsky explains, a meeting was then held at the Academy of Sciences under the chairmanship of Academician Mstislav Keldysh to formally discuss the subject and future of manned spaceflight, and the selection of suitable human subjects.

"The candidates might be chosen from the ranks of fighter pilots, submariners, rocket specialists, racing car drivers, or members of other physically challenging professions. The aviation physicians among us knew that fighter pilots would have a more relevant background than the others since their training includes exposure to hypoxia, high pressure, g-loads along various axes, and ejection by catapult. It seemed obvious to us that cosmonaut candidates ought to be fighter pilots, and this view was fully shared by Korolev and his colleagues. The Council of Chief Designers also insisted that the selection should be made by aviation physicians reporting to me under the supervision of the Medical Flight Commission and with the approval of the Chief Physician of the Air Force, Alexander N. Babiychuk" [1].

Developing the launch vehicle

The rocket that would carry the first cosmonauts into space was known as the *Semyorka* (Little Seven), or R-7, the Soviet Union's first long-range two-stage missile, which was originally developed for potential military use in delivering a warhead to any point on the territory of a potential enemy nation. The missile's design was fundamentally different from all the earlier Soviet rockets in terms of its configuration, loading patterns, dimensions, mass, propulsion system power, and systems functionality. A central core rocket was surrounded by four strap-on boosters, and the booster's 876,000 pounds of thrust was capable of lobbing a 5.3-ton warhead some 5,000 miles.

The origins of the R-7 rocket can be traced back to several key sources; most notably a proposal presented in 1947 by Tikhonravov, then the division head at RNII, the Scientific Research Institute of Jet Propulsion. His "rocket packet" concept was based on strapping several ancillary rockets around a core missile to provide sufficient launch and acceleration thrust, and then discarding them once their fuel was exhausted and they had become little more than excess weight that could be jettisoned. Korolev had seized upon this exciting concept and discussed it with Mstislav Keldysh, head of the V.A. Steklov Institute of Mechanics, U.S.S.R. Academy of Sciences (MIAN).

Keldysh agreed to provide a comprehensive study based on Tikhonravov's concept and design a powerful launch vehicle capable of hauling a nuclear warhead into space on a ballistic trajectory. Korolev could also envisage adding upper stages to the core vehicle, giving it the capacity to one day propel satellites and other massive payloads into orbit [2].

First authorized for development in early 1953 and developed at the Special Design Bureau No. 1 (OKB-1, now S.P. Korolev RSC Energia), the final line drawing of the rocket had been approved by Sergei Korolev on 11 March 1955. Following five failed attempts, the first successful launch of an R-7 Intercontinental Ballistic Missile took place at 3:25 PM Moscow time on 21 August 1957 from the newly completed Tyuratam launch site in Kazakhstan.

In an attempt to disguise its actual location from prying eyes, the Tyuratam site was officially called Baikonur, a real town located some 300 miles away. The test flight successfully carried a dummy H-bomb warhead 3,700 miles downrange, and ended with the rocket reaching its designated target area in the Kamchatka peninsula. The ICBM, destined to become the workhouse of the Soviet space effort, would later be added to the Soviet arsenal, on 20 January 1960.

The development, manufacturing and launching of the missile had involved dozens of companies and organizations throughout the country, including OKB-456 (V.P. Glushko), NII-885 (M.S. Ryazansky and N.A. Pilyugin), NII-3 (V.K. Shebanin), NII-4 (A.I. Sokolov), TsIAM (G.P. Svishchev), TsAGI (A.A. Dorodnitsin and V.V. Struminsky), NII-6 (V.A. Sukhikh), and the A.N. Steklov Mathematical Institute (M.V. Keldysh).

Key developers, designers and researchers for the R-7 at OKB-1 were B.Y. Chertok, P.I. Yermolaev, K.D. Bushuyev, S.S. Krukov, Y.F. Ryazanov, I.P. Firsov, A.I. Nechaev, G.S. Vetrov, G.N. Degtyarenko, Y.P. Kolyako, O.N. Voropaev, S.S. Lavrov, R.F. Appazov, P.F. Shulgin, P.A. Yershov, V.M. Udodenko, A.F. Kulyabin, V.F. Roshchin, A.F. Tyurikova, V.F. Gladky, O.D. Zherebin, S.F. Parmuzin, V.M. Protopopov, V.M. Liventsev, A.N. Voltsifer, V.A. Udaltsov, M.V. Melnikov, I.I. Raikov and B.A.Sokolov.

The manufacturing and testing of the rocket at factory No. 88 of OKB-1 was principally supported by R.A. Turkov, V.M. Klyucharev, N.M. Berezin, S.K. Koltunov, V.M. Ivanov, I.V. Povarov, N.A. Pshenichnikov, A.V. Kirov, D.M. Shilov, A.G. Zigangirov, Y.D. Manko, L.A. Medvedev, B.M. Afanasiev and A.N. Andrikanis. Within weeks of that first successful dummy payload test flight the way into space had also been cleared, and a specially modified R-7 would be used as the launch vehicle to place Sputnik, the world's first Earth artificial satellite, into orbit [3].

"We need many more"

Yazdovsky also acknowledges that during the 1950s there was a total lack of cooperation between the aviation and rocket communities, and even a degree of confrontation. One of the chief protagonists was the Commander in Chief of the Air Force, Pavel Zhigaryov, who was strongly opposed to the use of valuable ballistic missiles for what he saw as the mundane purpose of biomedical research. However, things would change dramatically with the appointment of a Marshal of the Air Force as his replacement. This well-respected man, Konstantin Vershinin, would provide a completely different and vigorously helpful attitude to the question of combining research and rocketry, although several of his more pragmatic officers were still vehemently opposed to what they viewed with disdain as radical ideas, and would interfere where they could, delaying and blocking work to the continuing frustration of the scientists and designers, making their tasks even more difficult to achieve.

Meanwhile, there was the vexing problem of selecting the right military men to ride the rockets from the many thousands who would be considered eligible, as Yazdovsky explained:

"The physicians were well aware that the fighter pilots in our Air Force were very similar in age, health, and flight experience. It was therefore unnecessary to go looking for cosmonaut candidates in the Urals, Siberia, and the Russian Far East. The search would be limited to the European part of our country. Korolev went before the team of physicians entrusted with the selection task and formulated his requirements regarding a future cosmonaut: Age maximum 30, maximum height 170 centimetres. He was asked how many cosmonauts should be chosen.

'A lot,' he replied with a smile. 'The Americans have chosen seven men—and we need many more.'

His answer created some bewilderment, but nobody made any comment. Everybody understood that it was not one or two flights that were being planned, but a much greater number" [1].

If the candidates were to be selected and trained in time to keep pace with Korolev's established plans, there was a definite urgency involved in setting things in motion. The preliminary selection of cosmonaut aspirants would take place under the auspices of the Central Military Scientific Aviation Hospital (TsVNIAG), created in 1959 in the Sokolini region of Moscow, and carried out by the hospital's Central Medical Aviation Commission (TsVLK). The preliminary selection, based solely on medical data, would be the responsibility of the so-called Chief Medical Commission (GMK), a board of leading doctors, all members of the Russian Academy of Sciences, who would be representing the Ministry of Defence. Those candidates surviving selection by the GMK would next undergo further evaluation by the State Interdepartmental Commission (GMVK), then known as the Credentials Committee, the top government commission for cosmonaut selection. If they successfully passed through this review board they could finally be recognized as cosmonaut candidates and begin space-related training.

The GMVK, according to Sergei Shamsutdinov, a research assistant at the Russian Centre for Space Documentation, was primarily a political committee which "selected candidates on the basis of their political reliability and both moral and human qualities" [4].

Seeking out the finest

Tasked with the initial responsibility for selecting candidates, the TsVLK physicians were paired off and despatched to selected locations in order to begin the process. Initially they were to select several dozen candidates who would undergo further screening. According to Yazdovsky, word soon leaked out to a number of Air Force units that teams of physicians would be travelling from Moscow to interview candidates for "special flights", Doctors at these units were apparently aware of the real purpose behind these visitations, and they quickly came up with a list of candidate pilots, well in excess of three thousand names. The physicians now had a massive job ahead of them, but they were well briefed and prepared.

The strict height, weight and age limitations would quickly eliminate a number of potential candidates, making the task of the physicians a little easier, while a study of individual flight logs and medical certificates narrowed the field even further. Diseases such as chronic bronchitis, angina, a predisposition to gastritis and colitis, renal and hepatic colic, and pathological shifts in cardiac activity as revealed by electrocardiograms—all were sufficient reasons for immediate disqualification from the dwindling list of eligible candidates. It was then time to begin the serious business of personal interviews.

COSMONAUT CANDIDATES

The weather at the Murmansk air base on 12 October 1959 was typically clear, with bright blue skies, light winds, a blinding sun and not a cloud to be seen. It was also frigidly cold, with frost on the ground, ice hanging from the roofs of buildings, and a temperature well below zero. Situated north of the Arctic Circle in the Kola Peninsula, the Murmansk air force base was strategically important in shielding and protecting Soviet submarine activity in the nearby Barents Sea, but its resident pilots and their families knew that they would soon have to endure several weeks of even worse weather and total darkness with the bitter onset of winter.

A mystery interview

However, there was also a distinct buzz in the air at Murmansk that day, as news had rapidly spread throughout the garrison that a special commission had arrived to conduct interviews with some of the airmen for a secret assignment. It was all very mysterious and puzzling. In all, 12 young pilots were ordered to trudge through the snow and ice from their squadron hut to the headquarters building. They had been selected to be interviewed, and two of those chosen were Senior Lieutenants Yuri Gagarin and Georgi Shonin. The 12 men made their way into their commander's office and saluted, after which he directed them into a small waiting room where, one by one, they were invited into another office.

When it was Shonin's turn he entered the office and sat opposite two grey-haired men who he came to understand were medical people. "This was a bit of a poser," he later wrote. "What did any doctors want with me? They asked me to sit down and began asking questions. We talked about the usual, perhaps I should say, boring things: how I was enjoying the Air Force, did I like flying, had I become adapted to the Far North, what did I do in my free time, what did I read, and so on" [5].

Yuri Gagarin undergoing flight instruction at Saratov.

Finally the two men looked at each other and Shonin knew the interview was at an end. As he stood to leave one of the men said that they might want to meet him again and continue the conversation. "I left the office and met my comrades' questioning looks," he wrote, "but as I didn't have any sensible answer, I could only shrug my shoulders."

The vagueness with which each meeting ended naturally gave rise to rumours, Shonin recalled. "Our town was a small place where nothing stayed secret for long, with everybody in full view of each other. A couple of days later a second round of talks began. This time far fewer of us had been called for." On this occasion the questions were far more specific and penetrating. "I was asked detailed questions about my flying experience starting from my first flights in air school. They listened to my answers very attentively even though they must have known it all: they had my flight log on the table before them."

All of a sudden the questions about his past ceased, and Shonin was asked how he would feel about flying more modern airplanes. He had been speculating to himself and others that they might be recruited to fly the brand new supersonic jets that were already being delivered to some air force regiments, and he promptly stated that he would certainly like to be involved. One of the men looked hard into his eyes. "And if it were a question of flying something of a completely new type?" he asked. Shonin said that he hesitated at this point, thinking that he might be talking his way into flying for a helicopter unit; something he definitely did not want to do. "I'm a fighter pilot," he blurted out. "I specially chose a flying school where I would be taught to fly jet fighters, and you ..."

The older of the two men quickly held up a hand and smiled. "No, no," he said reassuringly. "You don't understand. What we're talking about are long-distant flights, flights on rockets, flights around the world."

Shonin was taken back: this was certainly not what he had expected to hear. "Even though there were quite a few satellites in space by then, manned flights were still an idea from the realm of the fantastic. Even amongst us pilots, no one spoke seriously about such a thing" [5].

Only the best would go through

As Shonin had discovered, the interviews were necessarily thorough. The physicians wanted only the best candidates to go through to the next stage, so Yazdovsky had stressed to his team of physicians that they should quiz the pilots thoroughly about their health, achievements, work, moods, and quality of life. Following this, they were to cautiously ask how the men would feel if given the opportunity to pilot "a different kind of craft". If they expressed interest, they were made to understand that the vehicle they would be flying was not a conventional aircraft, but a large satellite that would be fired aloft from a launch pad and orbit the Earth. That news certainly came as a surprise to many of those interviewed, who were expecting nothing more than a transfer to another squadron flying a new type of jet aircraft or helicopter. According to Yazdovsky:

> "One of the Moscow physicians reported that 90 percent of their interviewees had asked whether they would be flying conventional aircraft. The candidates obviously enjoyed their profession and were proud of their rank as military pilots. Approximately three out of ten declined the offer immediately, not necessarily because they were afraid, but usually because they liked the Air Force, their teams, and their friends. They had a clear vision of their future military and professional careers. Many had a well-established family life which they were reluctant to give up in exchange for vague promises. There was a general rule that any candidate could decline at any stage without giving his reasons. Some asked to be allowed to consult their wives or family; others gave their agreement immediately or only after lengthy consideration.
>
> The most frequently asked question was how long they would have to wait to do whatever we were promising them. Would they have to wait until retirement? That would be okay, they said, except that meanwhile they had families to support" [1].

Later that evening, Shonin and the other candidates were told that only six of them would be receiving an invitation to travel to Moscow for a more comprehensive examination and evaluation by a medical selection board. He recalled that some of the original 12 candidates from his unit had decided to reject the offer, preferring to remain where they were. "United by our common wish, our common interests, beliefs and hopes for an unusual future full of romanticism, we waited to be summoned. We did not have to wait for too long. The first to leave were Yuri Gagarin and three others. The agony of suspense dragged on longer for me" [5].

An ever-shrinking list

As Sergei Korolev and his OKB-1 design team worked hard on building components of the future Vostok spacecraft, some parts were already undergoing tests and final adjustments were being made.

Meanwhile, the selection team had also been hard at work, winnowing down the list of candidates even further by going back through flight logs and medical records and being even more critical in their evaluations. Before long it was time to begin carrying out comprehensive medical examinations of the remaining candidates, who were brought to Moscow in groups of 20.

According to Alexei Leonov, one of those who travelled to Moscow, a total of 40 pilots out of the 3,000 interviewed would present themselves for evaluation, all with "experience flying the most modern aircraft, MiG-15s and MiG-17s, under all conditions" [6]. Once in Moscow, the pilots were closely scrutinized by a team of specialists, and those who failed were sent back to their units.

Research assistant Sergei Shamsutdinov maintains, however, that contrary to Leonov's recollection, documentation from the selection process reveals only 29 out of 154 preliminary candidates successfully passed their medical tests at the TsVNIAG in Moscow. "However, in the end only twenty of them were selected by the Credentials Committee," he added. "These twenty later became known as the Air Force's Gagarin Team." All of them made it to the stage of cosmonaut candidates [eligible for spaceflight] and later many of them went on to become cosmonauts. The names of the nine aspiring cosmonauts who were

Cosmonaut candidate Alexei Leonov.

rejected by the Credentials Committee are now known. They are N.I. [Nikolai] Bessmertny, B.I. [Boris] Bochkov, G.A. [Georgi] Bravin, G.K. (Grigory) Inozemtsev, V.A. [Valentin] Karpov, L.Z. [Leonid] Lisitz, V.P. [Valentin] Sviridov, I.M. [Ivan] Timokhin and M.A. [Mikhail] Yefremenko [4].

Some would make it, some not

Meanwhile, back at the Murmansk base, Georgi Shonin was waiting impatiently for news of Gagarin's group, as well as his own call-up to travel to Moscow. Eventually, one of the candidates returned to the base from Moscow; a friend of Shonin's from flying school days named Alik Razumov, who said with obvious disappointment that he had been rejected. Soon after, another two pilots returned with similar stories, and Shonin began to wonder if any of them would qualify. "Then Yura [as he was affectionately known] came back. The great grin with which he greeted me from afar told enough: everything was all right, he had passed!"

Gagarin had expected the examination to be rigorous, but even he was surprised at the extent of the medical tests. The medical screening process was carried out at the Central Aviation Research Hospital, and medical staff at the centre were exacting in their tasks, having been charged with selecting only the fittest candidates from the many dozens they would test. They would also reject those they felt were not completely up to the demands of spaceflight.

"The examination was really thorough," Gagarin later recalled. "It was not at all like our yearly pilots' medicals." He continued:

"We pilots were used to these and saw nothing terrible about them. But here, beginning with the very first specialist—the oculist as it happened—I understood how serious it was going to be. My eyes were checked very thoroughly. One had to have perfect vision; that is, to be able to read all the required letters and signs on the chart from top to bottom, from the biggest to smallest. I was carefully checked for a concealed squint, my night vision was verified, and my retina was also assiduously examined. Instead of going to the oculist only once (as is usually the case), I had to go seven times, and each time we went through everything from the start: the charts with letters and signs, colour sensitivity tests, look with your right eye, look with your left eye, look here, look there . . . He searched and searched but was unable to find even a hint of a fault in my eyes.

There was also a test of ability to work in unusual and difficult conditions. The task was to solve certain arithmetical problems with figures that first had to be found in a special chart. Both speed and the correctness of the answer were taken into account. At first glance, it seemed it would be relatively easy to solve the problem. Then suddenly a loudspeaker was turned on and a monotonous voice began to prompt answers. It became much harder to concentrate, and one had to force oneself to continue calculating without paying any mind to this "obsequious friend". It was tough. Incidentally, this was only the beginning; worse was yet to come.

There were a lot of doctors and each one was as stern as a state prosecutor. There was no appeal against their sentences. Doctors of all sorts, including therapeutists, neuropathologists, surgeons and ear-nose-and-throat specialists examined us. We were tested from head to toe: little hammers were tapped all over our bodies, we were twisted about on special devices, and the vestibular apparatus of our ears was checked ... Our hearts were the main object of their examination. The doctors could read our whole life history from them. One couldn't hide a single thing. Complicated instruments detected everything, even the tiniest cracks in our health.

The commission was headed by an experienced Air Force doctor, a very educated and knowledgeable man. He was a blue-eyed handsome man with a sense of humour; he immediately won over the whole of our group and even those who were rejected for health reasons went away with friendly feelings towards him" [5].

Yevgeny Karpov

The man of whom Gagarin was writing was Colonel Yevgeny Karpov. He was not only an experienced air force physician and chairman of the commission for the selection of cosmonaut candidates, but the future first director of the cosmonauts' training centre.

By this time the thoroughness of the medical examinations and other tests had enabled the physicians to assess each candidate's strength of character, will power, persistence, and their ability to cope in new and stressful situations. "The candidates were also placed in a pressure chamber and on a centrifuge," according to Yazdovsky, "in order to determine their resilience against hypoxia and various accelerations. Every day the group of candidates became smaller and smaller" [1].

The majority of those who failed were not overly upset at missing out, and were quite happy to return to their units, although others were quite disappointed. One rejected candidate is said to have turned to his friend, who had been successful, and said, "Well, you've made it. Good for you ... they'll turn you into another Laika!"

Alexei Leonov has revealed that only 8 of the 40 candidates in this first group were selected for further appraisal. More would be needed. A senior air marshal addressed the lucky eight in what Leonov described as "a fatherly way" and then offered the men an exciting prospect. "He told us we had a choice to make. Either we could continue our careers as fighter pilots in the Air Force, or we could accept a new challenge: space. We had to think it over. We left the room briefly and talked amongst ourselves in the corridor outside. Five minutes later we filed back into his office. We wanted to master new horizons, we said. We chose space" [6].

Beginning an incredible journey

Towards the end of November, a relieved Georgi Shonin was finally given his call to travel to Moscow, where he also found the examinations harder than he had expected, even after discussing them with Gagarin. "On average only one candidate

Alexei Leonov at the controls of a turboprop aircraft.

out of fifteen got through all the various stages," was his recollection of the medical evaluation. "Some of them were taken off flying duty altogether" [5].

Shonin returned to his Murmansk unit from Moscow on the last day of 1959. Once he had reported in to his commanding officer he sought out Gagarin so the two friends could once again compare notes on their experiences. He mentioned hearing that additional candidates had been sought earlier on and screened at different air bases, and then the best of these were sent to Moscow in order to better identify those not only possessing strong personalities but in excellent physical condition. Although they encouraged each other, both men were realistic about their chances with so many candidates to choose from. All they could do was be patient and wait. They did not know it, but as Yazdovsky revealed, they had already made the team.

"By the end of 1959 we were down to twenty cosmonaut candidates who had passed all the medical tests in 'Theme 6', the official code name. That group of twenty pilots—also known as Team 1—formed the first cosmonaut team in our country. Who could have predicted then that most of these young men in their warm coveralls would some day become not only the first cosmonauts, but also famous generals, national heroes, parliamentary deputies, and honorary citizens of many foreign cities? Others would disappear into oblivion, as is always the case in major undertakings. Yet none of these achievements would have been possible without the intensive efforts of thousands of scientists, engineers, and designers" [1].

On 14 January 1960, orders came through from naval air arm headquarters for Senior Lieutenants Gagarin and Shonin to be dispatched to Moscow. The orders did not specify how long they would be gone from their unit. Neither man could possibly know the incredible and challenging journey that would soon point them firmly on the road to the stars.

REFERENCES

[1] John Rhea (ed.), *Roads to Space: An Oral History of the Soviet Space Programme*, McGraw Hill, New York, 1995, pp. 228–233.
[2] Rex Hall and David Shayler, *The Rocket Men: Vostok and Voskhod, the First Soviet Manned Spaceflights*, Springer/Praxis, Chichester, U.K., 2001.
[3] Anon., *50th Anniversary of the Russian ICBM*, Energiya press release, Korolev, Russia, 23 August 2007.
[4] Sergei Shamsutdinov, *The Selection of Cosmonauts*, translated by Bart Hendrickx, from (1) *Vesti Meditsiny*, Nos. 4–5, 1994, p. 19, and (2) Nikolai Kamanin, *Skrytyi kosmos: kniga pervaya (1960–1963)*, Infortekst, Moscow, 1996, p. 206.
[5] Yaroslav Golovanov, *Our Gagarin*, Progress Publishers, Moscow, 1978.
[6] David Scott and Alexei Leonov, *Two Sides of the Moon*, Simon & Schuster, London, 2004.

3

Russia's future spacemen

Twenty suitably qualified cosmonaut candidates had been selected, and with training set to commence in March 1960 the rapid completion of a specialized but covert accommodation and training centre now became a priority. The site eventually selected was a former military installation in the middle of a dense birch forest in the Shchelkovo area, some 25 miles northeast of Moscow. Eight years later, in 1968, the facility would be given the official and appropriate name *Zvezdny Gorodok*, correctly translating to Starry Town, but more popularly known as Star City. In the meantime it would be known simply as the Cosmonaut Training Centre, or TsPK.

CREATING A TRAINING FACILITY

There were many crucial factors involved in selecting a suitable site for the new cosmonaut training centre, as determined by a special committee that had been established for this purpose. The committee, chaired by General Nikolai Kamanin, had been hurriedly formed after a government decision on 11 January 1960 to create the cosmonaut training facility, known as TsPK (*Tsentr Podgotovka Kosmonavtov*), and to select someone to command the new centre. The position would fall to someone who had already been involved in the process of selecting the first cosmonauts.

As Nikolai Kamanin would later write: "Who would head the group of future pilot-cosmonauts, to arrive at Zvezdny Gorodok as commander, tutor, and simultaneously bold experimenter? It appeared to us that there were few candidates for this post. It was given to the outstanding specialist in the field of aviation medicine, Colonel Yevgeny Anatolevich Karpov. For many years he had studied with pilots and knows their characters and the nature of flying well. From the first days [he] was set alight by the new tasks, perspectives and dreams."

General Nikolai Kamanin, the first Director of Cosmonaut Training.

Three men of considerable influence in Soviet cosmonautics: General Nikolai Kamanin, Sergei Korolev and Vladimir Yazdovsky.

Sergei Korolev with Yevgeny Karpov, head of the cosmonaut training centre, and (right) Colonel Nikolai Nikitin, head of cosmonaut parachute training.

Karpov's appointment as head of the cosmonaut training centre was approved on 24 February 1960. He quickly set to the task of working out how many specialists and workers would be needed to staff the facility. As Kamanin recalled in his diaries, Karpov submitted a proposal requesting a planned personnel of 250 people, but his application met with understandable scepticism. "The deputy chief of the Air Force, F.A. Agaltsov, smiled, appreciating the boldness of the 38-year-old colonel, and reduced the staff to 70 people. Marshal Konstantin A. Vershinin listened first to the colonel, then to the Colonel-General, and said to Agaltsov: 'You, Filip Aleksandrovich, don't know how they will train, and neither does he.' The Marshal pointed to Karpov: 'Look out. You must value this.' And he confirmed the 250 personnel."

The centre would be organized into a number of departments headed by Vladimir A. Kovalov, Nikolai F. Nikiryasov (in charge of political activities), Yevstafi Y. Tselikin (in charge of flight training), A.I. Susoyev and Grigori G. Maslennikov. Medical specialists included Grigori Khlebnikov, H.K. Yeshanov (optic physiologist), A.A. Lebedev (specialist of heat-exchange and hygiene), I.M. Arzhanov (otolaryngologist), M.N. Mokrov (surgeon), V.A. Barutenko (oculist-surgeon), A.S. Antoshchenko (hygiene systems, spacesuits, survival clothing), A.V. Nikitin (therapist, attached to the cosmonauts for constant medical monitoring), A.V. Beregovkin and others. The two senior trainers were Colonel Mark L. Gallai and Colonel Leonid I. Goreglyad, both Heroes of the Soviet Union.

Finding a suitable site

The criteria the selection committee worked to were formidable; these included locating a large and secluded area near an established military air base, with roads and a railway in reasonable proximity. It also had to be within handy enough reach of Moscow, but conversely remote enough to allow the training to be carried out in complete secrecy and security, as well as providing comfortable living quarters and facilities for the staff, trainers and cosmonauts who would occupy the centre.

Fortunately the ideal site they finally focused on, surrounded by a dense forest in the Shchelkovo region, was located just 25 miles from Moscow and within easy reach of the Chkalovsky Air Base, the largest military airfield in the Soviet Union. Already a mostly disused military installation, the site in question had once been used as a radio range and was still dotted with a few small buildings. It was also situated adjacent to the Yaroslavl railroad between Moscow and Monino, then home to the Monino Air Force Academy and nowadays the Central Air Force Museum. There was further advantage in the fact that Sergei Korolev's OKB-1 design bureau was located along the Yaroslavl railroad in nearby Kaliningrad.

The committee's selection of the site was approved, and the construction of some new buildings and facilities was begun.

Military physician Colonel Yevgeny Karpov, who had been fully involved in the cosmonaut selection process, was appointed the first chief of the new space training centre, while General Nikolai Kamanin would take up the post of chief of cosmonaut training within the High Command of the Soviet Air Force (VVS).

Initially, as the training centre and a small residential area for the military and civilian personnel serving the facility were being constructed, the cosmonaut trainees and their families would be housed at the Frunze Central Airfield in Moscow, located near the Zhukovsky Air Force Engineering Academy and the Central Sport Club of the Soviet Arm, and later in a Moscow apartment block, located in nearby Chkalovsky.

Although all 20 names were considered a state secret and several would not be officially revealed and recognized until April 1986, the new cosmonauts were an elite mix of highly trained professional pilots who now formed a very unique unit. As space journalist Yaroslav Golovanov wrote of them in the *Izvestia* article that finally revealed all 20 names, they were: "Fine fliers . . . each a leader in the group from which he had been chosen. Differing in temperament, pursuits, sympathies and antipathies, family status (some were married) . . . they formed a well-knit team in which mutual assistance and support was law."

IVAN ANIKEYEV

Although he was actually revealed as a potential candidate for the first cosmonaut group in Evgeny Riabchikov's book, *Russians in Space*, first published in the United States in 1971 under licence from the Novosti Press Agency, Ivan Anikeyev's place in that elite cadre was not officially revealed until April 1986, on the 25th anniversary of the history-making flight of Yuri Gagarin. In an officially sanctioned story on the first cosmonaut group written for the government newspaper *Izvestia* by Yaroslav Golovanov, all 20 cosmonauts were identified for the first time. The article told of the untimely demise of several from this group who left the cosmonaut corps through injury, death, or in Anikeyev's case, deep disgrace. The popular young senior lieutenant, a man fascinated by the theatre and music, was nevertheless also fond of a good time and drinking, which led to his early dismissal from the first cosmonaut team.

Ivan Nikolayevich Anikeyev was born on 12 February 1933, the son of railroad worker Nikolai Nikolayevich and Natalie Ivanovna Anikeyev, in Novaya Pokrovka, a large town in the Voronezh Region of Central Russia. In 1937 his birthplace, a major railway junction situated on the banks of the Don River, would officially become a city and was renamed Liski.

Affectionately known to his parents as Vanya, Anikeyev would complete ten years of grade education at Liski No. 12 School in 1952. On 11 July that year he was called to service in the armed forces of the U.S.S.R. by official decree of the Liskinskeye District Military Commissariat of the Voronezh Region, enrol-

Ivan Anikeyev.

ling by choice in the Russian Navy, or VMF (*Voyenno Morskoy Flot*). He would undertake flight training at the Stalin Naval Aviation School in the port city of Yeisk in southwestern Russia, graduating in 1955.

The following year Anikeyev completed an advanced course with the training fighter regiment (IAP) of the No. 12 Naval Air School in the city of Kuybyshev (now Samara), located on the Volga River, graduating on 30 July with the rank of lieutenant. Following this, on 18 August, he commenced service as a pilot with the 255th fighter aviation regiment of the 91st air division (IAD) of the Soviet Air Force's Northern Naval Fleet, stationed in the Kilp-Yarv region. From 25 October that year he flew with the northern fleet's 524th fighter regiment of the 107th IAD as a pilot-operator, and as senior pilot-operator from 30 November. Three weeks later, on 18 December, he was still flying Yak-25 fighters from his North Sea base, but now as chief pilot. The following year, on 24 April, he would attain the distinction of military pilot, third class. Anikeyev was subsequently promoted to the rank of senior lieutenant on 17 August 1958.

Late in 1959, and still a bachelor, the quietly spoken Anikeyev was one of a number of pilots asked to attend a special interview at his base, impressing the interviewing physicians sufficiently to be ordered to Moscow for further interviews and tests. Here, while undergoing the examinations and tests, he was billeted with a young senior lieutenant named Gherman Titov. Another pilot in that selection group they befriended was the dashing and impressive Grigori Nelyubov. In his 1971 book, author Evgeny Riabchikov recounted one evening when a friend of Titov's, only identified as "Volodya", washed out of the examinations and was immediately ordered back to his fighter unit in Leningrad.

"That night before going to sleep, Titov was sitting on the bed of his roommate, Ivan Anikeyev. 'It's too bad to lose a good man like Volodya,' he was saying. 'Do you suppose you and I will be going out the gate like that some day, with our suitcases? I couldn't take it!'

There was a light knock at the door, and Nelyubov stepped quietly in, looking worried. 'Not sleeping?' he whispered.

'Come here.' Titov got up to greet Nelyubov, and then flopped back on Ivan Anikeyev's bed. 'I was saying that Volodya was a great loss. Do you suppose they're driving us so hard for no good reason?' He ran a hand through his curly hair. 'The time will come when we'll remember this evening and all our fears and anxieties ... you know?'

'Right! Quite right. But let me get some sleep!' said Ivan Anikeyev, twisting around in the bed so that Titov almost fell off it. 'Don't go tearing your heart to pieces!'

Titov stood up. 'See you tomorrow, boys. We'll make it to the stars!' "

Despite their apprehension all three candidates—Anikeyev, Nelyubov and Titov—would make it through and be selected in the cosmonaut group. Sadly enough, only the latter pilot would eventually make it "to the stars".

Anikeyev browsing in a Moscow store.

PAVEL BELYAYEV

One of six eventual children, two boys and four girls, Pavel Ivanovich Belyayev was born on 26 June 1925 in the village of Chelishchevo in the picturesque Vologda region, northeast of Moscow, although he would spend his early childhood in the village of Minkovo in the Leninsky district. His father Ivan Parmenovich Belyayev, who had served in the First World War, was a physician's assistant at a local hospital, while his mother Agrafina worked on a collective farm. With both parents absent for the greater part of each day, Pavel (known affectionately as Pasha) and his siblings grew up having to learn how to fend for themselves and each other. Sadly, neither of his parents would live to share in the glory and pride of their older son's first and only spaceflight.

In September 1932, seven-year-old Pavel began his formal education at the Minkovsky secondary school, three miles from the family home. In winter time when the snow lay thick on the ground he would have to ski and trudge his way to and from school, where physics and geography would become his favourite subjects.

As he grew up, young Pavel also developed a fondness for playing hockey and started hunting. When he was still in sixth grade at school his father presented-him with a single-barrel hunter's shotgun, which became his pride and joy. At an early age he developed into an expert marksman, and would often sling the weapon over his shoulder and go game hunting in the nearby woods with adult companions.

Just before Pavel's 13th birthday in June 1938 his family moved to the Kamensk-Uralsky region in southwestern Siberia, a mining area and manufacturing centre for steel and aluminium. Here his education would continue at the Gorkogo secondary school in the Sverdlovsk region. Early in 1942, as savage fighting raged against the German invaders across his homeland, Belyayev learned to operate a lathe and took on temporary war-related work as a turner and later check operator in the Sinarsk pipes factory; but he was sure his destiny lay elsewhere. One day he read in the local newspaper that admissions were being accepted to a special air force school in Sverdlovsk and decided to apply, but missed out on selection. Undaunted, and armed with his experience in skiing and shooting, he then tried to volunteer for service in a fighting ski unit, but was rejected because he was considered too young. In May 1943, just before he turned 18, his persistence finally paid off when he received his call-up papers.

Pavel Belyayev.

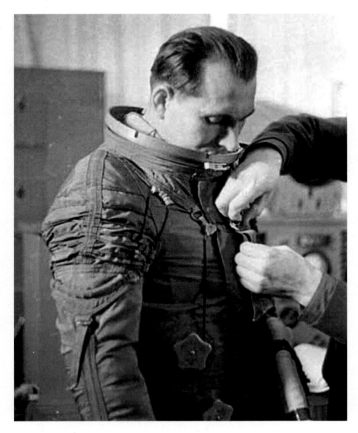

Dressing for space flight.

Belyayev entered military college at the 3rd Sarapul School on 20 May, where he was given initial cadet naval pilot training, passing out 14 months later on 2 July 1944. He was then assigned to the Stalin Naval Air School at Yeisk on the Azov Sea. Graduating as a military pilot on 9 May 1945 and receiving the rank of junior lieutenant a week later, he was too late to participate in the war in Europe, but was instead assigned to the Soviet Far East. Beginning on 16 June he flew aerial combat missions in the last days of the Pacific war against Japanese forces, piloting Yak, La and MiG fighters. Over the following 11 years Belyayev was stationed with various naval air squadrons attached to the Pacific Fleet Air Force, involved in guarding the Siberian frontiers. He rose steadily through the ranks, promoted to lieutenant on 13 May 1947 and senior lieutenant on 21 August 1950, while flying seven different aircraft types, including high-speed jets, and came to be regarded as one of the Soviet Union's best commander pilots. During this time, in 1948, he also met and married Tatyana (Tanya) Filippovna Prikazchikova, and they would eventually have two daughters: Irina, born 27 October 1949, and Ludmila, born on 20 March 1955.

On 11 November 1953 Belyayev was awarded the Distinguished Combat Service medal. Promoted to the rank of captain on 29 April the following year, he still enjoyed hunting whenever he could find the time, but was also fond of quieter pursuits such as reading and writing poetry, as well as playing the piano and accordion. He had a particular fondness for the music of Pyotr Tchaikovsky and Prince Michael Oginskiy, and the songs of Arkadiy Ostrovskiy.

In 1956, Belyayev passed his entrance exams and undertook advanced studies at the Red Banner Air Force Academy in Monino, graduating on 18 November 1959 with the rank of major and as a military pilot, second class. During the latter part of his studies at the academy he had also been interviewed and tested as a potential cosmonaut candidate. Douglas Hawthorne in *Men and Women of Space* states that: "Examiners were amazed at his ability to endure the high-gravity forces exerted by the centrifuge without adverse effects."

While waiting for news of his possible selection, Belyayev was dispatched to Air Squadron 661, 4th IAD regiment of the Black Sea Fleet Air Force, and the following month, December, he was appointed squadron commander with the regiment's 241st squadron, also attached to the Black Sea Fleet. By this time he had logged around 1,000 hours flying time in piston engine and jet aircraft, and had also completed around 40 parachute jumps.

VALENTIN BONDARENKO

According to his cosmonaut peers Valentin Bondarenko was a mild-mannered young man with a pleasant disposition, a good singing voice, a fascination for football, and almost unbeatable at table tennis. Sadly, he would die aged just 24 under tragic circumstances.

According to those who record such things, there is very little about the brief but remarkable life of Valentin Bondarenko held in the archives at the Star City training centre museum, which is otherwise home to around 20,000 exhibits. A single folder contains only a couple of photographs, a handwritten account of his life, schooling and military training prior to his selection; copies of his examination results; letters of commendation as well as diplomas from various flying schools and academies; and a photograph of his grave. There is precious little documentation with which to reconstruct and record the life of such a talented young man with so much unfulfilled potential.

Valentin Vasilyevich Bondarenko was born on 16 February 1937 in the city of Kharkov, then the capital of the Ukraine, and grew up as part of an ordinary working family. Both of his parents worked in a local fur factory where his father, Vasily Grigorevich, was employed as head tailor, converting animal pelts into fashionable clothing. Valentin would only have one other sibling—a brother named Anatoli, seven years his elder. Then, when Valentin was only two years old, his father suddenly volunteered to join a Soviet Army division to help repel the German advance, and later joined the resistance movement. His departure left Olga Ivanovna to raise and care for their two young children.

Valentin Bondarenko.

During the six years of the Second World War, Kharkov would become a place notorious in history, and not simply because it was the most populated Soviet city occupied by German forces. During April and May of 1940, prior to the German invasion of the Soviet Union and on the specific orders of the brutal Stalinist government, more than 22,000 Polish soldiers, prisoners, academics and civilians were rounded up and executed in what became known as the Katyn Forest Massacre. Around 3,800 Polish prisoners of Starobelsk camp were killed in the basement of Kharkov's NKVD (*Narodny Komissariat Vnutrennick*—secret police) building and later buried in mass grave pits deep in nearby forests.

In June 1941, Nazi forces invaded the Soviet Union and Kharkov became the site of several fierce military engagements. The city would be captured twice by Nazi invasion forces and in turn recaptured by the Red Army, before the final liberation of the city on 23 August 1943. Around 70% of the city was destroyed in the fighting, and tens of thousands of citizens were killed.

For two desperate years Valentin, his mother and Anatoli lived under the harsh German occupation of the area while his father courageously operated as a partisan scout with a guerrilla group in the resistance movement. Vasily Bondarenko would eventually survive the war and returned to his peacetime occupation in the fur factory. On special occasions or anniversaries he wore the seven medals of distinction he was awarded as a result of his bravery, and Valentin would often say how proud he was of his father. Vasily Bondarenko would later work on the city executive committee of Kharkov. He died in 1993 at the age of 85, sadly in the same year as his older son Anatoli, who passed away aged 63.

With the German occupation of the area at an end, young Valentin, who had meanwhile developed a childhood fascination for aviation heroes, was able to resume his studies at local grade school No. 115 until 1951. Later, while advancing his education in his final years at secondary school No. 93, he also took lessons at the Kharkov aero club, maintaining his boyhood dream of one day becoming a military aviator.

Bondarenko graduated from secondary school in 1954, was conscripted into the Soviet army and with his flying experience assisting his application, entered an air force college in the Ukrainian city of Voroshilovgrad (now Lugansk). The following year he was transferred to the air force college in Grozny (in what is now Chechnya).

Fierce street combat took place in Bondarenko's home town of Kharkov. (Photo: Associated Press)

In 1956 he married his sweetheart, a medical worker named Galina Semenovna Rykova (who preferred to be called Anya), and their first child, Aleksandr Valentinovich—affectionately known as Sasha—was born later that same year. Also in 1956 he was sent to the Armavir Higher Air Force Pilots School (HAFP), graduating with the rank of lieutenant on 2 November 1957, just a month after his nation had launched the first Sputnik satellite into orbit.

From 19 December 1957 Bondarenko flew MiG-15 and MiG-17 jets with the No. 868 unit of the 175 IAD (Fighter Regiment) of the 30th Soviet Air Force while based at Kaunus in the Baltic Union Republics. He would then operate as chief pilot with No. 43 unit, 263 IAD of the 30th SAF in Tukums, situated east of the Latvian capital of Riga. On 29 August 1959 he attained the status of military pilot, third class, and four months later the rank of senior lieutenant on 2 December 1959. Instructors at Bondarenko's earlier flying schools had been unanimous in their admiration for the skills and determination displayed by the young fighter pilot, praising him as industrious, tireless, bold, competent and confident. These written comments would cause Bondarenko to be the first from his unit to be interviewed by a selection panel on 26 December 1959, in the search for the nation's first cosmonauts. By this time, the records indicated he had flown and mastered several aircraft, including the Yak-18, Yak-11 UTI, MiG-15, MiG-15bis, and the MiG-17. He had flown 288 hours in these five aircraft types.

VALERY BYKOVSKY

Situated some 35 miles east of Moscow, and bisected by the Moscow–Vladimir train line, rests the ancient town of Pavlovsky-Posad. Renowned for textile mills that produce brightly coloured woollen shawls in the Russian style, Pavlovsky-Posad is also the place where Fyodor Fyodorovich Bykovsky, a former seaman, miner and

Valery Bykovsky.

railway worker met and married a young weaver named Klavdia Ivanova. This happy union would produce a future cosmonaut, Valery Fyodorovich Bykovsky.

Valery was born on 2 August 1934, a baby brother to Margarita, then aged three. The family lived a sparse life in a quaint unpainted wooden house so common to that area, with ornate wooden windows. World War II broke out when Valery was aged four, and the family made their way to the safer port town of Kuybyshev in the southeastern part of European Russia. Soon after, they moved again to the nearby port and rail centre of Syzran. In his free time young Valery could often be found on the wharves, dreaming of one day becoming a sailor like his father. Then the family moved back to Moscow, where his father had found a better paying job. It is rumoured that this work was with the KGB, but it was only ever reported that he worked for the Ministry of Traffic.

In 1941, Fyodor Bykovsky was transferred to a government office in Teheran, Iran, where they would spend the next seven years and where both of the children continued their education in a special Soviet school.

It would be the summer of 1948 before the family returned to Moscow. Now 14 and nearing the end of his secondary schooling, Valery told his father he would like to attend naval school. But Fyodor wanted something better for his son and insisted he stay on at school to better his chances later in life.

One day three years later, now studying in tenth grade, Bykovsky heard some fellow pupils talking about a free lecture at school that afternoon on aviation and flying lessons, and decided to attend. The spokesman, Nikolai Yerofey, was an air force pilot instructor, and as he spoke Bykovsky grew interested as it was explained how they could learn to fly and later join the Soviet Air Force. At the end of Yerofey's talk he asked if anyone would like to enrol in a flying club. Several hands, including Bykovsky's, shot up. However, on the day Bykovsky presented himself for the medical examinations and admission tests he was one of only two attendees from his school. He passed all the tests and was soon undertaking flight theory lessons at the Moscow City Aviation Club.

In 1952 Bykovsky not only graduated from aviation school but achieved his sports pilot's licence. He then enlisted for advanced flight studies at the Kachinsk Military Aviation Academy, and flew solo for the first time on 27 September, touching down twice in a Yak-18. His instructor that day was the person who had

Bykovsky undergoing
training.

first inspired him to fly, Nikolai Yerofey. He would graduate from the academy on 28
November 1955 with top marks in flying and combat training, and the rank of
lieutenant.

On 23 January 1956 Bykovsky was assigned to a jet fighter unit populated by
several veterans of the war. These men were only too eager to share their amazing
stories of aerial combat, as well as assisting the eager young pilot by physically
demonstrating actual air battle tactics and manoeuvres. While studying aeronautics
in his spare time, Bykovsky made sure he flew every day in order to maintain his
proficiency. "He has always been courageous and exciting, and dangerous professions attracted him," his father would later proudly reflect of his son's determination.

Late the following year, on 17 December 1957, Bykovsky began operating as
chief pilot and certified parachute instructor with the 23rd Interceptor Flight Regiment (IAP) of the 17th Air Division (IAD) in the Moscow military district. He was
promoted to senior lieutenant on 28 January the following year, and achieved the
rating of military pilot, 3rd Class, on 8 December 1959. By the time he underwent
testing as a cosmonaut candidate late in 1959 he not only had 72 parachute jumps to
his credit but had also met his future wife, a history student named Valentina
Mikhailovna Sukhova.

VALENTIN FILATYEV

The agricultural village of Malinovka is located in the Ishimsk district of Tyumen Region, situated east of the Ural Mountains in the steppes of Western Siberia. It would be the birthplace of future cosmonaut Valentin Ignatyevich Filatyev, known to his family and friends as Valya, who came into the world on 21 January 1930 as one of five children, with three brothers and a sister. Tragically, his career as a cosmonaut was a surprisingly short one of just three years for a man once described by Soviet space journalist Yaroslav Golovanov as "calm, well balanced". It was brought to an abrupt halt when he became embroiled in an alcohol-fuelled escapade one night with two cosmonaut friends that grew rapidly out of hand and led to his dismissal—perhaps unfairly—from the cosmonaut corps for disciplinary reasons, along with his two colleagues.

Very little is known of Filatyev's early family life, except that he undertook his first taste of education in 1938 at the Shablyninsky secondary school, and his childhood, like that of so many other Russian families during the war, was one filled with anxiety and profound grief. His father Ignatius and one of his brothers perished at the front while fighting during World War II and a second brother would also pass away after treatment for severe wounds failed.

Quiet, introspective and well-read, a youth who enjoyed basketball and nature walks and collected specimens of fungi, he graduated from secondary school in 1945 after completing seventh grade. It then seems that he or his mother decided he should pursue a career as a schoolteacher, and he duly enrolled at the Ishimsk State Pedagogical Institute (also known as the Ishimsk State University of Pedagogy). In 1951, now aged 21, he graduated and was awarded his Diploma of Education.

Having deferred his compulsory military service due to his studies, Filatyev was seconded on graduation into the armed services, electing to undertake flying lessons the following August as a student at the Stalingrad (Kacha) Higher Air Force School for pilots. He would remain there until October 1955, learning to fly incrementally more powerful aircraft; eventually specializing in the "operation and the combat employment of aircraft and their equipment", and obtaining qualification as a pilot-technician. On 29 November the rank of lieutenant in the Soviet Air Force (SAF) was awarded to the promising young fighter pilot.

Valentin Filatyev.

Two months later, on 11 January 1956, Filatyev began full-time service in the SAF as a pilot with the 472nd fighter regiment (IAP) of the 15th Anti-Aircraft Defence Force, flying in the defence of the city of Orel. It was a tense time; on 4 November that year other elements of the Soviet Air Force bombed part of the Hungarian capital of Budapest, with Russian troops then pouring into the city during a massive dawn offensive. This was in response to a

Always a keen sportsman, Filatyev enjoys a relaxing game of table tennis.

national uprising led by the anti-Soviet Prime Minister Imre Nagy, who wanted to withdraw from the Warsaw Pact, while promising the Hungarian people independence and political freedom; this was deeply concerning to Eastern Bloc countries.

Six weeks later, on 15 December, Filatyev transferred to the 3rd IAP of that same air defence division, and 12 months later, on 12 December 1957, was further promoted in rank to senior lieutenant. The following day he was also promoted to chief pilot with his regiment. It is known that in this time he served with two future cosmonauts from the first group: Valentin Varlamov and Mars Rafikov. He also met and married his first wife, Larissa.

Filatyev eventually became a regimental parachute instructor, and would have some 250 jumps to his credit when he was first interviewed at his base for a place in the cosmonaut team.

YURI GAGARIN

In 1934 spring came early to the small rural village of Klushino, in the Smolensk Region, some one hundred miles west of Moscow. There was a lot to do, and with her skilled carpenter husband already working from morning to night, Anna Timofeevna

Yuri Gagarin.

Gagarina returned from the hospital in nearby Gzhatsk with her new baby son Yuri Alekseievich, born 9 March, and quickly resumed her normal routine in a small house on the collective farm (known as a kolkhoz) where she worked as a dairy maid.

The newest addition to the Gagarin family had two older siblings—a brother named Valentin (born 1924) and a sister Zoya (born 1927), while a younger brother named Boris would arrive in 1936 to round out the family. In later years Anna Gagarina, described as a voracious reader, would often be asked if there had been any special events or influences that would lead her younger son on the path to world renown. "No, I can't recall any," was her usual response. "He grew up like all the other boys in the village."

Yuri entered school in 1941, but his education was abruptly interrupted for two years by the Nazi invasion, which drove his family out of their home, forcing them to live in a dugout. In 1943 the Germans retreated from the Smolensk area, taking Valentin and Zoya with them to work as slave labourers.

When liberation from the German occupation finally came, the Gagarins moved to the town of Gzhatsk, where Yuri continued his secondary school studies. In later years Gzhatsk would be officially renamed Gagarin in his honour. Meanwhile, to their profound relief, Valentin and Zoya had been released by the Germans and were soon reunited with their family. The years of war had caused extreme poverty and shortages; at school the children used to write their lessons on scraps of newspaper and wallpaper. Around this time Yuri's uncle Pavel would also inspire in him a youthful interest in reading science fiction books by such well-known authors as Jules Verne and H.G. Wells.

In 1949, aged 15, Gagarin completed his regular schooling and entered a trade school in Lyubertsy. Following his graduation in 1951 he took on a temporary job as a foundry-shop moulder. At the same time he also completed an evening school for young workers, but Gagarin was already looking to a better future and contemplating a career in aviation. In October 1954, while undertaking further studies at the Saratov Specialized School of Industrial Technology, he and four friends enrolled for flying lessons at the local aero club. According to a college friend, Aleksandr Shikin, "Many people joined the flying club, but only Yuri completed the course."

"I did not become an air force pilot by chance," Gagarin would later declare. "During the war we boys felt powerless. Certainly, we did what we could to hurt the

Nazis; we would sprinkle nails and broken glass on the road to puncture the tyres of their cars ... but when we were older we realised how important our country's security is. And that was what led me to make the choice I did—I dreamed of becoming an air force pilot."

In addition to studying the theory of flying, aircraft and engine design and high-speed aerodynamics, Gagarin would also make his first parachute jump on 18 May 1955 from 800 metres. Two weeks later, on 2 July, he completed his first flight with instructor Dmitri Martyanov in a Yak-18. "That first flight filled me with pride," he later stated, "and gave meaning to my whole life." In June that year he graduated with honours from the industrial school as a founder-technologist, and three months later received his ground school diploma from the Saratov Aero Club, having accumulated a total of 196 flights, with 42 hours and 23 minutes in the air in his flight log.

By now Gagarin knew where his future lay; much to his father's chagrin he had abandoned any thoughts of becoming a carpenter like him, or a foundry labourer. Armed with recommendations from his instructors and a supporting document from the Oktyabrsky District Military Commissariat he successfully applied to enter the 1st Chkalovsky Higher Air Force Pilots School, close to the Karavanny state farm outside Orenburg, as an aviation cadet. Initially he was flying the familiar Yak-18s, but would later graduate to the MiG-15 as a cadet-sergeant (promoted 22 February 1956). Short in stature, standing at just five foot two inches, his height nearly caused him to wash out of flying school.

An article in *Pravda* on 12 April 2004 (on the 43rd anniversary of his historic spaceflight) showed just how close he came to being unable to demonstrate his proficiency to a flight instructor in a dual-seat jet trainer, which would then qualify him to fly a single-seat fighter. According to the article in *Pravda*, pilots from a stand-by fighter regiment at Orenburg were used for the examinations. When it came Gagarin's turn he took off successfully, but coming in to land the jet aircraft suddenly dipped and the examiner/instructor had to quickly take over the controls to rectify the situation. The same thing happened two weeks later.

The rules were inflexible: any flaw, however minor, in piloting techniques was enough to cause the dismissal of any underperforming students, and papers were to be drawn up for Gagarin's dismissal from the school. However, one day the commander of the regiment, Ivan Polshkov, was walking by the sports ground and noticed a solitary cadet hard at work, training in the rain on a fitness apparatus. It was Yuri Gagarin. Polshkov summoned Gagarin's instructor to his office and bluntly asked, "Where are the documents disqualifying Sergeant Gagarin?" The instructor replied saying they were not ready, as he didn't have the heart to bring himself to do it. Gagarin, he said, was keenly disappointed and had told him he couldn't live without flying.

Fortunately, Polshkov decided to look into the young cadet's case, and after noting his otherwise excellent record discussed the problem at length with Gagarin's instructor. They agreed there might be two reasons for poor landing skills. First, the pilot might lack a sense of the ground—a sense which helps him to determine altitude with an accuracy of plus/minus ten centimetres. No instrument at that time could substitute this feeling. Second, the viewing angle might be wrong, particularly for shorter pilots such as Gagarin. It was decided to give him one more landing attempt—his third—but if he failed, he was out. This time, however, the instructor gave Gagarin a cushion to sit on, bolstering his height by a few centimetres and increasing his overall viewing angle. The landing was still a little rough, but this time it was within the set limits. He would be permitted to solo, and if there were no further problems, to graduate. Gagarin improved rapidly, and in his second year at the school was given the added responsibility of assistant platoon commander.

While at the flying school Gagarin had met his future wife in Moscow's Red Square. She was an attractive young woman named Valentina Goryacheva, the youngest of six children. "When I first met Yuri I was a nursing student," she would later recall of their meeting in her memoirs, "and strange to say I was part of a nurse's gymnastic brigade participating in the Moscow May Day celebrations. What do I remember from our first meeting? Let's see; first of all an overexcited cadet and my friend Helena giggling, 'If you want to get a general you must start off with a lieutenant'." For Gagarin, the chase was on. "My priorities then were my hair, flying school and chasing Valentina." Eventually they became engaged, and would be married at a registry office on 27 October 1957.

Gagarin during parachute training.

Earlier that month, while completing some flight training at the airfield, Gagarin learned of the successful launch of the first orbiting Soviet artificial satellite, Sputnik. The following day all the cadets were eagerly reading newspaper accounts of this momentous event. "I drew this spaceship in my notebook," Gagarin would later write, "and again felt that familiar, somewhat obsessive and not yet recognised urge; that same attraction to space, which I was afraid to acknowledge, even to myself."

On 5 November, having completed 166 hours and 47 minutes of flight time, Gagarin was promoted to lieutenant by order of the U.S.S.R. Minister of Defence. He graduated the following day at a solemn ceremony, during which he formally received his lieutenant's shoulder straps.

Assigned to air force service with the Northern Fleet, Gagarin then began a two-year basing at Luostari air base in the Murmansk Region of the Arctic Circle, near the Norwegian border. He was appointed to Senior Lieutenant Leonid Vasilyev's flight. Meanwhile, Valentina was completing her nursing course, after which she joined him at the base. Their first child, a daughter they called Yelena (affectionately Lena) was born on 10 April 1959. Three months later, on 7 July, Gagarin became a military pilot, third class.

Following the launch of the unmanned probe Luna 3 on 4 October 1959, by which time he had accumulated 265 hours of flight time, Gagarin is said to have written a report to his commanding officer in which he indicated he would like to volunteer to be a part of the thrust into space. "In connection with the expansion of space exploration going on in the U.S.S.R. people may be required for manned spaceflights," he wrote. "I request you to take note of my own ardent desire, and should the possibility present itself, to send me for special training." The letter was passed on in turn to the appropriate SAF authority by his commanding officer, Lt. Colonel Babushkin, who had marked it with own endorsement.

On 6 November 1959, just three weeks after being interviewed by the medical commission who were seeking out cosmonaut candidates, Gagarin was promoted to the rank of senior lieutenant. His next promotion, to major, would occur during his historic spaceflight.

VIKTOR GORBATKO

Born to Vasili Pavlovich and Matrena Aleksandrovna (nee Shmayunova) Gorbatko on a collective state farm in the Northern Caucasus settlement of Ventsy-Zarya in the Gulkevich district, Krasnodar Region, on 3 December 1934, Viktor Vasilievich Gorbatko spent his childhood and school years living and studying at school No. 30 on the Voskhod stud farm for horses in the Novokubanski district, where his father worked as a veterinary surgeon. Vitya, as he was known, had a 12-year-older brother, Boris, and two older sisters, Elena and Valentina. Another sibling, Ludmila, would be born five years later.

Viktor's first teacher was Nadezhda Karaulova, and with her determined influence he learned to study from an ABC book, censored by the occupying Nazi forces to exclude anything pertaining to Soviet heritage, although Karaulova would

Viktor Gorbatko.

help him to fill in those gaps. She would speak to her pupils about Soviet history, and secretly read to them from books she had carefully hidden away. He would complete his seventh-grade education at the school in 1949, before spending a further three years at secondary school No. 2 of the Progress settlement, also in the Novokubanski district.

There was no early incentive to a flying career, although "I remember with what ecstasy we, the village boys, watched our aircraft shooting up the German cavalry." What he had observed was a dogfight over the stud farm between six red-starred Yak aircraft and a number of German fighters, resulting in the shooting down of an enemy aircraft before the combatants flew out of the area. To his horror he also witnessed a Soviet officer being callously gunned down in front of him by a German soldier, and lived in mortal fear of the same thing happening to his two older sisters, who were devout members of the outlawed Young Communist League. His mother would tell him some years later that their names had been on a list for execution, but a timely Russian advance through the area, repelling the Germans, had spared their lives. His older brother Boris had returned wounded from Stalingrad, and used to tell Viktor how the fighter pilots had fought so bravely in the skies. One of his sisters had also married a pilot, and Viktor listened in awe to stories of the man's adventures in the air. With any sort of clothing in short supply in the area after the Nazis had been driven out, he used to wear his brother-in-law's old uniform.

When it came time to join the Soviet army late in 1952 Gorbatko reported to the recruiting station and requested an assignment to flight school. He was subsequently dispatched to the 8th Military Aviation School of Pilot Basic Training in the Ukrainian town of Pavlograd, Dnepropetrovsk Region, where his instructor taught him and other shy, unskilled cadets the rudiments of flying, leading to their first solo flights. One of those cadets was another future cosmonaut he befriended, Yevgeny Khrunov. In a letter he sent home some time later, Viktor wrote: "We have become real pilots and we fly real jet planes! If Boris [his brother] could see us now ... Tomorrow we go flying again. We get up early, at five in the morning. And now

Gorbatko (right) with
Alexei Leonov and Valery
Bykovsky.

it is already twelve! I am glad that I am here. Lieutenant Baskavov, our instructor, is
an excellent flier and a likable man. True, he has a difficult temper: on the ground he is
very polite, but as soon as we take off he starts to grumble. When you land you think
he will scold you, but he calmly analyses the flight.''

Over the next three years he and Khrunov would hone their skills considerably,
flying increasingly high-performance aircraft and jet fighters at the Serov Higher Air
Force School in Bataisk (named after Soviet Air Force fighter ace, General Anatoly
Serov) before graduating as lieutenants in the Soviet Air Force on 23 June 1956.
Around this time Gorbatko also married his first wife, a pretty young gynaecologist
named Valentina Pavlovna Ordynskayar.

Gorbatko and Khrunov were then assigned to the 86th Guards Fighter Regiment
of the 119th Fighter Division (IAD), attached to the 48th Air Army operating out
of the Odessa Military District in Moldavia. They began flying with the unit on
22 August, and Gorbatko became a senior pilot with his regiment on 22 June 1957.
Four months later the first of his two daughters, Irina, was born. A second daughter,
Marina, would be born in April 1960, two months after he began his cosmonaut
training in Moscow. Late in 1957, when the Soviet Union issued postage stamps to
mark the launch of the first *Sputnik* satellite, he found himself drawn into the hobby
of philately which, together with his love of the cinema and literature, would become
a lifelong passion. He also loved the outdoors life, with a particular fondness for
hunting, fishing and tennis.

Gorbatko rose in rank to senior lieutenant on 6 August 1958, and then along
with his friend Yevgeny Khrunov achieved military pilot, third class, two months
later on 16 October. Despite operating with a crack fighter unit, he would

nevertheless later categorize the majority of his flying in those days as "routine" and spoke of "the sheer repetitiveness from day to day".

ANATOLI KARTASHOV

There are certain parallels that can be drawn between the interrupted spaceflight careers of Anatoli Kartashov and NASA astronaut Donald "Deke" Slayton. Both men were highly regarded fighter pilots selected to their nation's first space team who were under serious consideration to make the first spaceflight. Early in his cosmonaut career Kartashov impressed his training superiors to the extent that he was assigned as one of six pilots to receive advanced training for the first Vostok missions. Both men were in superb physical condition, but it would be the discovery of a minor medical ailment during training that would see them abruptly grounded by over-cautious doctors. While Slayton would eventually be cleared to fly, and became part of the Apollo-Soyuz Test Programme (ASTP) flight crew in 1975, for Karshatov his vision of seeing the Earth from orbit was at an end. During their early training, Yuri Gagarin would actually describe the popular Kartashov as "the best amongst us. He will be the first man in space."

The engineering and richly agricultural Voronezh Region in southwestern Russia, close to the Ukraine, occupies the central part of the Russian plain. The region's administrative centre is the city of Voronezh, located on the Voronezh River, a tributary of the Don, some 370 miles from Moscow. It is an ancient and historic area, occupied in turn over many centuries by the Scythians, Alans, Huns, Khazars, Pechenegs and Polovtsians.

On 25 August 1932, two years before the Voronezh Region was formed as part of the Russian Federation, Anatoli Yakovlevich Kartashov (nicknamed Tolya) was born to Yefrosinya Timofeyevna and Yakov Prokofyevich Kartashov in the village settlement of Pervoye Sadovoye, in the Sadovoye district. He began attending junior school but his education, like that of so many young children, would be interrupted by the war years.

In 1941, Anatoli's father Yakov was one of those unfortunate individuals who fell foul of a local organ of the dreaded NKVD (*Narodny Komissariat Vnutrennikh Del*), the Soviet secret people charged with carrying out Stalin's brutal political repressions. He was arrested

Anatoli Kartashov.

and imprisoned, but with the onset of the German invasion was released to serve in the Soviet Army. From 28 June 1942 until 25 January 1943 Voronezh was the scene of fierce, bloody fighting between German forces and the tenacious Soviet Army, which resulted in the near-total destruction of the city. One of those who gave his life early in this 212-day battle was Yakov Kartashov.

Yuri Gagarin once described Kartashov as "the best amongst us."

In 1948, Anatoli Kartashov finally completed seven years of secondary schooling. He then took on further studies at the Voronezh aviation college, aspiring to become a technician-mechanic in the field of aircraft engine production. At the same time he was also learning to fly at the Voronezh Aero Club. On finishing his course at the technical college in 1952 he was called into military service, and for the next two years was a student at the Chuguyev Higher Air Force School for military pilots. He received the rank of lieutenant on 1 November 1954. Following his subsequent graduation he was assigned to serve as an Air Force fighter pilot, taking up his first posting on 25 December. He became a senior lieutenant 14 months later on 19 February 1957, and from 21 June that year flew as chief pilot with the 722nd Fighter Regiment, attached to the 26-1 Air Defence Division of the 22nd Air Force based in the Petrozavodsk defence district of northwestern Russia.

He later became an instructor in parachute jumping, and prior to being interviewed as a potential cosmonaut had completed around 200 jumps, more than any other member of the first cosmonaut team. Married to Yuliya Sergeyevna, they would eventually have two daughters: Ludmila, born in 1960, and Svetlana, born in 1967.

YEVGENY KHRUNOV

As part of a large peasant farming family that would eventually comprise two girls and six boys, Yevgeny Khrunov grew up in the village of Prudy, south of Moscow in the Volovsky district of Tula Oblast. Born on 10 September 1933 as the second child (and second son) to collective farmers Vasily Yegorevich and Agrafena Nikolayevna Khrunov (nee Bakunova), he would be given the nickname Zhenya by his family.

As a young boy, Yevgeny loved nothing better than to ride beside his father on their tractor. Inquisitive and eager to learn, he began school when he was six years old. However, even though he occupied a desk in the classroom, he could not be enrolled as a student because he was too young. His official status as a student was

Yevgeny Khrunov.

granted in the fall of 1941, the year in which the Nazi invasion of his homeland began and his father joined the fight against them as a brigade leader.

Living in a nation torn apart by war was difficult and confusing, and Yevgeny would often witness savage aerial dogfights being fought out in the skies over his village. He did not comprehend the totality of war, but the sight of aircraft streaking through the skies nevertheless excited the youngster, and he decided that flying was something he would like to pursue later in life. But he would also witness the ugly side of humanity when retreating Nazi occupation troops callously sacked and burned the small village as they passed through. One of the consequences of war for Yevgeny and a number of local children was that they were late in beginning their formal education. However, he would soon be praised by his teachers as a good, eager student, and he quickly made up lost ground.

In April 1948, Yevgeny was still in seventh grade of middle school when his 42-year-old father unexpectedly passed away. This meant even greater hardships for the family and their struggling mother, then four months pregnant with her eighth child. Yevgeny's 19-year-old brother Vladimir would subsequently assume responsibility as the male head of the Khrunov family, and he had some words of advice for his four-year-younger sibling, telling him that much was now expected of him, and he had to become the scholar in the family.

Having graduated from the seventh grade, Yevgeny was keen to take up a trade involving farm machinery. He decided to join a local polytechnic where he would receive a small but helpful scholarship, and was enrolled at the nearby Kashira Agricultural Secondary School. As part of his hands-on studies he became involved in machine work on a collective farm, and nearing graduation could easily take apart and reassemble a grain harvester. His teacher Boris Shago would later recall the eager young student. "I remember Zhenya very well. He was an active, ingenious and inquisitive lad. He studied perseveringly. This is proved by the high marks entered into his personal papers, which have been preserved in the polytechnic. He loved to tinker with agricultural machinery and said that in this way he could better absorb the material he had studied. Zhenya also took an active part in social life."

While studying at the polytechnic, Khrunov also had the opportunity to read a lot, and would keenly devour any book he could get his hands on that dealt with aviation heroes and aircraft. On graduating from the Kashira school in 1952, he was drafted into the Soviet Army and immediately applied for admission to a pilot school.

To his delight the application was accepted and he was sent to a military aviation school at Pavlograd in the Ukraine. Despite all of his recent studies, the young man from the Tula region now knew that he wanted more from life than to simply work with farm machinery.

Khrunov soon became proficient at flying training aircraft, leading his instructor Lieutenant Baskakov to report that steady progress was being made. "He flies competently, intelligently," he wrote. His commanding officer later added: "Fighter pilot Yevgeny Khrunov flies excellently." While at the school, Khrunov befriended another young pilot who would figure prominently in later years, Viktor Gorbatko.

The following year, 1953, Khrunov and Gorbatko applied for a transfer to the Serov (Bataisk) Higher Air Force School in Rostov Oblast, southwestern Russia. Both men were successful in their applications, and it was here that he nurtured and developed his passion for flying. On 16 June 1956 he graduated from the school and a week later was awarded the rank of lieutenant. On 22 August he and Gorbatko commenced a posting as fighter pilots with the 86th Guards Fighter Regiment, 119th Fighter Division of the 48th Air Army, located in the Odessa Military District of Moldavia. He would be promoted to senior lieutenant on 6 August 1958, attaining the rank of military pilot, third class, at the same time as Gorbatko, on 16 October.

While assigned to the fighter division, Khrunov met and married Svetlana Sokolyuk, a teacher, with the whole regiment attending the wedding. Their first son Valery was born on 13 July 1959. By this time Khrunov had been giving serious thought to applying for a transfer to the Air Force Academy, but he was in two minds

Khrunov (left) shares a humorous moment with Pavel Popovich.

about making the transition as he knew it might mean less flying, which he loved. However, fate would step in and make the decision for him just two months later.

It was September 1959, and the regiment was quartered in their summer camp when Khrunov unexpectedly received a summons from his commanding officer. Arriving at the office, he was introduced to a stranger and told that he should answer the man's questions honestly. Those questions, according to Khrunov, came thick and fast. "Do you want to study? How would you like to fly aircraft of unheard-of models? You wonder what it would be like: how shall I put it? If we compare the plane you're now operating with a hand scooter, then we might say that the aircraft I have in mind is like a racing motorcycle. Grasp the difference?"

Khrunov was a little bewildered, and hesitated. "But I don't feel fit as yet ..." he began. The stranger patiently held up his hand. "Of course you don't. But you will, with time, if you study and try hard." He would later find out that Gorbatko had also been interviewed, and later they were both subjected to a medical examination, which they passed.

VLADIMIR KOMAROV

Fellow cosmonaut Pavel Popovich would later remark of Vladimir Komarov that: "He was one of the oldest in our group [and] was already an engineer when he joined us, but he never looked down on the others. He was warm-hearted, purposeful and industrious. Volodya's prestige was so high that people came to him to discuss all questions—personal as well as questions of our work."

Vladimir Komarov.

Vladimir Mikhailovich Komarov was born in Moscow on 16 March 1927, growing up in an old house in that city, on the Third Meshchanskaya Street. The only son of Mikhail Yakovlevich and Kseniya (Sigalyeva) Komarov, he had a 12-year-older sister Matilde as his only sibling. His father was a labourer who supported his family in a variety of low-paid jobs as a yard-keeper, sidewalk sweeper, janitor, porter, storekeeper, fitter and turner.

In 1935, at the age of eight, Vladimir began his education at local secondary school No. 235, achieving good marks in mathematics, but doing poorly in writing essays. His schooling was suspended for a time in 1941 due to the war, so he worked as a labourer on a collective farm. Fascinated by aviation and aeroplanes from the age of seven, as a youth he had several self-made model aircraft resting on top of an old wardrobe in his bedroom. Inside the wardrobe he also kept a much-loved collection of flying magazines and pictures of different aircraft, as well as a propeller he had carefully fashioned from a tin can. He loved nothing better than to climb up the stairs of his house, through a narrow door on the top landing, and cross a dark, silent attic to a dormer window. From here he enjoyed a spectacular view of Moscow, but this did not fascinate him as much as watching the war planes that flew over the city. At an early age he could even identify the different fighter or bomber aircraft before they came into view, simply by the sound of their engines. As he grew older, his infatuation with aircraft also continued to grow.

In July 1942, 15-year-old Komarov graduated from school and enrolled in the 1st Moscow Special Air Force School for potential air force pilots. Around this time the family found out that Mikhail Komarov had been killed in an unknown war action. Due to the war, the piloting school would be temporarily evacuated to the Tyumen Region in Siberia. Here, apart from learning the elementary principles of flying, students also studied such diverse subjects as geometry, zoology, physics and learning a foreign language. Later the cadets would move on to actual flight instruction and make their first solo flights at the 3rd Sasov Air Force School in the summer of 1945. Graduating as a pilot with honours in July that year, he was just too late to take part in the action as the Second World War came to an end. His concerned mother Sigalyeva hoped that he would now give up his thoughts of a career in aviation, but it was already in his blood and he wanted desperately to become a fighter pilot and later a test pilot. The following year he completed the first-year course at the Chkalov Higher Air Force School in Borisoglebsk, on the Vorona River in Voronezh Region. His training then continued at the A.K. Serov Military Aviation College in Bataisk, receiving his lieutenant's wings at a graduation ceremony on 10 December 1949. Sadly, his mother would not live to see her son graduate; she had passed away seven months earlier, on 30 May.

Komarov once said of flying, "Whoever has flown once, whoever has piloted an airplane once will never want to part with either an aircraft or the sky."

From 31 December 1949 the young graduate served as a fighter pilot with the 383rd Regiment (IAP) of the 42nd North Caucasian Fighter Air Division, operating in the Grozny military district. He married his sweetheart Valentina Yakovlevna Kiselyova, a graduate of the history and philology department of the Grozny Teachers' Training Institute, on 21 October 1950. The following year, on 21 July, their son Yevgeny was born. Komarov was made a senior pilot with his regiment on 28 November 1951. Promoted to senior lieutenant on 17 April 1952, he was next assigned as chief pilot to the 486th Regiment of the 279th Fighter Air Division based in the Prikarpate Region on 27 October that year. He would remain in that post until August 1954, when he was enrolled for the next five years at the N.E. Zhukovsky Air Force Engineering Academy, undertaking higher education studies in the faculty of air armament. On 10 December 1958, while he was still studying at the academy, Valentina gave birth to their second child—a daughter they named Irina.

Gagarin and Komarov enjoyed hunting together.

The years of study were paying off: on 31 August 1959 Komarov's rank became that of senior engineer-lieutenant. Following his graduation from the academy he took on work, beginning 3 September 1959, as an assistant leading engineer at the Central Scientific Research Institute of the U.S.S.R. Ministry of Defence, in the town of Chkalovsky, testing aircraft equipment. During this time he was a test pilot, third class, in the Institute's Department 5. On 3 September that year he received a further promotion to engineer-captain, and was also invited to undergo examinations for enrolment as a cosmonaut candidate.

ALEXEI LEONOV

Alexei Arkhipovich Leonov was born on 30 May 1934 into a working family living on the outskirts of the remote village of Listvyanka, in the southwestern Siberian Tisulsky District of what is now known as the Kemerovo Region. He would be the eighth of nine surviving children—three boys and six girls—born to Yevdokia (nee Sotnikova) and Arkhip Alexeievich Leonov, an electrician, miner, zoo technician and railroad worker. Three other children had died in infancy. Alexei's maternal grandfather had originally been exiled to Siberia by the Czarist regime for taking an active part in the 1905 revolution, and for organizing illegal workers' strikes in his Ukrainian home city of Lugansk.

Political turmoil would cause Alexei's family to become destitute and virtually forced out of their home in Listvyanka by vindictive neighbours when he was aged just three. The children were forced to quit school, and his father was thrown into gaol without so much as a trial. "He was sent to prison on the strength of the false testimony

Alexei Leonov.

of a corrupt co-worker," Leonov wrote in his co-authored 2004 autobiography, *Two Sides of the Moon*. "He was not alone; many were being arrested. It was part of a conscientious drive by the authorities to eradicate anyone who showed too much independence or strength of character. These were the years of Stalin's purges. Many disappeared into remote gulags and were never seen again."

For the next few years the large family would live under cramped circumstances with one of Leonov's older married sisters in Kremerovo, several hundred miles away. Fortunately Arkhip was eventually absolved and even compensated for wrongful imprisonment, and was able to re-join his family in Kremerovo, where he took on work in the local power plant.

As he grew up, young Alexei found he had a talent and liking for drawing. He used to earn a little extra bread for the family by carefully drawing pictures of flowers on the white ovens in neighbourhood houses, and would later paint canvasses depicting mountain and woodland scenes that people would buy to hang on their bare walls.

But, while he loved painting and drawing, Alexei had already decided what he wanted to be: a pilot. When he was six years old he had met a Soviet pilot staying at a neighbour's house, and the man greatly impressed him with his stories and demeanour. "I remember how dashing he looked in his dark-blue uniform with a snow-white shirt, navy tie and crossed leather belts spanning his broad chest," Leonov would later recall. Meanwhile his older brother Pyotr, an aviation student, also helped inculcate an interest in flying in him. He began building model aeroplanes, as well as studying aircraft engines and the theory of flight.

In addition to being a superb pilot, Leonov was an accomplished artist.

In 1948 the family heeded a plea from the government for more families to settle in Soviet-occupied East Prussia and moved to the city of Kaliningrad, situated on the Baltic coast between Poland and Lithuania. Here, in 1953, Leonov would finish his secondary education at school No. 21 with excellent marks. During the spring holidays he applied for admission to the Academy of Arts in the Latvian capital of Riga, but after visiting the academy he reconsidered (mostly due to the unexpectedly high cost of tuition), enrolling instead at a preparatory flying school in Ukrainian Kremenchug. In January 1955 he flew for the first time with an instructor and in May that year made his first solo flight. He then undertook a two-year advanced course for training as a fighter pilot. Meanwhile, he also took up part-time art studies at the Riga academy.

Leonov graduated with an honours diploma and the rank of lieutenant from the Chuguyev Higher Air Force Pilots School on 30 October 1957 and then served as a fighter pilot, first with the 113th Parachute Aviation Regiment, 10th Engineering Aviation Division of the 69th Air Army operating in the Kiev military district, and from 14 December 1959 as chief pilot with the 294th Reconnaissance Regiment of the 24th Air Army, operating out of Altenburg in East Germany.

Prior to taking up his basing in Germany, Leonov had proposed to his girlfriend Svetlana Pavlovna, an editor who had graduated from teachers' college, and they were married just two days later—the day before he was sent to Germany. In a very short time Leonov had logged an impressive amount of flight time (he had 278 hours to his credit when selected as a cosmonaut), and was highly thought of by his senior officers. Prior to his interview in October 1959 he had made 115 parachute jumps with varying degrees of difficulty, and earned the title of Instructor of the Military Air Forces for paratroop training.

GRIGORI NELYUBOV

Grigory Nelyubov's life—once so full of zeal and daring—ended in irony and ignominy just 31 years after it began. Selected as a member of the first cosmonaut detachment in March 1960, he was a young man who might have blazed a trail of glory across the firmament. He brought the skills and daring of his MiG-19 piloting days to the cosmonaut corps, quickly becoming a star pupil and a leading candidate for one of the first flights. Brash and fiercely competitive, Nelyubov made a point of excelling at all exercises and academic tests laid before him. But he died in disgrace, killed when hit by a locomotive, spurned by the same nation that might once have embraced him as a colossus of the Space Age.

Grigori Nelyubov.

Although his official date of birth is given as 9 April 1934, Grigory Grigor-yevich Nelyubov was actually born a few days earlier on 31 March 1934—the date given on his gravestone in the Ukrainian town of Zaporozhe. The anomaly is not an uncommon one in Russia, and generally stems from con-fusing the birth and birth registration dates. His birthplace was the Crimean city of Porfiryevka, in the Yevpatoriya Region (now the Krym Republic), 55 miles north of Sevastopol. His father, Grigori Makarovich Nelyubov, had served in the border forces as a captain in the NKVD, while his mother Darya was a housewife. At the time Grigori (nicknamed Grisha) was their only child; his younger brother Vladimir would not be born until 1945.

In 1953, in his ninth year of secondary school, 19-year-old Grigori enrolled for flying lessons at Zaporozhe Aero Club, and a year later he not only completed his formal school education, but had likewise graduated from the aero club with 50 hours of flying in a Yak-18 training aircraft under his belt. From October 1954 he attended the Stalin Naval Aviation School No. 12 in Yeisk, southern Russia, undertaking a year's preparatory course.

After further studies in seaborne naval operations, Nelyubov advanced to the second phase of his flight training at the Order of Lenin Yeisk Naval Aviation School, named at the time after I.V. (Joseph) Stalin. Today, the city of Yeisk is home to an Air Force Academy and the V.M. Komarov (formerly Stalin) Higher Military Avia-tion College. Grigori Nelyubov would graduate as a naval aviator on 2 February 1957, with the rank of lieutenant.

From March to December 1957, Nelyubov served as a senior pilot in the 639th Fighter Regiment of the 49th Fighter Aviation Division, an arm of the Soviet Air Force's Black Sea Naval Fleet, first in Crimea's Sevastopol region and then on the

Nelyubov and his wife enjoy a quiet game of chess.

Kerch peninsula, also in the Crimean Ukraine. Following this, from 13 December, he was assigned as a senior pilot to the 966th Fighter Regiment of the 127th Fighter Aviation Division, again part of the Black Sea Naval Fleet, flying a MiG-19—the first Soviet jet fighter capable of supersonic speeds in level flight. It was a MiG-19 pilot who had first reported seeing an American Lockheed U-2 spy plane flying over Soviet territory that same year, but the pilot could not make up the 7,000 foot difference in altitude in order to try to shoot it down.

Nelyubov was promoted to senior lieutenant on 4 March 1959, and on 24 August that year would achieve the status of military pilot, third class. He remained with the 966th Fighter Regiment as a navy pilot until his selection as a cosmonaut candidate. This would require a service transfer to the Soviet Air Force, but that did not pose any problems for Nelyubov, although it was later discovered he did harbour thoughts of one day resuming flying with his MiG-19 unit. He would undergo medical screening along with Gherman Titov and Ivan Anikeyev, passing with ease.

The author of a 1990 biographical study of the entire cosmonaut team, Gordon Hooper, wrote that Nelyubov was regarded as "a remarkable person, an excellent pilot, a sportsman, and also stood out for his amazing vivacity, lightning reactions, charming nature, and his ready ability to make general conversation with people."

At the time of his selection, Grigori Nelyubov was married to the former Zinaida Kostina, a sports parachutist.

ANDRIAN NIKOLAYEV

One of his early ambitions was to be a lumberjack, and Nikolayev's short but stocky frame would certainly have been well suited to that profession. Instead he would realize a cherished boyhood dream: that of flying. Unlike many swaggering pilots of high-performance aircraft, Andrian Nikolayev (Andrei to his friends) was a quiet and reserved man. During their cosmonaut training, Gherman Titov would come to admire the man who later became his back-up pilot for Vostok 2. He would characterize Nikolayev as "amazingly calm and unruffled in any situation, and modest to the point of shyness," as well as "the embodiment of composure, which is necessary for the commander of a spaceship—a man of iron endurance and courageous determination."

Andrian Grigorievich Nikolayev was born into a peasant family on 5 September 1929 in peaceful Shorshely, a small village in the Marinsky-Posad Region of the Chuvash Autonomous Republic. His father Grigori was a stable hand on a local collective farm, where his mother Anna Alekseyevna also worked as a dairymaid. When Andrian was quite young his parents inexplicably changed their family name to Zaitsev. It was under this surname that Andrian and his older brother Ivan would grow up in Shorshely, as well as their younger siblings Pyotr and Zina, born and registered under that name. The situation seems to have involved a tax agent by the name of Nikolayev who had unexpectedly come to live in the village, but the reason behind the name change has never been publicly revealed. A few years after his father's death in 1944 Andrian would formally apply to have his surname revert back to Nikolayev and this was granted.

At the age of seven, Andrian's life changed forever one day when his father took him to a nearby airfield. He became completely enthralled by the sight and sound of fighter aircraft taxiing back and forth, and decided on the spot to become a pilot when he grew up. His fascination with aviation grew even stronger over the years, but following his father's death he and his older brother Ivan had to help his widowed mother and their two baby siblings as the family struggled to survive crippling post-war poverty. Things were so tough that Andrian told his mother he wanted to leave school and take up work to support his family, but Anna would not hear of this, admonishing her son to finish his education and make something of his life.

Andrian Nikolayev.

On completing secondary school Nikolayev entered medical school in Tsivilsk but soon decided he would rather be like his older brother and take on forestry as a career. He subsequently enrolled at the Marinsky-Posad Institute, graduating as a forest technician in 1947. Over the next three years he worked as a lumberjack and foreman in the Karelia area before being drafted into military service in April 1950. He decided to become a fighter pilot and entered the Air Force, where he completed courses in air gunnery and radio operation. He later served as deputy commander of a platoon of gunners and radio operators at the Kirovobad Higher Air Force School, flying on Tupolev Tu-2 bombers. The urge to fly now grew even stronger, so following a period operating as an operational air gunner he applied for a transfer to pilot training. This was approved and Nikolayev attended a flying school based in Chernigov, later transferring to the Frunze Higher Air Force Pilots School of Air Army 73 in the Turkestan military district.

Late in Nikolayev's training he demonstrated his skill and coolness under pressure when the jet engine flamed out while he was flying a single-seat MiG aircraft. His attempts to restart the engine were futile. Rather than taking the recommended option of bailing out, he elected to stay with the fighter and brought it down to a safe crash-landing in a field. He was uninjured, and the expensive aircraft was able to be repaired and returned to service. The trainee's calm proficiency under extreme pressure would receive high praise from his superiors, which later aided him in his application for cosmonaut training. After graduating on 29 December 1954, Lieutenant Nikolayev was assigned to the Moscow Air Defence District, flying MiG-15 fighter jets with the 401st (Smolensk) Air Regiment.

On 15 December 1957 Nikolayev became a senior pilot and squadron adjutant with the 401st Regiment, and was promoted to senior lieutenant on 30 April 1957. That same year he became a full-time member of the Communist Party. He achieved the rank of military pilot 3rd Class on 25 June the following year, and on 31 October 1959 was awarded his 2nd Class ranking.

Nikolayev (right) with
Valery Bykovsky.

Although a bachelor when he applied for cosmonaut training, Nikolayev was said to have been engaged when he was accepted. Unable to tell his fiancée what he was doing, and absent for long periods, she finally broke off the engagement in frustration.

PAVEL POPOVICH

Pavel Romanovich Popovich entered the world on 5 October 1930, the son of Roman Porfirevich and Feodosia (Kasyanovna) Popovich. The family lived and worked in the small Ukrainian village of Uzin in the Kiev Region.

"Ours is a working family," his father would later recall. "Since 1927 I have been a stoker at the furnace [in a sugar refinery]. My parents and all my relatives were farmers. Pavel's mother is also from a working family. I was twenty years old when I married Feodosia ... We have five children—three sons and two daughters. We tried to bring them up to have a liking for work, and not afraid to do any kind of job; to be ready to overcome difficulties of every sort." Roman Popovich could never have imagined that "any kind of job" would one day have special significance for his tousle-haired son Pavel, who, while normally cheerful, would grow up enduring the brutal German occupation of Kiev during World War Two.

There are defining moments in most people's lives, and for 11-year-old Pavel this occurred one day when he and his father witnessed a solitary Ilyushin 2 airplane being attacked overhead by several superior German fighters. The IL-2 was shot down, slamming into the ground a short distance away. Out of respect Roman and other villagers buried the pilot's remains, but that brutal act would remain a potent memory for Popovich. He wanted to fly, and exact revenge on the despised German invaders.

By this time the Germans had begun forcing children to attend school lessons they had organized, but in defiance Pavel would stuff cotton wool into his ears so he could not hear the words of their German tutors. As a consequence of this and other acts of rebellion he was dismissed from the school. With German troops rounding up local youths suspected of sabotage activities he soon faced the prospect of being sent to a slave labour camp in Germany. Alarmed, Feodosia took the unusual step of dressing Pavel in old frocks, passing him off as a girl until the immediate danger had passed.

Pavel Popovich.

Popovich in his study.

With Nazi forces in full retreat after catastrophic losses in the battle for Stalingrad early in 1943, life slowly resumed in the battered Ukraine. Pavel would assist his family financially by working as a herdsman while taking on part-time studies at night school to make up for his missed education. He would save precious money by not wearing shoes on his tough herdsman's feet until the first snows dusted the hillsides. He would later join his father at the sugar mill, performing manual labour.

In 1946, having graduated from the Belaya Tserkov vocational trade school, he moved east and took on further studies at the Magnitogorsk Industrial Polytechnic School in the Ural Mountain area. He would receive his diploma with high marks as a building technician in 1951 despite the lessons being given in Russian—an added difficulty for a young man whose native tongue was Ukrainian.

While studying at the polytechnic school an old passion was rekindled when he discovered there was a flying club nearby. He joined up, managing to fit in early-morning flight instruction before hurrying off to school. Graduation from flight school with high marks came in September 1951.

Drafted in the Red Army when he turned 21, Popovich cited his graduation from flying school in requesting service in the Soviet Air Force. He was assigned to a first-year course at the Stalingrad Military Pilot School near Novosibirsk, and from 26 September 1952 was a cadet at Military Pilot School No. 52 in the Far East village of Vozzhayevka. When this school was disbanded he was sent to the famed Myasnikov Air Force Flight School in Kacha, near Sevastopol. Popovich was awarded his lieutenant's wings on 30 October 1954 and would remain at Kacha until 25 December 1954, when he began service as a fighter pilot with the 265th Fighter Regiment of the 336th airborne division, flying in the Far East and Siberian regions.

One evening while on regimental duty in Siberia, Popovich attended a graduation ball and found himself conversing with an attractive, spirited 20-year-old pilot named Marina Lavrentevna Vasilyeva. She was from a working family—her father was a woodcutter—but she was also a proficient stunt flyer. They got on famously, and would be married in 1955. A daughter named Natalya was born on 30 April 1956.

Popovich was promoted to the rank of senior lieutenant on 24 April 1957. From 19 June that year he was assigned as chief pilot and squadron adjutant to the 113th regiment of the 22nd Air Force, then operating in the Karelia region.

On 4 October 1957, the Soviet Union launched the first Sputnik satellite into Earth orbit. "Beginning that day," he recalled, "I got the dream of becoming a cosmonaut." He next served as chief pilot with the 772nd Fighter Regiment of the 26th airborne division, also with the 22nd Air Force, and from 31 May 1958 flew as chief pilot with the 234th Fighter Regiment, 9th air force, operating over the Moscow region from an airfield in Kubinka. From 31 January until 10 October 1959 he also served as the squadron adjutant, during which time (on 30 March) he was promoted in rank to captain, and (on 12 May) achieved the status of military pilot, 3rd Class. From there, the stars awaited Pavel Popovich.

MARS RAFIKOV

In Roman mythology the god or war was Mars, the son of Jupiter and Juno. Sharing that stellar name, Mars Zakirovich Rafikov, the son of Zakir Akhmetovich and Marziya Galyautdinovna Rafikova, could not boast such an illustrious ancestry, but when it came to being at war with his superiors he certainly appeared to uphold the tradition. More than any other candidate, Rafikov blew away any chance he might have had of flying in space through a carefree insubordination, defying rules, drunken misbehaviour and womanizing.

Rafikov was born a Tartar national in the village of Begabad, in the Jalal-Abad region of Soviet Kyrgyzstan on 29 September 1933. His father was the supervisor of a large collective farm until he was killed in 1943 while trying to repel German forces in the battle for Kharkov. He left behind three children: Mars, his sister Clara (born 1935) and younger brother Rinat (born 1939). Mars apparently carried out some vital

Mars Rafikov.

work against the occupying German forces; in 1947 he was awarded an individual medal "for the valiant labour in the World War II, 1941–1945" in the Timurovtsev area of Kharkov.

After the war, Rafikov's mother found work as a teacher at a local school to support her three children, and would later become a nurse. In 1948 he would complete his seventh-grade studies at No. 3 Frunze School in Jalal-Abad (the school named after war aviation hero Mikhail Frunze) where one of his fellow pupils, 11 months younger Valentin Varlamov, also studied. By sheer chance, both would be selected in the first cosmonaut team in 1960. Rafikov would then attend an elementary air force flight training school in Leninabadsky, Tajikistan, graduating in 1951. Called into military service, he subsequently took on a two-year student course training on Yak-11 and Yak-18 aircraft at the Chuguyev Higher Air Force School for military pilots, which he completed at the Borisoglebskom base. Following this he would attend Air Force School 151 in Syzran, Kuybyshev Region, graduating on 29 December 1954 with the rank of lieutenant in the Soviet Air Force.

From 12 March 1955, Rafikov served as a pilot with the 472nd Fighter Regiment of the 52nd Aviation Division, where he flew MiG-17 jet aircraft. During that time he met and married the first of his three wives, Ludmila Voizechovskaya. The marriage would not last the distance, but it survived long enough to produce a son, Igor Marsovitch, in 1956.

Transferred on 18 January 1956 to the Soviet Air Force's 3rd Guard Fighter Regiment, attached to the 15th Fighter Aviation Division of the Soviet Anti-Aircraft Defence Force, Rafikov would continue flying, now patrolling the city and region of Orel in western Russia. On arrival at the air base that day he was delighted to be reacquainted with an old classmate from their days at the Jalal-Abad School, Valentin Varlamov. They would subsequently serve together in two regiments of the15th Fighter Aviation Division, and later in the year they would be joined by yet another fighter pilot also destined to become a member of the first cosmonaut team, Valentin Filatyev. Promoted to the rank of senior lieutenant on 30 April 1957, Rafikov would soon divorce his first wife and begin courting Olga Borisovsyaya, whom he would later marry. This second marriage would eventually produce two daughters; Zhanna Marsovna (born 1960) and Elmira (born 1965).

Andrian Nikolayev, Mars Rafikov and Pavel Popovich.

Becoming a military pilot, second class, on 12 February 1959, Rafikov would remain with the 15th Air Division until his interview and subsequent selection as a cosmonaut on 20 February 1960.

GEORGI SHONIN

Georgi Stepanovich Shonin was born on 3 August 1935 in the Ukrainian village of Rovensky, in the Lugansk (Voroshilovgrad) Region, but spent his childhood in the town of Balta. Its name coming from the Russian word for "pole-axe", Balta, founded in 1797, is situated on the Kodyma River nearly 130 miles north of the city of Odessa, and is home to factories involved in manufacturing processes for clothes, sheep skins and cheese, in addition to tile and brick kilns and other industries. During the Second World War partisan fighters would make their hideouts in the nearby forests.

"Zhora", as Georgi became affectionately nicknamed, was the second child born to Stepan Vasilyevich and Sofia Vladimirovna Shonin, an accountant. His older brother Oleg had been born a year earlier, and his younger sister Dzhuletta (Juliette) would arrive some six years after him, in 1941. That same year, a month after the outbreak of war, German invaders swept into their village, looting, burning and shooting local villagers. The Shonin family was one of many forcibly evacuated from

Georgi Shonin.

their home, but they had to do this without Georgi's father, who had been sent to fight on the front during the first days of the war. He would never return, being killed in action in 1941. The family then lived with Georgi's grandmother Mariya Petrovna in her Balta home. For a time they also shared the house with a Jewish family that the sympathetic grandmother was hiding, at great personal risk.

On 29 March 1944, the town was finally liberated and life could slowly begin returning to normal. Georgi was attending school in Balta, but for the first few years he was said to have been a mischievous child and his marks were poor. It wasn't until seventh grade that there was a marked turnaround in his attitude; he began knuckling down to his schoolwork and reading books, particularly those about ships and sailors. One day in 1950 he saw a small notice in a local newspaper announcing the formation of a special air force school in Odessa, and despite his earlier interest in becoming a sailor he decided to apply. It would lead him on a path to realizing both ambitions in a sense, when he became a naval pilot.

In August 1952, aged 17, Shonin successfully concluded a two-year course at the air force school in Odessa, following which he was sent to the Stalin naval aviation school in Yeisk where he also completed a preparatory course. He then attended Naval Aviation School No. 93 in Leningrad, and in the summer of 1954 was sent to the nearby seaport of Kronstad, the base of the Russian Baltic fleet. Here he would undertake interactive service exercises aboard the ships *Sedov*, *October Revolution* and *Admiral Makarov*. After passing his exams he was returned to complete a second course of studies at the naval aviation school in Yeisk.

Years later, Shonin would describe 18 April 1954 as one of the most memorable occasions of his life. "It was the day I went up in the Yak-18 aircraft for my first solo flight. The sun was shining brightly, and everything inside me was rejoicing. I worked accurately like an automatic machine. But I had to sweat a lot before this day of my dreams dawned. Once my grandmother, a person who had suffered a lot, said about pilots; 'They do not earn their bread easily.' It is not simple to become a pilot. Experienced and clever pilots can transform a young flier into a real aviator. I shall remain thankful throughout my life to Senior Lieutenant Polyakov, who taught me to understand aviation and flying."

Sometime in this period, Shonin married Lydia Fedorovna Shumilova, and their first child, a daughter Nina, was born on 9 September 1955.

On 2 February 1957, Shonin was awarded his lieutenant's wings after graduating from the naval aviation school. He was then posted to the 935th Fighter Regiment of the Baltic fleet, a naval arm of the Soviet Air Force, transferring on 27 December that

Shonin in early training.

year to the 789th Fighter Regiment of the 237th Air Division, based in the Murmansk area. From 17 March 1958 he served as the chief pilot of the 768th Fighter Regiment of the Northern Fleet's 122nd Air Division, based at Luostari in the Murmansk Region. Becoming a military pilot, third class, on 30 November 1958, Shonin was further promoted to senior lieutenant on 22 February the following year. It was here, while serving beyond the Artic Circle, that he would befriend another young senior lieutenant from another regiment named Yuri Gagarin.

"I often met the young, jovial energetic pilot in the officers' club and at the sports meets," Gagarin would later reflect. "We got to know one another closer and I learnt that this chief pilot also dreamt of outer space. I was able to pass the medical examination before him. When I came to know that he had also passed I was happy for my comrade from the bottom of my heart."

GHERMAN TITOV

As a young boy growing up in the southern Siberian Altai district of Russia, some 2,000 miles east of Moscow, Gherman Titov did not initially dream of becoming a pilot. Instead it was the heavens and the stars that held the greatest fascination for him, and he often wondered what it would be like to travel beyond our planet. He could not know that one day those childhood dreams would come true, and that he came tantalizingly close to being remembered as the very first person ever to be launched into space.

A young Gherman Titov.

Gherman Stepanovich Titov was born in a log cabin in the Koshikha district village of Verkhneye Zhilino on 11 September 1935. His father Stepan Pavlovich Titov was the local schoolteacher, and it is said (although later denied by him) that he named his only son Gherman as a salute to a lead character in his favourite Pushkin play, *The Queen of Spades*. The young boy learned to ski by the age of three, and would also grow to love ice skating, although at one time he ventured too far out onto thin ice and nearly drowned when the ice shattered beneath him.

Despite the fact that his father was also his schoolteacher, no favouritism was shown; if anything he was under more pressure than the other students to study and do hours of homework. The war eventually took his father away from the family for three years, which taught young Gherman the meaning of responsibility as the male head of the family, supporting his struggling mother Aleksandra Mikhailovna and baby sister Zemfira, born in December 1941.

One post-war spring day, aged 14, Titov was riding his father's bicycle when a hen unexpectedly ran onto the road and under his front wheel. "There was a collision," he later recalled. "I flew over the handlebars and landed on the ground. There was sharp pain in my left arm. I could not get up." His arm was broken, and he would leave hospital with it encased in a heavy plaster-of-Paris cast. The break would knit slowly, and when the cast was eventually removed Titov began exercising his arm, "thinking up all kinds of physical exercises that would completely prevent the break from having any permanent effects." The fracture should have prevented him from ever becoming a cosmonaut and flying into space, but it was a secret he kept close to his chest for many years.

Called up for service in the armed forces when he turned 18, Titov attended the Military Registration Office, where he was asked if he wished to go into the navy, or aviation. He quickly elected to undertake elementary flight training at an aviation school, but was worried that if anyone found out about his previously fractured arm it might preclude him from the service. Luck was on his side; he was passed fit and assigned to basic training at the 9th Military Aviation School in the Kazakh city of Kustenai.

Graduating from the aviation school in 1955 (albeit with a disturbing record for insolence and poor attitude to his studies) he was then sent for a further two years of training at the Stalingrad (now Volgograd) Higher Air Force School. His instructor

Titov in discussion with General Nikolai Kamanin.

was a man named Kiselev, a strict and uncompromising pilot. Under his guidance Titov learned to fly the Yak-18, then moved on to the faster Yak-11 and finally the MiG-15. There were many times during his training that Titov's outspokenness and occasional hot-headedness nearly caused his dismissal from the aviation academy. On one occasion his defiant attitude went beyond acceptable limits, causing a staff officer to formally demand his dismissal from the academy. He only just managed to survive this episode through the insistence of his instructor, Major Vladimir Gumennikov, who reported to the academy hierarchy that Titov was too good a student to lose. But it had been a salutary lesson for the young aviator. He would graduate from the academy on 11 September 1957 with a first-class honours degree, the status of military pilot, third class, and the rank of lieutenant. Coincidentally, graduation took place on his 22nd birthday.

From 5 November 1957 Titov would serve as a combat pilot, based at Siverskaya village with the 26th Guards Fighter Regiment of the 41st Air Division, Air Army 76, Leningrad Military District. From 28 October 1959, as a senior lieutenant (promoted 12 days earlier on 16 October), he became chief pilot with the 103rd Guards Regiment of the same airborne division. While serving at Siverskaya in the spring of 1958 he began dating a Ukrainian cook at the air base named Tamara Vasilyevna Cherkas, and they would be married just two months later.

VALENTIN VARLAMOV

His fellow cosmonauts would unhesitatingly describe Valentin Varlamov as a very experienced and talented man, with obvious technical capabilities, and one who was distinguished by his impeccable health. He loved sport and drawing, played the guitar, and was noted both for his success at mastering the precise sciences and his stubbornness. According to his cosmonaut colleague Georgi Shonin, Varlamov was easily one of the most promising members of the first team, and the only one who had no trouble at all with higher mathematics.

Valentin Stepanovich Varlamov was born into a working family in the village of Sukhaya Tereshka in the Talchinskyon district, Penza Region, on 15 August 1934. The son of Klaudia (nee Prokofevna) and Stepan Nikiforovich Varlamov, he would have a sister Olga and a brother Dmitri as his siblings.

Post-war, and for an unknown reason, the Varlamov family uprooted themselves and travelled to Jalal-Abad in southwestern Kyrgyzstan, situated a little north of the Uzbekistan border, in the foothills of the Babash Ata mountains. Jalal-Abad had earlier become renowned as a popular resort area because of its mineral and spa springs. Valentin, then aged 15, would continue his studies at School No. 3 in Jalal-Abad. By a remarkable coincidence one of his fellow adolescent scholars at the school was Mars Rafikov, 11 months older than Valentin. A little over a decade later, both men would become part of the first cosmonaut group.

At school, Valentin excelled at both his studies and on the sports field, but it seems the family made yet another move, as he would finish his tenth-grade education at School No. 1 in Solz-Iletsky in 1952. Solz-Iletsky is located in the Orenburg Region, at the southeastern end of the Southern Ural mountain range. He then undertook preliminary flight instruction at an Air Force introductory school in Western Siberia. From here he moved on to the Kacha Higher Air Force School and in 1954 progressed to the Stalingrad (now Volgograd) Military Aviation School for Pilots, graduating with a commission as a fighter pilot and was awarded his lieutenant's wings on 29 November 1955.

From 18 January 1956, Varlamov served as a fighter pilot with the Soviet Air Force's 3rd Guard Fighter Regiment, attached to the 15th Air Division, operating in the defence of the city and district of Orel. He had arrived at the station the same day as his old classmate from Jalal-

Valentin Varlamov.

Varlamov with his wife
Nina.

Abad's Frunze school, Mars Rafikov, although Rafikov had already been an operational fighter pilot with another regiment for the best part of a year. Later that year another future cosmonaut, Valentin Filatyev, would also join their fighter unit at Orel.

On 13 December 1957 Varlamov was made a chief pilot with the regiment, and 11 days later was promoted to senior lieutenant, some eight months after the more experienced Rafikov. He would achieve the status of military pilot, third class, on 10 October 1958.

At the time of his selection, Varlamov was married to the former Nina Federovna Dmitrieva, and would list as his hobbies drawing, playing guitar and singing.

BORIS VOLYNOV

A short, scenic drive from the spectacular beauty of Lake Baikal—the world's largest and deepest freshwater lake—the city of Irkutsk, the capital of East Siberia, was originally founded as a Cossack garrison in the mid–17th century. It would also develop into a setting-out point for expeditions to the far north and east. In the early 19th century Irkutsk became a primitive home in exile to a large number of rebellious but gentlemanly army officers formerly based in St. Petersburg, who had taken part in a poorly conceived coup against Tsar Nicholas I on 26 December 1825, thus earning for themselves the now-popular sobriquet of Decembrists. Once the ringleaders of the coup had been executed the 121 surviving mutineers were promptly tried and sent to serve out sentences in labour camps and prisons across Siberia, where many were later joined by their wives and families.

Prominently positioned on a tree-lined avenue named in his honour, a bronze bust of Yuri Gagarin proudly gazes out over the fast-flowing Angara River that sweeps through Irkutsk, but one would seek in vain for any similar monuments or tributes to the locally born Boris Volynov. By the time he eventually flew his first space mission in 1969 some eight years had elapsed since Gagarin's epic flight, and the

Boris Volynov.

Soviet space effort to fly to the Moon was rapidly waning ahead of America's impressive efforts. As well, the heroic status once accorded pioneering cosmonauts had greatly diminished, along with the public's interest in what was perceived as their less noteworthy exploits in space. It is therefore not surprising that Volynov is still not formally recognised in Irkutsk, but he is also a man who through temerity and persistence overcame many prejudices in order to achieve his goals, not only in his youth and as a cosmonaut, but in his capacity as the first Jewish person to fly into space.

Boris Valentinovich Volynov was born into a middle-class Jewish family on 18 December 1934. His birthplace of Irkutsk was situated within a sprawling, desolate area known as the Jewish Autonomous Region that had been created by the dictatorial Joseph Stalin, who harboured a particular hatred for Russia's Jewish population. The region was 5,000 miles east of Moscow, spartan and poor, but even at that great distance still not remote from Stalin's relentless purges, arbitrary murders, arrests and imprisonment of political leaders who might pose a perceived threat to his brutal leadership.

Tragically, his father Valentin Spiridonich Volynov died when Boris was only three years old. While he hardly got to know his father, his mother Yevgeniya Izrailyevna was a qualified paediatrician, and so they were able to live in relative comfort. Before the outbreak of World War Two, he and his mother left Irkutsk and moved to the coal-mining and industrial town of Prokopyevsk in the Kuznetsk Basin region, where she had taken on work as a traumatologist. Yevgeniya would later train in emergency medicine and become a surgeon in the village where, because of the harsh and often dangerous and improper work practices that were part of everyday life for the inhabitants, her services were always in demand. Post-war, Yevgeniya would remarry, and Boris became the stepson of war veteran Ivan Dimitrievich Korich.

Initially, young Boris had resigned himself to the fact that his destiny lay in the grim local coal mines, but he soon became fascinated in books about Soviet aviation heroes and even began designing small propeller-driven engines, which he would successfully use on lightweight wooden aircraft models. His mother, on the other hand, patiently dismissed any romantic dreams he harboured about a career in aviation, urging him instead to consider enrolment in medical school. Then, with ten years of schooling behind him, a determined Volynov took the first step in his future plans by secretly writing to the regional Komsomol committee, asking for a letter of recommendation to an air force school. To his delight, this was granted. On the appointed day, and knowing that his mother would never agree to letting him go, a guilt-ridden but excited Volynov stole out of his house and caught a train carrying several other young recruits to the flight training school. At the next rail station the locomotive's water tank had to be refilled, so he disembarked during the lengthy stop and sent his mother a telegram, begging her to understand and pleading that flying was the only career he wished to pursue. Much to his relief his mother would later forgive him for his deception. As his training continued he not only wrote to her often, but would also send long, affectionate letters to his childhood sweetheart back in Prokopyevsk, Tamara Fedorovna Savina, who was studying to be a metallurgical engineer. In the course of this long-distance correspondence their affection deepened and they decided to marry.

After graduating from flying school in 1952, Volynov was enrolled at the Stalingrad Military Aviation School for advanced training, serving under decorated squadron commander Major Viktor Ivanov. Another trainee at the school was a brash young pilot named Gherman Titov. After four years' training Volynov received his wings, as well as a glowing commendation from Major Ivanov, and was despatched to the 133rd Air Division of the Soviet Air Defence Forces at Yaroslavl in northern central Russia. Tamara would join him here in 1957, and while she sought out a local job they set up their first home in a small rented attic near his air base.

Volynov with Yuri Gagarin.

Volynov would be promoted to full lieutenant on 14 March 1958, and soon after Tamara told him she was pregnant. They were married on 9 May, following which his new wife would travel to the comforts of her mother's home in Prokopyevsk to await the birth. Together with her new baby she would later return to the air base and introduce Volynov to his son Andrei.

In the fall of 1959, now a highly-regarded and popular senior lieutenant in the Air Defence Forces, Volynov was asked to attend a mystery-shrouded interview at the base, following which he was sent under equally secret orders to Moscow, where he would find out that he was under serious consideration for the first cadre of Soviet cosmonauts.

DMITRI ZAIKIN

Dmitri Alexeyevich Zaikin was born on 29 April 1932 in the small agricultural town of Yekaterinovka, situated 14 miles northeast of Salsk in Rostovskaya Oblast, in southeastern Russia.

Dmitri started school in 1940, before the Great Patriotic War reached Russia, but the conflict subsequently caused his family to move several times. His father, Alexei Gavrilovich Zaikin, a senior lieutenant in an artillery unit, was killed during fierce fighting against the Germans at Stalingrad on 12 December 1942. His mother Zinaida Vasilyevna (nee Grinenko) would later remarry, to Anatoli Kononovich, a tractor operator and chauffeur.

Dmitri Zaikin.

After he'd completed seventh class at a school near Salsk, Zaikin was involved for a time in horse breeding before attending the 10th Rostov Air Force School to learn the basics of flying, graduating on 18 August 1951. A keen gymnast, he then entered the Chernigov Air Force Pilots School at Armavir, a rail junction town on the line from Rostov-on-Don to Baku. At Armavir he had learned to fly the dual-seat Yak-18 trainer aircraft, and was then selected from 50 students to attend another Air Force Pilots school in the city of Frunze, named after Bolshevik military leader Mikhail Frunze. The city, renamed Bishkek in 1991 following the break-up of the Soviet Union, is now the capital of Kyrgyzstan. Zaikin would graduate as a lieutenant from this school on 29 December 1954, together with another future cosmonaut, Andrian Nikolayev.

From 24 February 1955 Zaikin served as a pilot with the 439 IAP Regiment of the 50th Soviet Air Force division, and from 26 May 1957 with the 163 IAP of the 144th SAF division. On 5 November 1957, now with the rank of senior lieutenant, he served as chief pilot with the 95th IAD of the 26th SAF in western Belorussia. He achieved the distinction of military pilot, third class, on 16 April 1958.

One day in the fall of 1959 some dignitaries from Moscow paid a visit to his unit, and Zaikin, together with a fellow fighter pilot, was requested to attend an interview. While the other pilot waited outside, he was ushered into a room. What he was told by the mysterious visitors came as a complete surprise, as he recalled during an interview at Star City with Soviet space researcher Bert Vis on 13 August 1993.

"Initially they asked us if we wanted to become test pilots. I said that every pilot dreams about that. Then they asked me, 'and if we asked you to fly into space?' I said I only knew about the first satellite, and Laika, and that Tsiolkovsky had written about space. They said, 'We want you to be a cosmonaut, we will ask you to come'."

He must have impressed the interviewers, because he was then directed to attend a further meeting of the State Commission in Moscow.

"We were asked not to tell anyone about why we had been asked to come; we were asked to invent some story why we had gone there. There were three colonels doing the meeting and we were senior lieutenants." Zaikin then chuckled as he

Zaikin undergoing
fitness checks.

recalled the excuses he gave at the time. "I invented the story that I had had some disciplinary problems during my vacation, but nobody believed me."

"The meeting was in September 1959 and when I was asked to come again in January 1960, my squadron commander asked me what I was going to do ... did I want to be a cosmonaut? He knew about it because some other people who had been to Moscow too, but who had declined the offer to become cosmonauts, had told everybody about it ... about the commission ... everything. So the commander knew where I was going.

"He asked me what I would do if I [was not] selected, as he was certain I would be dismissed from the Air Force altogether. I told him that I would leave the army to enrol at the polytechnical institute in Novocherkassk [in Rostov-on-Don]."

Dmitri Zaikin was not identified as a cosmonaut until 1977, when Georgi Shonin published his book *The Very First Ones*. His full name and other details were not released until *Izvestia* published a series of articles in April 1986 commemorating the 25th anniversary of Gagarin's flight. He and his wife Tatyana (nee Sukhoryabova), who was born 10 August 1941, have two children; Andrei Dmitrivich (born 5 September 1962) and Denis Dmitriyevich (born 13 March 1971).

REFERENCES

The biographies above were principally sourced from many publications, and the authors wish to acknowledge them equally. In alphabetical author order they are: *Gherman Titov: First Man to Spend a Day in Space* by Pavel Barashev and Yuri Dokuchayev; *Cosmonaut Yuri Gagarin: First Man in Space* by Wilfred Burchett and Anthony Purdy; *Gherman Titov's Flight into Space* also by Wilfred Burchett and Anthony Purdy; *Who's Who in Space (The International Space Year Edition)* by Michael Cassutt; *Starman: The Truth behind the Legend of Yuri Gagarin* by Jamie Doran and Piers Bizony; *Cosmonaut No. 1* (serialized in *Izvestia*, 2–6 April 1986, extracts translated by Jonathan McDowell); *Into that Silent Sea* by Francis French and Colin Burgess; *Our Gagarin* by Yaroslav Golovanov; *The Rocket Men: Vostok and Voskhod, the First Soviet Manned Spaceflights* by Rex Hall and David J. Shayler; *Russia's Cosmonauts: Inside the Yuri Gagarin Training Center* by Rex D. Hall, David J. Shayler and Bert Vis; *Men and Women of Space* by Douglas B. Hawthorne; *The Soviet Cosmonaut Team* (1990 edition) by Gordon Hooper; *Sons of the Blue Planet (Syny Goluboi Planety)* by L. Lebedev, B. Lykyanov and A. Romanov; *Soviet and Russian Cosmonauts 1960–2000* by I.A. Maranin, S. Samsutdinov and A. Glushko; *Russians in Space* by Evgeny Riabchikov; *Two Sides of the Moon* by David Scott and Alexei Leonov; *Cosmonauts of the USSR* by V.A. Shatalov and M.F. Rebrov; *Challenge to Apollo: The Soviet Union and the Space Race, 1946–1974* by Asif Siddiqi, and *First Man in Space: The Life and Achievements of Yuri Gagarin* edited by Nikolai Tsybal. All of these fine works have been fully detailed in other references within this book. Much appreciation is also extended to Dr. Vladimir Golyachovsky, author of *Russian Doctor* (St. Martin's/Marek), New York, 1984.

4

Training days

Early in March 1960 the first of 20 cosmonaut candidates began arriving in Moscow. The first to enter the temporary training facility at Frunze airfield were Pavel Popovich and his wife Marina. Popovich would then offer to serve as an unofficial guide and quartermaster for the other trainees as they arrived. An ebullient character given to bursting out in song, he was actually a near-perfect person to greet and quickly familiarize his new colleagues with their temporary training facility and living quarters.

SETTLING IN

Valery Bykovsky—then still a bachelor—recalled that the day before he left his regiment, Chief Air Marshal Konstantin Vershinin had communicated with him, reminding him (and the other potential cosmonauts ready to leave for Moscow) of their responsibilities as spaceflight candidates, while warning them that the programme of preparation, developed by the Academy of Sciences, would be very complex. He gave them a realistic view of their new assignment, said the training would be difficult, even exhausting at times, and it would require diligence and discipline on their part. Vershinin gave them a clear picture of the expectations being thrust upon them, and left them in no doubt as to the difficulties of what they were now facing in their new role. As Bykovsky prepared to leave his regiment he followed tradition by giving away some personal possessions that his colleagues could use, although they only knew that he was being transferred to an undisclosed location. They knew better than to press for details. He therefore gave away to friends his air rifle, motorcycle and radio receiver.

Reporting for training, Bykovsky would find that things had been done in such an obvious rush that permanent housing had not been completed. He found himself quartered in a small, two-storey gymnasium building belonging to the Central Sports

Marina and Pavel
Popovich.

Club of the Army at the Khodynskoye Field, within the Frunze Central Airfield on
the northern outskirts of Moscow. Here he would be met and briefed by Popovich,
and that evening the two of them got together so Popovich could give his new
colleague what details he could of the training facilities and programme.

Wife of a cosmonaut

Meanwhile, further candidates began arriving. Beyond their own cosmonaut group,
one person who impressed a lot of them quite early on was Popovich's wife, Marina.
She was a renowned and admired, record-breaking aviator in her own right, and this
would lead to later speculation that she was about to become the first Soviet woman
to travel into space.

Marina Lavrentrevna Vasilyevna was born on 30 July 1931, the daughter of a
woodcutter. As a child, Marina's family had moved to Novosibirsk ahead of
advancing German forces, and she said her desire to become a pilot came about
when she learned of the bombings of Smolensk and how German troops were
rumoured to be shooting unarmed Soviet pilots parachuting to Earth after being
shot down. Having completed her schooling in 1947 she would apply several
times over the next three to four years for a commission in the Soviet Air Force.
Meanwhile, she joined the parachute section of an amateur flying club in
Novosibirsk.

At that time, she recalls, the Soviet Air Force had only three female pilot
regiments; one for fighters, one for bombers and one for transport aircraft. Finally,
in desperation after all her applications had been flatly rejected, Marina made her

way to Moscow and appealed directly to Marshal Kliment Voroshilov, chairman of the Presidium of the Supreme Soviet. He was greatly impressed by her drive and enthusiasm, and wrote to Lt. General Kamanin, then the Chairman of the Army, Navy and Air Force Voluntary Aid Committee, asking him to give her application a far better hearing, and if he found it to be suitable, to find a place for her on a flying course.

Kamanin agreed, the application was approved, and Marina was finally permitted to become a pilot. She later served in what would become the Seregin Regiment, where she piloted Ilyushin-14 and Antonov-26 aircraft, the type from which the first cosmonauts would carry out parachute jumps. She and her future husband Pavel Popovich first met at her graduation dance near his air force station in Siberia in 1951, and while there was a definite attraction they somehow

Marina Popovich was an accomplished pilot in her own right.

drifted apart. By the time she met him again in 1955 she was a major in the Soviet Air Force, a test pilot, and holder of 13 aviation world records. This time they would stay together, and were married that same year.

Marina Popovich recently recalled for the authors that when her husband was selected for cosmonaut training the news came very suddenly and unexpectedly. They had to pack in a matter of minutes and were transported to a new barracks complex [1].

Later arrivals

Pavel and Marina Popovich would spend the first three uncomfortable days waiting to be properly accommodated. Then Valery Bykovsky turned up, followed in turn by Anikeyev, Volynov, Gagarin, Gorbatko, Leonov, Nelyubov, Nikolayev, Titov, Khrunov, and Shonin. The others would arrive some weeks later.

The new arrivals were disappointed with their temporary living quarters, especially those with wives and young children. Alexei Leonov described the accommodation given to him and his wife Svetlana as "bunk beds in the corner of a volleyball court. We had to drape newspapers over the net in order to get some privacy," he recalled, "because another pilot and his wife were sleeping at the other end of the court" [2]. They were told things would soon improve. One of the new arrivals, Gherman Titov, also felt somewhat let down at the lack of actual spaceflight training facilities. "We might as well have moved into a training camp for the Olympics," he later dryly observed.

According to Bykovsky, their first formal function was a briefing at 9:00 AM on the morning of 14 March, headed by Dr. Vladimir Yazdovsky, Deputy Director of the Institute of Aviation and Space Medicine, who gave the introductory lecture. In essence, this described some of the physiological and stressful aspects of launching into and flying in space, and all the known factors affecting living organisms during spaceflight. Then the physicians took over, and on subsequent days explained in an abundance of detail what they knew about the effects of *g*-forces and weightlessness on the human body.

"Listeners began to be bored," reflected Bykovsky. They were straining at the leash, and had quickly tired of the seemingly endless talks on the known and possible effects of space travel. They were pilots, impatient and ready to fly. Then there was a change of direction, as specialized lecturers from Korolev's design bureaus began to discuss thematics involving the rockets and spacecraft they would fly one day. These included rocket engineering, the dynamics of flight and the construction of the spacecraft and its separate systems. The lectures were given by top designers and scientists including Konstantin Tikhonravov, Mikhail Bushuyev and Boris Raushenbakh, as well as experienced design bureau engineers, some of whom would one day become civilian cosmonauts themselves—Konstantin Feoktistov, Oleg Makarov, Vitaly Sevastyanov and Alexei Yeliseyev [3].

Gagarin's wife Valentina would later record in her memoirs for an article (penned by an unnamed ghost writer) that her husband's curiosity was insatiable at these lectures, and he would ask innumerable questions. "Vitaly Sevastyanov was an engineer at Sergei Korolev's design office who lectured the cosmonauts on the 'hardware.' And he recalls that the largest number of questions were asked by a stocky fair-headed young man with blue lively eyes and a winsome smile and by a solid and reticent captain with dark thoughtful eyes. The dark-eyed man was Volodya Komarov and the blue-eyed was Yuri" [4].

Among those now beginning their physical training were Anikeyev, Nikolayev, Nelyubov, Popovich and Bykovsky.

Popovich performing some
basic training exercises.

Before a proper training regime had been set up, including extensive parachute training, training centre commander Yevgeny Karpov had a simple philosophy directed at preventing the cosmonaut candidates from becoming idle in between lectures—physical drills and more physical drills. Each morning there would be outdoor exercises including cross-country runs carried out in all types of weather.

"We began with physical training," Karpov once explained. "Not all the flyers understood why. Some were puzzled, even irritated. Why all the suddenly intensified setting-up exercises every morning and all the gymnastics and track-and-field work? And why so much less attention to the centrifuges, vibrostands and other special equipment?"

"We had to do a lot of convincing to bring them around to seeing it our way. Our prospective cosmonauts had to change many of their notions. There are some people, young flyers especially, who regard any doctor with suspicion. To them he is the fellow who is cantankerous about trifles, always looking for something to latch on to.

Anikeyev also participated in
individual and team exercises.

Several of the latecomers to our group felt this way, annoyed at the medical checks
before and after training. 'More and more indicators! Endless checkups! You'd think
we were a bunch of guinea pigs!' "

Inevitably, Karpov said, there were constant checkups, and the doctors patiently
explained to their charges that they could not be avoided. "But the grumblers were
not convinced. The medics, they felt, were out to cook up some non-existent defects."

"Take Gherman Titov, for example. He would hear the doctors out, then often
do things his own way or ignore their instructions altogether. When they insisted that
he follow instructions, Gherman became irritated. But they would not give him an
inch. They argued with him calmly, even with a little humour. 'No one takes medicine
because he likes it. The best medicine for you is athletics.' Gherman agreed, but he felt
imposed upon by what he considered unnecessary medical supervision of his training
and personal life, limitations on his freedom."

"But then he underwent a change. It had nothing to do with his mood and spirit,
although they had risen a couple of degrees. There was something else that surprised

Wearing sensors to monitor his
condition, Popovich endured
the centre's fitness campaign.

him—he felt stronger. He seemed to weigh less; he walked and ran much more
easily" [5].

What the doctors did not know at the time was something Titov would only
smugly reveal many years later—that in his boyhood he had taken a bad tumble from
his bicycle and broken his wrist. Feeling ridiculous, he decided not to tell his mother,
and though the wrist eventually mended it would continue to cause him pain for years
to come. Through a strict exercise regime he had managed to strengthen the wrist
sufficiently that it was not detected when he applied to enter the Air Force, and he
also kept it secret during all the cosmonaut medical tests. He knew that if he'd
revealed the untreated break to the doctors he would have been immediately
disqualified. It was for this reason that he was more noticeable than others in
questioning the amount of exercise they had to perform as cosmonaut candidates [6].

"Not all cosmonauts were in perfect shape when we began their physical training," Karpov continued. "A bit of running or volleyball tired some of the men very
quickly. They complained of muscle pains. But they were soothed by the physical
training instructor [Boris Legonkov]. 'Don't worry about it,' he said. 'In a short while

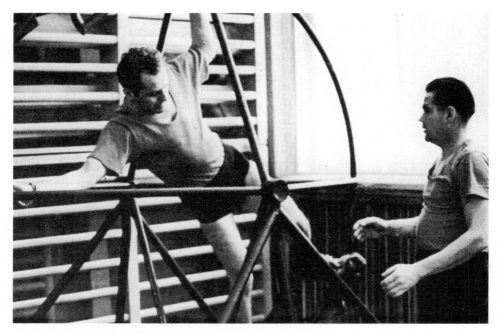

Titov, seen here with Nikolayev, was one of those initially arguing against the tough physical programme.

you won't know yourselves.' And one of the doctors added, 'Just note down your present blood pressure and pulse. You'll find it interesting to make comparisons after a while.'

"The suggestion was duly followed. In a month their pulse rates had slowed down, even though they were exercising harder. They began to look upon the doctor as an older comrade, a well-wishing friend or, what was more to the point, an experienced athletic coach. He was not only checking pulses, they felt, but helping the instructor to build them up. That broke the ice, and the cosmonauts and medics became close friends" [5].

Apartments at last

Shortly after their training intensified, those with families (and some senior managers) were moved to an apartment block located on Ultisa Tsiolkskaya in nearby Chkalovsky, which they were informed was only a temporary arrangement while construction of the Cosmonaut Preparation Centre was being completed at TsPK. They would actually remain in two apartment blocks in Chkalovsky for nearly six years.

Here, the Popovichs initially had to share a two-room apartment with the Titovs. There were no beds, Marina remembered, nor any other furniture. She smiled

Titov in training.

when recalling that Yevgeny Khrunov managed to get hold of two nails which he hammered into the wall of their apartment so he could at least have a place to hang his jacket! Grigori Nelyubov had also arrived there with his fiancée, a parachutist who at least had a job. They were married soon after.

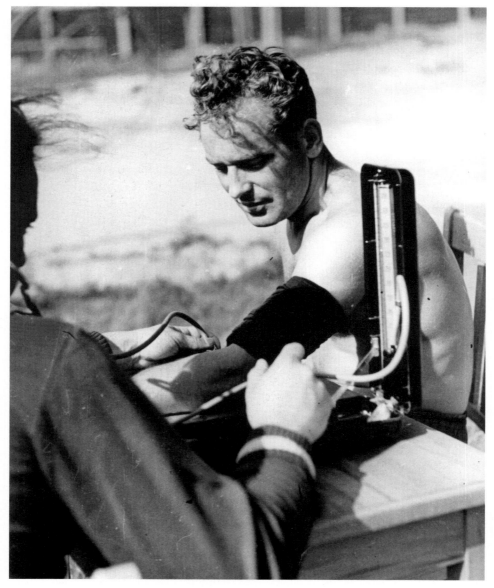

Measuring Titov's improving physical condition.

"I had just gotten an apartment when my family joined me in Star City," Boris Volynov told journalist Yuri Dokuchayev. "Tamara, my wife, was at a loss about what to do. We didn't have one piece of furniture; only a few carpets that were lying rolled up in the corner of the room" [7].

Bykovsky would marry
Valentina Sukhova within
months of their first meeting.

Later that year, as Gherman Titov told Dokuchayev, he and his wife would suffer a profound sadness. "Friendship is tested by deeds," he said. "When we future cosmonauts came together for the first time, it happened that the Gagarins and my family were next-door neighbours. Their oldest daughter, Lena, a toddler at the time, was born up north where Yuri had served with an Air Force unit. And my wife Tamara was expecting the first child. Our child [a boy they named Igor] died shortly after it was born. Yuri offered genuine support. Without being mawkish or heaving sentimental he simply behaved like a truly close and real friend. I was grateful to him and, though I did not know him very well at the time, began to like him a great deal" [7].

Valery Bykovsky, meanwhile, would meet his future wife Valentina Sukhova within weeks of his arrival at the training facility. Theirs would be a whirlwind romance and they would marry early in 1961 with several cosmonauts in attendance and his good friend Andrian Nikolayev acting as best man. Bykovsky's wife would later become a historian at the museum in Star City.

Basketball was a favourite training aid for the future cosmonauts. Shirtless in this photo are Filatyev (in cap) next to Nikolayev; at front are Nelyubov and Volynov.

Chief Designer Sergei Korolev would occasionally visit the construction site for the new training centre that would later become known as Star City. He wanted to check the layouts of the buildings, and requested a draft be drawn up for a new laboratory building which he would ensure was funded and fully equipped with the latest equipment. He also told them to include plans for an indoor swimming pool and outdoor gymnasium. "We've got to have some effective mock-ups," he also told Karpov. "And don't be bashful. If there's something you haven't got, ask for it. I'll find it."

"After a couple of months our cosmonauts were ready to admit that the athletic training was indispensable," Karpov observed. "They felt stronger and had much more endurance. The programme expanded when our sports field and outdoor gym

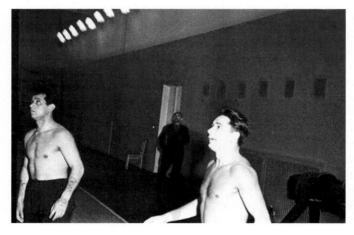

Nelyubov (left) and Bondarenko keep an eye on the action.

Bondarenko in white top
during gym training.

were equipped with jumping nets of the kind circus acrobats use for their incredible stunts. In a few weeks the men were doing some very complicated exercises on these trampolines, not up to circus standards, perhaps, but their performance was neat, bold and certain."

For the cosmonaut trainees, their day began at 7:00 AM with some light gymnastics, following which they would have breakfast at 8:30 AM. After this they would attend lectures until midday, and then sit down to lunch at 12:30 PM. The afternoons would be given over to physical training directly related to spaceflight, as well as vestibular exercises. They would also carry out exercises in a simple mock-up of a spacecraft, under the direction of their Air Force instructor, Colonel Mark Gallai. On these exercises they would sometimes carry out a simulated three-day flight in the mock-up. The men would also learn the principals of radio communications, the operation of survival equipment on land and in the water, and learn to carry out various functions while wearing a spacesuit.

The men would also be introduced to the high diving board above the swimming pool. Most were tentative at first; anxious about the height as they walked out onto the edge of the diving board, but after a few dives they had conquered their fear and were eager participants.

Despite the gruelling schedule, these was also some time given over to relaxation, as shown in top photo by (from left) Shonin, Rafikov, Gagarin and Zaikin. In the photo below, Rafikov, Gagarin and Shonin are joined by Yevgeny Karpov and an unknown woman—possibly Karpov's wife.

Initially, there was no spacecraft simulator for use in training the future cosmonauts, so the academic load was increased to include courses in radio and electrical engineering, spacecraft telemetry, guidance and navigation. By this time the candidates for the first flights knew that parachute recovery at the end of their mission would be an essential part of their training. "Nor were our cosmonauts enthusiastic about parachute jumping," Karpov added. "Like most flyers, they were accustomed

Parachute training begins.

to carrying the item as standard equipment, but they hated to have to use it. The first jumps, as expected, were the hardest." Many of the cosmonaut group were novices insofar as parachute jumps were concerned, although some of the younger members including Gagarin had already completed five mandatory jumps from the early phase of their Air Force flying careers.

The parachute jumps were carried out under the watchful guidance of a highly experienced and accomplished sky-diver, Colonel Nikolai Nikitin. Three years later, in May 1963, he died during a jump when he and another parachutist from the drop team, Alexei Novikov, collided heads at high speed, killing both men. The cosmonaut training jumps were carried out at an airfield in the steppes at Engels, with each man making several jumps each day. As Georgi Shonin would recall, "Parachute jumping in the course of six weeks was, I think, one of the most difficult aspects of the training" [8].

Those who made rapid progress were soon permitted to bail out over water. "Full-kit diving has no resemblance whatsoever to the kind you do in a swimming pool," said Karpov. "In spite of that, the whole group came through this baptism successfully; all their timidity washed away before they were through" [5].

Andrian Nikolayev takes the plunge.

All of the men had to learn how to land on the ground and on water, in both day-time and night-time jumps from low and high altitudes. "There were some critical situations in the air," observed space journalist Yaroslav Golovanov, "but being in full command of the parachute, the men were able to handle these situations well. For instance, Ivan Anikeyev was able to come out of a deep spin unscathed, and Gherman Titov did not lose his nerve when the main parachute did not open. Another important thing was that the strenuous training made the men close friends" [8].

Nikolayev soars to the ground.

Unfortunately for Belyayev, he broke his ankle during parachute training in August 1960, which would delay his training for almost a year, ruling him out of an early Vostok mission.

While their husbands went off to Engels to carry out initial parachute training, Marina Popovich and Tamara Titova managed to obtain a floor-polishing machine that they used to clean other people's floors for 20 roubles a time. In this way they were able to make a little money for domestic necessities, and prove their

A beaming Gherman Titov after making a successful jump.

Commemorative plaques indicate the early residences of two pioneering cosmonauts. The left one reads: In this house, from 4.10.1960 until 13.4.1966, lived Pilot-Cosmonaut of the USSR Hero of the Soviet Union Colonel Yuri Alekseyevich Gagarin who on 12 April 1961 made the world's first flight in space on board of the satellite ship "Vostok". The right plaque reads: In this house, from 9.10.1961 until 28.5.1966 lived Pilot-Cosmonaut Twice-Hero of the Soviet Union, Engineer-Colonel Komarov, Vladimir Mikhailovich.

independence to their husbands when they finally returned. Marina would also take the opportunity to enrol in a Moscow flying club and renew her proficiency in aviation [1].

Nikolai Kamanin would later write that Marina Popovich's flying seemed to cause some problems for her husband, especially later on when he began specific mission training for Vostok 4. Distracted by what she was doing, he failed an important test and asked to be allowed to sit it again, but this was refused. As a result, Marina was asked to ground herself from flying activities until her husband had flown his space mission. There would be other conflicts resulting from her flying activities and his role as a cosmonaut.

As a footnote to this story, the cosmonauts would continue to reside in two apartment blocks in Chkalovsky for some years while more suitable, comfortable accommodation was being constructed within TsPK. After his flight in 1961 Gagarin was accorded the privilege of having the walls between adjoining two-room apartments in Ulitsa Tsiolkovskaya knocked down to create a more spacious home. The following year the other cosmonauts moved into another nearby apartment block in Ulitsa Lenina which offered them larger, three-room apartments. Gagarin, however, was quite content to stay where he was.

"They made us live in the flats, not far from the [N.E. Zhukovsky Air Force Engineering] academy," recalled Dmitri Zaikin, "and didn't allow us to leave the flats before we passed our exams. We did the exams together. Gagarin passed, majoring in aerodynamics; Khrunov majored in engine systems, and I passed, majoring in construction and integration." But he has very little evidence of those early training days. He and the other unflown cosmonauts were a State secret, and though photographs were taken of them, especially when Zaikin was later training for the Voskhod 2 mission, they were refused copies, and those photographs have nearly all vanished without trace (although a crew photo of him with Khrunov still exists). After the men's existence was made known many years later he asked officials of space museums for prints of these photographs, but they were nowhere to be found [9].

All of the cosmonauts from the 1960, 1962 and 1963 selections would finally move out of the flats in Chkalovsky and take up their new residences in the early spring of 1966 when the two training centre tower block apartment buildings known as *Dom* [House] *2* and *Dom 4* were completed.

Weightless training

Initially, the trainees could only undertake weightless training in the back of a two-seat MiG-15 flying parabolic curves, but these flights were regarded as unpleasant, uncomfortable and not a true experience in weightlessness, as they were tightly strapped in. It was known that America's astronauts were conducting their own training within the padded fuselage of a large, modified transport aircraft, evaluating the processes of eating, drinking and waste systems during brief periods of weightlessness.

Following complaints about the ineffectiveness of this MiG training, especially in comparison with that of the Americans, the cosmonauts were soon carrying out

Boris Volynov revels in the delights of weightlessness during a parabolic flight.

suited and unsuited parabolic flights aboard a specially modified Tupolev Tu-104 operated by the Gromov Flight Research Institute out of the Zhukovsky Air Base, 21 miles southeast of Moscow. These dizzying flights, generally lasting for around 90 minutes, would consist of a number of rapid dives. Once a speed of around 400 miles an hour had been reached, the pilot would pull the aircraft into an ascending curve until an altitude of 24,000 feet had been attained. The Tupolev's controls would then be set to idle, with inertia sending the aircraft on a parabolic arc, during which time weightlessness would be achieved for up to 30 seconds.

Prior to one such flight, Yuri Gagarin was given a fountain pen and asked to fill out a form attached to a clipboard before boarding the aircraft, which required him to write his full name, the date and place of his birth, and then append his signature. He would repeat this during one episode of simulated weightlessness, and again immediately after. Post-flight, he would comment to the flight leader that writing in weightlessness did not really cause any problems, describing it as just like in a movie. "You can't feel your fountain pen, nor can you feel your hands, but the letters run" [10].

Other training aids used by the cosmonauts at this time included a treadmill, which they called a "running path". One of those who showed the most athleticism and endurance on this was Grigori Nelyubov.

A LIFE-CHANGING INJURY

In July 1960, separate incidents occurred that would eventually lead to the loss of two members of the cosmonaut team. The first of these, on 16 July, would cause the removal of Kartashov from the cosmonaut team, while the second, occurring just eight days later, would similarly end the cosmonaut career of one of its most promising candidates.

End of a dream

When the cosmonauts were first introduced to the centrifuge as part of their training programme, no one seemed eager to volunteer. Ivan Anikeyev would be the first of the group to ride the g-producing device, known to the group as the "devil's merry-go-round". He first had to undergo an exhaustive medical examination involving checks on his temperature, pulse and respiration.

The gondola of the centrifuge was in the form of a Vostok ejection seat, and the trainees would be required to wear a prototype spacesuit. Everything about the test was presented as preparation for an actual launch, including a countdown and the word "Ignition!" as a button was pressed to set the centrifuge in motion. Leonov recalls being tested at around 14 gs. Ivan Anikeyev finally stepped forward and was strapped tightly into the seat. Sensors were then attached to his body to record his physiological reactions to the test. On the first run, the g-forces were kept to a minimum, but as he seemed to have encountered little discomfort the tempo was lifted. Again he was checked, but said he was fine.

"The third strain was a tough one," Karpov later observed. "Ivan could hardly move. It was no fun—the man weighed 130 pounds, with the centrifugal load he weighed 1,300. Your blood grows heavy then, your body feels as though it were full of

An unnamed candidate is shown strapped aboard the centrifuge.

lead, and the whole world around you seems to have sunk to the bottom of the ocean. When the chair came to a stop, Ivan stayed there for a while. He had to rest, waiting for the physical and nervous tension to ease. The doctors watched while his body accommodated itself. A quarter of an hour after the test the trainee feels as if a mountain has been lifted from him. Walking and moving about seem so easy that he thinks the added strength comes from the centrifuge. Actually, he's only reverting to normal" [5].

For 28-year-old Anatoli Kartashov, the future seemed not only assured but truly celestial as he was strapped into the gondola of the TsPK centrifuge in mid-July 1960. As one of the last of the cosmonaut trainees to join the group along with Varlamov and Rafikov just 11 weeks earlier, he had been

A medical condition would abruptly end Anatoli Kartashov's hopes of flying into space.

undergoing initial training, including parachute jumping, as the entire group prepared to move to the new, purpose-built cosmonaut training centre, TsPK.

Then a senior lieutenant, Kartashov was not only a superb pilot, but he had already been awarded a highly prized place in Sergei Korolev's advanced training group, or Vanguard Six as they were known. Under the specific command of Colonel Gallai these men would undergo an accelerated training programme as a special sub-group to the cosmonaut team, allowing them primary access to the training facilities, including a Vostok simulator which had been delivered to TsPK in early May. As Gherman Titov recalled in 1996: "There were 20 of us in the first team, and during the first month it was clear that all of us couldn't be properly trained because there were not enough training machines there. The equipment was still in the stage of preparation, and we had to study things from scratch. It became clear that only one person would be able to go. Six people were short-listed" [11].

Those initially selected in that advanced group of six were Gagarin, Kartashov, Nikolayev, Popovich, Titov and Varlamov. It was fully expected that from these six candidates would come the first Soviet cosmonaut (and possibly first person) to fly into space.

While one would logically feel that this elite group of six men had been chosen for their ability, determination and temperament during training, there is compelling evidence to suggest that the criteria for selection was less educational and more genetic. As weight was of critical importance for the Vostok craft, it was noticed years later that all six were the shortest and lightest men in the cosmonaut unit, as confirmed by Alexei Leonov.

"When they defined the group of six people ... they were identified in a very simple way. The first group were people shorter than 170 centimetres. Those longer than 170 centimetres were the second group. The first group was shorter than medium height: 160, 165, like Gagarin and Titov ... 167 Nikolayev ... 168 Popovich. The second group were taller and started working on a different type of spacecraft. I'm 174 centimetres and I was an instructor, and we taught each other. Bondarenko was an instructor for the Vostok ship."

Leonov explained that he was paired with Kartashov for certain medical examinations, including one that would test the fragility of their blood vessels. He later learned there was a suspicion among the medical staff at the time that Kartashov was prone to bleeding.

"A kind of suction cup is attached to the body and a vacuum is created by means of a special pump. And all of a sudden I could see how [Kartashov's] suction cup was filled with blood. I myself had three kinds of small dots, as if someone had stuck me with a needle. You could hardly see it at all—probably a magnifying glass would show it. But it was obviously clear that this person had very fragile blood vessels" [12].

Alerted to a possible problem, the medical staff would carefully supervise the centrifuge tests. Under the ever-watchful eye of Dr. Adilya Kotovskaya, these tests began at the Central Aviation Centre on 16 July. A specialist in pathologic physiology, she had earlier participated in pioneering biological investigations carried on rocket flights and Earth satellites, and had been instrumental in the medical examinations and selection of the cosmonauts.

Eventually it was Kartashov's turn, and he patiently submitted to all of the pre-run examinations, the securing of sensors, and then being dressed in the suit and helmet. Undoubtedly it was an exhilarating and very real preview of what it would be like to actually be a cosmonaut.

Everything went smoothly, and as the countdown ended the centrifuge lumbered into life, sweeping clockwise around the training facility as it rapidly built up speed. Unbeknown to Kartashov, a malfunction had developed in the mechanics of the centrifuge, causing it to exceed by several g's the permissible test load for this run.

When the centrifuge finally slowed and came to a halt, Kotovskaya's medical team unstrapped Kartashov and routinely conducted a post-test visual examination to check for any possible physiological trauma. To their relief everything seemed okay at first, until a vivid red area was noticed in skin above Kartashov's spine. A closer examination showed that this discoloration was in effect a mass of pinpoint haemorrhages. Meanwhile all centrifuge training was cancelled until the faulty apparatus had been checked and repaired.

The doctors then conducted more extensive tests on Kartashov, and it was finally determined that his body had become overstressed as a result of the excessively high g-load, causing internal damage leading to minor haemorrhaging down his back. After considerable discussion it was felt that Kartashov might have a weakness in his blood vessels. Once the centrifuge had been repaired and tested he would be subjected to further low-g rides and post-run tests. These somehow seemed to corroborate their original diagnosis.

The doctors decide

Kartashov was a strong and vigorous man in extremely good health, and he struggled to come to terms with the implications of this problem. Unfortunately, his condition was regarded as sufficiently problematic to have him immediately suspended from the Vanguard Six on medical grounds and replaced by Grigori Nelyubov. Devastated at this turn of events, Kartashov would nevertheless resume basic training with the other cosmonauts. Gagarin was one of those who pleaded with Kotovskaya for his friend's reinstatement to the group of six prime candidates, saying that Kartashov was "the best amongst us. He will be the first man in space." Sadly, his pleas fell on deaf ears.

However, there was worse news to come the following year when a final medical report resulted in Kartashov's dismissal from the cosmonaut team on medical grounds on 7 April 1961. In order to cover any possibly awkward questions, he was said to have left because of a "mild heart irregularity".

Some years later, fellow trainee Gherman Titov told journalist Yaroslav Golovanov that he felt the medical people were wrong, and had placed far too much emphasis on what was essentially a very minor problem. "With Toley Kartashov the medics overreacted," Titov stated. "He was a splendid pilot and he could have been an excellent cosmonaut. If he had undergone the tests which they carry out today, certainly he would have passed them" [8].

Following his dismissal from the cosmonaut team, Kartashov decided to keep flying and was assigned as a military pilot in the Far East. "According to the requirements that are expected from everyone, he wasn't fit," Leonov commented. "We did our best to get him settled . . . adjusted in life." Initially based at an airfield eight miles west of the Volga River city of Saratov, Kartashov subsequently served as a test pilot in a State-run scientific research institute of the Soviet Air and would retire from military service with the rank of colonel on 28 March 1985. That same year he took on a new career as a civilian test pilot in the Antonov aircraft design bureau in the Ukrainian city of Kiev, test-flying An-2 aircraft. It is obvious that the problems the medical team perceived as serious enough to be career-ending in TsPK had no influence at all in his subsequent work as a test pilot, first class. "An experienced test pilot," Leonov added. "He was involved in testing until the end of his career" [8].

In addition to his test-flying, Kartashov would also complete over 500 parachute jumps. Despite losing his place in the cosmonaut team so early on in his training he nevertheless remained in touch, visiting Star City on a number of occasions, most notably on the 40th anniversary of the group in 2000.

Anatoli Kartashov died in Kiev after what was described as a lengthy illness on 11 December 2005, aged 73. He left behind his wife Yuliya and two daughters: Ludmila, an economist born in 1960, and Svetlana, a dressmaker born seven years later.

THE TRAINING PROGRAMME

In addition to physical exercises and familiarization with the Vostok 3A spacecraft and its systems, other training procedures undertaken by the cosmonaut trainees were both complex and arduous, and would comprise the following physiological and psychiatric evaluations:

— Pressure chamber training at artificially created "heights" of 3–4 miles and 8–9 miles
— Centrifuge training from $3g$ to $8g$ for up to 40 seconds in a "head-waist" direction, then seven times from $7g$ to $12g$ for up to 13 minutes in a "chest-back" direction
— Training on staying alone for an extended period (10 to 15 days) in an isolation chamber in complete silence
— Familiarization and training in a thermal chamber at a temperature of up to 70°C from 30 minutes up to two hours
— Vibration stand training with a vertical vibration of $f = 50$ Hz, at an amplitude of 0.5 mm
— A number of parabolic flights simulating micro-gravity conditions
— Parachute training [13].

Georgi Shonin would be the first of the cosmonauts to spend a full day dressed in a spacesuit heated internally to a temperature of 55°C and with humidity set at 40%. Volynov would be the second to undergo this "steam bath" test, as he called it. After that it was the turn of Mars Rafikov, who spent three days in the suit and later declared: "At the end of the third day I became terribly sleepy. I dreamt of fountains, waterfalls and the sea" [8].

A simple accident

Medvezh'i Ozrea, or Bear Lakes, are shallow stretches of water located in pine forests 22 miles northeast of Moscow, located by the road between Balashinkla and Shchelkovo, and close to the cosmonaut training centre.

On 24 July 1960, three of the cosmonauts decided to spend a break in their training with a day at the lakes, relaxing, swimming and sunbathing. On arrival they found a picturesque spot on a grassy bank and stripped down, ready for a swim. The sport-loving Varlamov then began to kid around, suggesting they dive into the shallow water from the river bank instead of simply wading in. He was obviously

A simple accident would eventually rule Varlamov out of the cosmonaut team.

A few moments of levity between a bearded Pavel Belyayev and Grigori Nelyubov.

Life continued for the cosmonauts in training, with the marriages of (this page) Valery Bykovsky and (opposite) Dmitri Zaikin.

in a good mood, with a lot to look forward to as the first manned spaceflight moved ever nearer.

Varlamov had been promoted to the rank of captain the previous month, on 30 June, and knew he was firmly in line for one of the early Vostok missions—perhaps even the first. He was certainly up to the challenge. Bykovsky, meanwhile, was up to the challenge set by Varlamov as he dove head-first into the water, but grazed his head on the sandy bottom of the lake. When he stood up he cautioned the others that it was quite shallow and they should be careful.

Shonin then plunged in without mishap, also touching his head on the lake bed before surfacing. Varlamov was the next into the water, but he misjudged his dive and struck his head heavily on the sandy bottom. When he surfaced he was grimacing and cursing, and made his way to the river bank. He complained that his neck was really sore, and the others became increasingly concerned that he might have sustained a serious injury. They quickly gathered up their belongings and drove back to TsPK, where they took Varlamov straight to the hospital.

An examination showed that Varlamov had displaced a cervical vertebra in his neck and it was decided to place him in immediate traction. He spent several weeks patiently but morosely undergoing this treatment, for which he had to remain motionless. His misery was only relieved by visits from his fellow trainees, who brought him gifts, including a guitar so he could more easily pass the time.

Meanwhile, training continued for the remaining cosmonaut candidates. Here Dmitri Zaikin undergoes further tests.

Upon his release, and eager to catch up with his colleagues, Varlamov wanted to regain his former fitness and recommence light training, but this was not permitted until he had faced a medical commission.

Meanwhile, as Alexei Leonov told interviewer Bert Vis, extensive examinations and tests had shown up further anomalies and weaknesses in Varlamov's bone structure. The commission's report would devastate the young man—he was medically disqualified and was to be immediately removed from the cosmonaut group. The commission felt it could no longer risk having a trainee undergoing strenuous training exercises, including potentially violent centrifuge and parachuting activity, especially with any type of spinal weakness that might result in further and more permanent injury. They knew from training reports that he had obvious technical ability, had previously enjoyed impeccable good health, played sport, and possessed a strong determination to succeed. In fact, he had been one of the most promising candidates of the group.

However, there would be no stand-down period for Varlamov, and no possibility of rejoining the group at a later stage. He was formally discharged from the group on 6 March 1961, with no chance of appeal. Leonov told Vis that even if Varlamov had not injured his spine, the deficiencies in his bone structure would eventually have been discovered. "Sooner or later it would have shown," he stressed [14]. Varlamov's place in the core training group of six cosmonauts had already been assigned to Valery Bykovsky.

"Pulled by the stars"

Despite his profound disappointment, Varlamov did not leave the cosomonaut training centre after being forced out of the cosmonaut team. He grudgingly accepted that the doctors who'd prepared reports on his condition were probably correct, and were just doing their job, so he went to work in the TsPK. Up until the time of Gagarin's flight he served as deputy chief of the command point for controlling spaceflights. Later, beginning on 16 January 1963, he became a senior instructor for the cosmonauts at TsPK, specializing in astronavigation.

His cosmonaut colleagues did not forget that he was once a member of the group, and he would often be included in their free-time activities and holidays. As his fellow trainees began to fly missions of increasing length and complexity, he would always wistfully follow their progress, knowing that but for one rash moment at the lakes he should have been one of the first cosmonauts into space. But as his friends became world-famous, so they began to drift apart, and he subsequently retreated into the background and his own activities. Gherman Titov once reflected that his former colleague never really got over losing his place as a cosmonaut, saying "The stars above pulled at him."

Varlamov and his wife Nina Federovna (nee Dmitriev) would celebrate the arrival of a baby girl, Yelena, on 28 February 1964. He continued to work within various departments of the training centre, and was promoted to the rank of major on 18 July 1964. He would eventually achieve the rank of lieutenant colonel on 12 November 1968.

On 2 October 1980, aged just 46, Valentin Varlamov died suddenly as the result of a domestic accident, suffering a cerebral haemorrhage after slipping on the glazed tile floor of his bathroom and hitting his head. He is buried in the cemetery at Leonikha, a village near Star City. Following the loss of her husband, Nina Varlamov continued living and working in the mail room at Star City until her own death on 7 December 1988. Their daughter, and only child Yelena took on work as a telegraphist at Star City, and like her mother would also become a postal worker.

A near disaster

Valery Bykovsky also came perilously close to losing his place on the cosmonaut team in the early days of that first summer as a cosmonaut. By this time he and Andrian Nikolayev had become good friends. One day the two men were enjoying a break along with Georgi Shonin. Alcohol may or may not have been involved, but after a good-natured argument about determining visual compass bearings, Bykovsky threw off his jacket, discarded his shoes, and climbed to the lower branches of a pine tree. He described what he could see—a highway and buses—and then indicated what he thought were the main compass points. He was about to descend when Shonin chided him about being so low on the tree. Bykovsky was always one to accept a challenge,

Cosmonaut training took many forms, including games of ice hockey. Here (from left) Nelyubov, Titov, Filatyev, Khrunov and Komarov prepare for a match.

Cross-country skiing was also a fun physical activity. From left: Bondarenko, Shonin, Volynov and Komarov.

and climbed even higher. In a rash move he began to step out onto one of the branches, which suddenly cracked and gave way. Bykovsky fell, grasping another branch on his way down, which also broke off in his hand. He landed heavily on his upper back, and ominously lay there without moving as Nikolayev and Shonin rushed over.

Still clutching the broken branch, Bykovsky began to moan. Shonin dropped to his knees and began lightly slapping his cheeks, asking if he was okay. Soon after Bykovsky looked up at the white-faced Shonin and with a hint of a smile asked his friend why he had not caught him when he fell. He appeared to be only winded, and shortly after asked to be helped to his feet. His shoulders were sore, but nothing seemed to be broken, so later that day he asked the two men to give him a strong remedial shoulder massage, which seems to have helped. Much to his relief at a subsequent medical examination he passed as fully fit and nothing was ever mentioned of a shoulder injury [3].

As well, Alexei Leonov nearly lost his life towards the end of 1960 when he and his wife were passengers in a chauffered car that skidded on a patch of ice near the entrance of the training centre and plunged into a lake. Leonov not only managed to scramble free from the sinking car, but also rescued his wife and their driver. From that time on the stretch of water became known to everyone as Leonov's Lake.

At long last—Vostok

Winter would bring with it some different and more enjoyable methods of training for the cosmonauts—skiing, skating and high ski jumps.

The cosmonauts would also visit Korolev's design bureau to see an actual and heavily guarded Vostok spacecraft for the first time. They had studied drawings and trained in the simulator along with their test-pilot instructor, Mark Gallai, but this was their first glimpse of the real thing. Sergei Korolev was present, and he would allow the trainees to clamber into the confines of the Vostok 3A craft, touch the

Gagarin and Korolev: they would both soon make history.

controls and toggle the switches. "Who wants to get in and try it?" he asked. Almost immediately Gagarin stepped forward. "Allow me," he said, greatly impressing Korolev by removing his shoes before he stepped up the ladder and climbed into the craft. The men's excitement was evident, and they were full of questions for Korolev's designers and engineers as they familiarized themselves with the spherical descent craft. At the conclusion they were asked if they had any suggestions for changes to the craft. Several, particularly in regard to the rigidity of the seat, were made and noted and later actioned by Korolev.

Prior to one subsequent visit to the design bureau, Korolev mentioned that it was time to begin testing the men on their knowledge of the Vostok spacecraft, and he had arranged for a board of examiners to check what they had learned. The cosmonauts arrived on time, but the board of examiners was running late. To relieve the tension, Gagarin suggested, "Why waste time? Suppose I talk, and you correct me if I'm wrong." He once again removed his shoes and carefully clambered into the Vostok cabin and began a recital of the craft's construction, its equipment, and the conditions the pilot would be experiencing in orbit. Towards the end of his recital the board of examiners arrived, headed by Chief Air Marshal Vershinin. A nervous Gagarin was then asked to continue his recital for the benefit of the examination board, which he did with an impressive competence [5].

As far as Korolev was concerned, he had known since his first day with the cosmonauts who he wanted to be in charge of the first Vostok spacecraft in orbit, and that man was Yuri Gagarin.

REFERENCES

[1] Interview with Marina Popovich conducted by Rex Hall and Bert Vis, Star City, Moscow, 14 April 2007.

[2] David Scott and Alexei Leonov, *Two Sides of the Moon*, Simon & Schuster, London, 2004.

[3] Grigori Reznichenko, *Cosmonaut No. 5, Ch. 1 (1959–1962)*, Politizdat, Moscow, 1989.

[4] Valentina Gagarina (and an unnamed ghost author), "108 Minutes and an Entire Life," *New Moscow Times* (weekly English magazine), serialized from Issue 4 (April 1981) to Issue 1 (January 1982).

[5] Yevgeni Karpov (writing as Colonel Yevgeni Petrov), *Cosmonaut*, published c. 1962, unknown Soviet publication, from Rex Hall archives.

[6] Francis French and Colin Burgess, *Into that Silent Sea*, University of Nebraska Press, Lincoln, NE, 2007.

[7] Yuri Dokuchayev, "Yuri Gagarin," *Soviet Life*, April 1981, pp. 6–26.

[8] Yaroslav Golovanov, "Cosmonaut No. 1," *Izvestia*, 2–5 April 1986.

[9] Dmitri Zaikin interview with Bert Vis, Star City, Moscow, 13 August 1993.

[10] Olga Apenchenko, *Truden Put' Do Tebya, Nevo! (A Reporter's Account of Cosmonaut Training)*, Gosudarstvennoye Izdatel'stovo Politcheskey Literatury (State Press for Political Literature), Moscow, 1961.

[11] CNN, "Cold War Experience," interview with Gherman Titov, March 1996.

[12] *Pravda*, "Alexei Leonov, pioneering cosmonaut looks back," 26 June 2002.

[13] Gagarin Cosmonaut Training Centre (GCTC) website article: *http://gctc.ru/eng/gagarin/podgit_polet.htm*

[14] Bert Vis, interview with Alexei Leonov, Association of Space Explorers Congress, Edinburgh, Scotland, 17 September 2007.

5

Selecting the first cosmonaut

By the middle of January 1961, the elite group known as the Vanguard Six or primary cosmonauts—Bykovsky, Gagarin, Nelyubov, Nikolayev, Popovich, and Titov—had satisfactorily completed their three-day simulator tests, while at the same time concluding a series of exhaustive parachute and recovery training exercises. General Kamanin then informed the six men that over the next two days they would be undergoing a final and comprehensive "state examination" in order to judge each man's abilities and preparedness to fly. In less couched terms, this was it—the gold ring on the cosmic roundabout. The results of the exercise would be carefully evaluated by a special interdepartmental commission under the auspices of Kamanin. They would review all the results and from these apply a ratings list numbering the men from one to six, thus determining which of the cosmonaut candidates best deserved the unrivalled honour of being the Soviet Union's first man in space.

THE VANGUARD SIX

Those serving on Kamanin's selection committee were Major General Aleksandr N. Babiychuk (Chief of the Soviet Air Force Medical Service), Lieutenant General Vladilen Y. Klokov (Institute of Aviation and Space Medicine), Vladimir I. Yazdovsky (Institute of Aviation and Space Medicine), Colonel Yevgeny A. Karpov (Director of the Cosmonaut Training Centre), Academician Norai M. Sisakyan (Department of Biological Sciences of the U.S.S.R. Academy of Sciences), Konstantin P. Feoktistov (OKB-1), Semyon M. Alekseyev (Chief Designer of Plant No. 918) and Mark L. Gallai (test pilot at the M.M. Gromov Flight Research Institute and a senior cosmonaut trainer at TsPK). Additionally, Flight Research Institute Director Nikolai S. Stroyev would be present at the tests [1].

Five members of the Vanguard Six (in white overalls) undergoing medical examinations at the Air Force hospital near Moscow in 1960. Yuri Gagarin is absent from this photo.

Topping the list

Tuesday, 17 January 1961 was a big day in American history. President Dwight D. Eisenhower would prepare for a handover of the U.S. presidency to the dynamic and ambitious John F. Kennedy just three days later by giving a final State of the Union address to Congress, in which he warned of the increasing power of "a military–industrial complex". That evening, with a viewing audience of millions glued to their black-and-white television sets, he made his impassioned farewell address to the American people.

On the other side of the world, with freezing temperatures and deep snow drifts outside adding to their misery, the six cosmonauts most in contention for the first flight also faced one of the most challenging and concerning days of their lives as they prepared to begin the state examinations. In their first test, each of the six men would take turns spending up to 50 minutes inside a spacecraft simulator at the M.M. Gromov Flight Research Institute (also known as LII) in Zhukovsky, 25 miles southeast of Moscow. During this time they had to describe in detail the operation of the spacecraft's systems, instruments, and discuss the different phases associated with a space mission. They would then be grilled by members of Kamanin's committee on various facets of the craft and its operation, with particular emphasis on the question of orbital orientation.

Later that day the committee assembled behind closed doors and offered up their individual appraisals for discussion. Once the results had been promulgated, four of the candidates had received a rating of "excellent" (Gagarin, Nikolayev, Popovich and Titov), while Bykovsky and Nelyubov were rated as "very good."

The following day, 18 January, the harried candidates sat down to a comprehensive written examination. All of the scores were later checked and then incorporated into the previous day's results, after which the committee drew up its all-important priority list. The name at the top was Yuri Gagarin. Then, in order of priority, came Titov, Nelyubov, Nikolayev, Bykovsky and Popovich.

Gagarin, Titov and Nelyubov
photographed in Moscow's
Red Square.

Six becomes three

Now armed with the names of the top three candidates, this proficiency list effectively
gave Kamanin his first Vostok crew—prime pilot, backup and second backup—but
not necessarily the final flight order. It did, however, go a long way to determining
who might be the most qualified candidate. From that time on, judgements on
suitability could be made based on far less technical issues, such as each man's
personality, speaking and conversational skills, their background, political and
cultural ideals, and an ability to represent the Soviet people and Communism in
an ambassadorial role at all levels. It is not known for certain whether these results
were revealed to the six contenders at the time, but subsequent events must have
ensured that Gagarin, Titov and Nelyubov were aware they were the three candidates
most under consideration for the first flight.

According to space journalist Yaroslav Golovanov, while Gagarin had topped
the list and was considered by his peers as easily the most likely to be awarded the first
Vostok mission, he had yet to prove himself the best of the best, rather than just one
of the leading contenders for the first flight. Everyone had formed their own opinions,
according to Golovanov, and these opinions would grow stronger over the months as
the men, their work and behaviour, remained under close scrutiny.

> "Karpov valued in Nelyubov the speed of his mind, temperament, and skilled
> command of words, although he saw a deficiency in him too: he had a not always
> justified tendency to superiority over everybody, and a nearly complete absence of
> self-criticism.
>
> Karpov told me that at different periods of training, he gave his preference to
> first Popovich, then Titov. He liked Titov's frankness very much. Gherman, if
> coming under question, never twisted things, never invented excuses for himself.
> On the other hand, Karpov was alert to Titov's impulsiveness; if he failed, he
> became practically unmanageable. Gallai, too, valued Titov highly, and spoke of
> him to Korolev. Korolev himself, evidently, also gave some preference to Titov,

Nikolayev, Nelyubov and Gagarin photographed at Chkalovsky Air Base.

but still also to a large extent to Gagarin. Leonov thinks that Gagarin pleased Korolev as early as the time of the cosmonauts' first trip to the construction bureau. Yazdovsky related that Korolev said to him one day about Gagarin: 'I like that brat' " [2].

Golovanov added that Boris Raushenbakh, a vibrations specialist at OKB-1, had also expressed a preference for Nelyubov. But that particular candidate would often demonstrate one very complex personality.

"Grigori Nelyubov very much wanted to be the first. And possibly it was this outspoken thirst for leadership which prevented him from being first. To judge from the recollection of witnesses of all these events, Nelyubov was a remarkable character. A good pilot, a sportsman; he also stood out in his general scope, amazing vivacity, lightning reactions, charming nature, his ready ability to make general conversation with people. In [Georgi] Shonin's words, this was a 'through-passage' guy. No one knew as well as Nelyubov how to 'deal' with the doctors, instructors, and trainers. He possessed a talent for process, sometimes even in spite of his interlocutors, of introducing himself to their own circle of concerns and transforming himself into their ally and assistant. This man was a joker, a racconteur, 'the life and soul of the party', lover of noisy booze-ups; in short a 'Hussar'. However, the psychologists noted in him a constant desire to be the centre of universal attention, egocentrism, which prevented him reconciling personal interest with the interest of the project" [2].

KORABL-SPUTNIK 4

On 7 March 1961, with the tempo of specific flight training rapidly intensifying, Yuri Gagarin received some wonderful news on the home front. Valentina had given birth in hospital to their second child; another little girl they named Galina, or more affectionately Galya. As proud and happy as he might have been, Yuri nevertheless understood that in the present scheme of things there would be very little time for him to get to know his newest baby daughter before he flew down to the Baikonur cosmodrome at Tyuratam, in the steppes of Kazakhstan, east of the Aral Sea.

Chernushka

Gagarin was still in Moscow two days later on 9 March, quietly celebrating not only his 27th birthday and the fact that he was a father for a second time, but that a human-rated Vostok 3A spacecraft, designated Korabl-Sputnik 4 (or KS-4, and also identified in the west as Sputnik 9) had achieved a successful single-orbit flight carrying a small, dark-haired street dog named Chernushka, whose name translates to "Blackie".

The first of two precursory orbital flights to carry dogs, Korabl-Sputnik 4 was principally a dress rehearsal for the upcoming manned flight, then scheduled for the early part of April. The Vostok 3A craft that had been used was almost identical to the one designed to carry a cosmonaut, although Chernushka occupied her own pressurized sphere within the spacecraft, accompanied on her flight by 40 black and 40 white mice, guinea pigs and reptiles, and other biological experiments.

"Ivan Ivanovich"

Representing a cosmonaut onboard Korabl-Sputnik 4 was a life-sized, experiment-packed mannequin cheerfully christened Ivan Ivanovich, which had been tightly strapped into the fully functional ejection seat, wearing a Sokol SK-1 orange pressure suit and helmet.

Vladimir Suvorov was a distinguished chief documentary cinematographer who had recently arrived at Tyuratam in order to record events surrounding the launch. In his diaries, later published, he told of the earlier delivery of "Ivan" at the Assembly Testing Complex (ATC).

> "Having arrived, we got deeply engaged in the work. The next day we got acquainted with Ivan Ivanovich: a dummy pilot. In a spacious clean room of the ATC three men in white overalls opened a big, sealed box which arrived via the special delivery service. They lifted the dummy carefully from the box and put it into the cosmonaut's seat. "He" was dressed extraordinarily: bright-orange suit, white helmet, thick gloves and high, laced boots ... His head, the "skin" of his body, arms, and legs were made from synthetic material with durability, elasticity, and resistance mimicking that of the human skin. His neck, arms, and legs had gimbal joints so they could be moved ... Dressed in a complete

cosmonaut spacesuit he looked somewhat unpleasant with his fixed false eyes and a mask for a face" [3].

After the planned single orbit had been successfully accomplished, the Korabl-Sputnik 4 spacecraft was automatically aligned for re-entry braking and brought back to Earth. As scheduled, the hatch was explosively jettisoned and "Ivan" was automatically ejected from the spacecraft after a fiery passage through the atmosphere. The mannequin separated from the ejection seat soon after and landed by parachute. Meanwhile Chernushka and the other animals also descended by parachute to the ground, although still within the confines of the spacecraft, landing in an open field near the village of Krasny Kut, about 165 miles northeast of Kuybyshev.

Freezing, snowy conditions severely hampered the recovery effort led by General Kamanin, who was eventually first on the scene. Vladimir Suvorov was also on hand to film the recovery, and he later wrote that it looked "creepy" to see the mannequin lying motionless in the snow with eyes painted wide open in a fixed gaze. "He looked precisely as if a real cosmonaut had been killed during the landing," was Suvorov's first impression.

Eventually, the animals, mannequin, spacecraft, ejection seat and other components were all gathered up and triumphantly returned to Moscow amid great jubilation. It was just the sort of successful mission everyone had wanted, giving a huge boost in confidence to those connected with the Vostok programme, as well as the cosmonauts waiting eagerly in the wings.

Mounting a precursory flight

Following the tremendous success of Korabl-Sputnik 4, a second, similar test flight was scheduled for later that month. If this was carried out and completed without any significant problems the first manned attempt could proceed.

The waiting was nearly at an end. With all three frontrunners still basically in contention for the first manned flight, Gagarin, Nelyubov and Titov were pumped up and ready to fly, although Bykovsky, Nikolayev and Popovich were still hoping that some minor miracle might yet place them in the pilot's seat. In order to give all six candidates the opportunity to observe first-hand the vital pre-launch preparations, as well as conduct a number of tests, carry out some additional training and take part in suiting-up procedures, they would all fly down to the cosmodrome on 17 March. Korolev was also at Tyuratam to observe the launch, and this gave him the chance to conduct his own final and comprehensive evaluations of all six candidates at individual question-and-answer sessions. Meanwhile, with Varlamov and Kartashov now ruled out of contention, the 12 remaining cosmonauts would stay in Moscow to continue their own, less flight-specific training.

Just six days later, on 23 March, a horrifying tragedy would strike at the very heart of the Soviet space programme and especially the men of the cosmonaut team.

DEATH OF A COSMONAUT

These days, Russian émigré Dr. Vladimir Julievich Golyakhovsky is an associate professor in New York's Department of Orthopaedic Surgery. The son of a chief surgeon general in the Soviet army, he was just 30 years old on the morning of 23 March 1961 and living a spartan life with his wife, mother-in-law and baby son in a tiny, wooden two-bedroom house on the eastern outskirts of Moscow. That day, Golyakhovsky was performing an early morning shift duty at the sprawling Botkin Hospital in northwestern Moscow as their chief surgeon-traumatologist. What had started out as a fairly routine day at the hospital would change dramatically when he received a frantic phone call from a military surgeon, reporting the imminent arrival of a severely burned patient who required emergency aid.

Mystery patient

In his 1984 book, *Russian Doctor*, Golyakhovsky identified the harried caller as a Colonel Ivanov (most likely a pseudonym) from the Central Aviation Institute of Medicine (TsVNIAG) in Moscow. This facility was an auxiliary of the Institute for Medical and Biological Problems (IMBP), which was then responsible for conducting medical, physiological and psychological tests on aspiring cosmonauts. The TsVNIAG institute was located near Petrovsky Park, only a ten-minute drive from the hospital. Ivanov said he would be accompanying the patient.

"Such a warning was out of the ordinary," Golyakhovsky recalled, "so I guessed that there was a good reason for the doctor's anxiety" [4].

Shortly after, a military ambulance raced through the hospital gate, closely pursued by a convoy of five or six official-looking black Volga sedans. Once these cars had pulled to a halt, several senior military officers hurriedly spilled from their doors and ran up to the ambulance where they assisted in unloading the blanket-covered body of a person lying on a stretcher. The stretcher was quickly manhandled up a short flight of stairs into the admission area, where Golyakhovksy was waiting. The sickly, unmistakeable smell of burned flesh rapidly permeated the air.

Golyakhovsky, already realizing the prognosis would not be good, directed the bearers to carry the stretcher into a shock treatment room, where a nurse and an intern helped him to carefully remove the blanket and a sheet from the hideously burned body. Even as they worked, Golyakhovsky could not help but wonder who the patient might be to merit such urgent attention from so many high-ranking military officers, a number of whom were wearing medical insignia on their epaulets.

"The body was totally denuded of skin, the head of hair; there were no eyes in the face—everything had been burnt away," he recalled. "It was a total burn of the severest degree." Despite this, the patient was somehow conscious and trying to say something through his swollen, burnt lips. Golyakhovsky placed his ear near to the man's mouth and heard him whisper, "Too much pain ... do something, please ... to kill the pain."

An immediate intravenous injection of glucose and sodium chloride was needed, but a vein could not be located. The only unburnt part of the entire body was on the bare soles of the man's feet, which had been protected by thick-soled shoes. Golyakhovsky managed to insert the needles into veins in the feet and began injecting the solution. He then gave the man an injection of morphine, which seemed to ease his breathing a little. As a traumatologist, however, he knew the man would not survive much longer: hypovolemic shock and myocardial contractility would soon end his suffering.

After he had done all that was physically possible to help to subdue his patient, Golyakhovsky asked Colonel Ivanov how such a terrible accident could have happened. He also enquired after the man's identity, and in his book recalled that the name he was given was "Sergeyev".

The Chamber of Silence

It had begun ten days earlier as a routine, supervised exercise in sensory starvation, but the young man's name was actually Bondarenko—not Sergeyev. It may be little more than an ironic coincidence, but Sergeyev was the pseudonym used at that time to identify the Soviet Union's chief spacecraft designer, and a man well known to all the cosmonauts, Sergei Korolev. Twenty-four-year-old Bondarenko had been enclosed in a soundproof isolation chamber, and like earlier confinements of his colleagues he had no idea how long the exercise would last. That was the pivotal point of the exercise; to see how an individual would cope in solitude and silence over several sessions for an indeterminate time ranging from several hours to more than a week, with very little outside stimuli. In order to simulate spaceflight to a degree, the oxygen-enriched chamber was maintained at a reduced pressure, similar to that within a high-flying jet aircraft; however, the pressure could be raised or lowered by the duty doctor to test individual subjects' physiological reactions.

All 20 cosmonauts would undergo this daunting exercise, and the first member of the group to enter the chamber was Valery Bykovsky, who readily put his hand up when a volunteer was called for, saying "I'll try it!" [5]. While his series of isolation tests was taking place in Moscow during March 1960, the rest of the cosmonauts would be in Engels near Saratov carrying out parachute training. A contemporary report on cosmonaut training, purportedly written by a Colonel Yevgeni Petrov, "a commander of the Soviet Cosmonauts Group" (which was the *nom de plume* used by the first director of the training centre, Yevgeni Karpov) told of Bykovsky's experiences in the soundproof chamber:

"How Valery felt I didn't know, but we were all a bit worried. This was the first time one of our men was to enter that unknown realm of silence. There was no danger; watchful researchers would be there all the time. What troubled us was something else—how would the first man react? The behaviour of the rest hinged largely on that. Registering instruments of the latest type kept constant watch. Valery behaved strangely at first. He seemed to be in an unreasonable hurry. Finished with one thing, he moved to another after only a moment's thought. He reached suddenly for the telegraph key and hurriedly began to send the message, '... air temperature ... pressure ... humidity'.

Sporting a brave smile, Vladimir Komarov enters the dreaded isolation chamber for an unspecified period of days to test his endurance and adaptability.

But the initial excitement subsided as soon as he adjusted to the situation. He began to work according to schedule, rationally and without undue haste. He kept an eye on the instruments and reported from time to time. He skilfully repaired one of the devices that went out of order. When he was not busy at assigned tasks, he wrote and sketched.

The pressure was changed several times and the routine disturbed by bright shots of light and harsh sounds, but he reacted rationally and in good time to all disturbances. He slept soundly, awoke exactly on schedule, and moved promptly to his assigned tasks. Valery came through the test with flying colours and returned from his 'cosmic' voyage in perfect health and good spirits" [6].

Bykovsky later reassured his fellow trainees, probably a little boastfully, that his experience with the isolation test was "nothing extraordinary".

Yuri Gagarin would be the 5th of the 20 cosmonauts to spend time in the chamber, while the records show that Bondarenko was number 17 in line. Some of them would later state that this was one of the most gruelling facets of their training, leaving them mentally exhausted and lacking in concentration. "It was often a trying experience," Gagarin would later report, "especially since we were never told how long our stay in the cell would be. Several hours? A day and a night? Several days?" [5]. Popovich would only admit "it was tough," while Andrian Nikolayev said he "would have been happy to hear even the chirping of a little bird or to see any living creature."

The chamber, set in the middle of a large laboratory, was an elaborately constructed facility suspended on several thick-rubber shock absorbers, with thick walls capped by an airlock. There would be no outside sound or even the vibration from footsteps coming from beyond the chamber. Two small covered portholes, each made of extremely thick glass, could be uncovered from the outside to allow an occasional viewing of the interior and subject. A single door led into the isolation chamber; it was a massively heavy brute more than 16 inches thick—the same thickness as the chamber walls. Inside, the furniture was sparse and basic, comprising a small steel bed, a wooden table, and a large seat identical to the one planned for the Vostok spacecraft. There were also toilet facilities and a small hotplate with a heating ring and a saucepan for warming their meals. Mounted on the wall was a large board containing numbered squares, alternately coloured red and black, for mental agility tests [7].

Popovich bides his time within the isolation chamber.

Titov takes his turn in the so-called Chamber of Silence, while (bottom photo) his progress and reactions are monitored.

At first, each cosmonaut would undertake short visits to the chamber so doctors could assess their suitability for much longer periods of silent isolation and sensory deprivation. The subject would never know how long each test would be, but once the doctors were satisfied the longer test could proceed the door would be closed for periods lasting anywhere from 10 to 15 days. No other cosmonauts would be present; only a team of medical support staff and technicians.

Early morning each day, usually around 5:45 AM, a light would illuminate in the chamber. This was the signal for the subject to apply medical sensors to his body, using a special paste to ensure good contact with the skin. Once this had been done, a green light would allow the doctors and technicians outside to know that physiological readings could begin.

In his book *700,000 Kilometres through Space*, Gherman Titov wrote of conditions in what the cosmonauts balefully called the Chamber of Silence:

"It is quiet, very quiet. But the word does not really describe the situation. Complete absence of sound. Not a tap, not a rustle, not a splash, not a sigh. Such absolute silence takes some getting used to; one must acclimatise oneself to it, preserve what the doctors call one's neuropsychic equilibrium. A glance at my temporary residence and its scanty furniture. A small armchair at a table. A special switchboard and, beside it, a television camera. Everything needed for a long-duration flight lies ready to hand—food, water, living utensils, books to read, a notebook.

I make entries in the log-book and perform a number of other tasks. It is just like a real flight. I know I must keep my watch with faultless accuracy, and not so much because I am being observed by a television camera as to get used to maintaining a steady rhythm of life in such conditions" [8].

Learning to cope

On his first extended stay in the chamber Titov would use the small stove-top cooker, but even in this he displayed a pleasing sense of lateral thinking. Food for the exercise was supplied in tins, and when it came time for the first meal several of the men simply removed the top from a can and dumped the contents into the saucepan, which they placed onto the hotplate. That was fine until they realized afterwards that there was only a small supply of water for drinking and personal hygiene, and they would have to cook all their meals in the same food-encrusted saucepan. Using his ingenuity, Titov poured a little water into the saucepan, allowing it to boil before he popped in the unopened tin. He was then able to open the tin and enjoy a hot meal without all the mess, and he could also use the water in the saucepan several times.

The test subjects were given a programme of daily activities that lasted for four hours; over the remainder of each day they had to amuse themselves, rest and eat. For the sake of comparison some were allowed to read, and others were not. Those that were not permitted books were instead supplied with coloured pencils, paper, small blocks of wood and a carving knife. To pass the time, Titov recited some favourite verses of Aleksandr Pushkin aloud, and over several days also mentally stripped down and rebuilt a car, while during his confinement the always-ebullient Pavel Popovich sang Ukrainian songs with great gusto. Others wrote poetry or made small models from the paper and wood [9].

Boris Volynov would occupy the isolation chamber in December 1960, during which time he would celebrate his birthday in enforced solitude, as he told journalist Olga Apenchenko.

"At first it was strange to be without people. Then I got used to it. I started to do a lot of work. You know, it is a good thing that the day in the chamber is so completely filled that there is no time to be lonely. There is always something to do. The doctors are wonderful in this respect. I checked on the thermometers, the hygrometers, I kept check on the instruments, and prepared my own food . . . but the day comes when you finally want to talk to someone. I had such a day . . . my birthday. Imagine—I spent my 25 birthday on Earth, and my 26th had to be spent in the 'cosmos.' Have you any idea what it means to sit alone, enclosed in four walls, for so many days? And take into consideration that I am a pilot . . . the whole sky was mine. And suddenly, instead of sky, a small narrow chamber.

On my birthday I wanted very much for someone to be with me . . . to hear some good kind words . . . a live human voice. A human word, only a single word, what would I not have given for a single word.

When I came out, something improbable happened to me. Everyone gathered around me, motion-picture cameras began humming, and the girls were shouting 'Well, tell us something; tell us.' And I can do nothing but stand there and remain silent. I am so happy that I look at the people with blank eyes and can't say a word; almost as if I had forgotten how to speak. I had longed so for human speech that I am waiting until someone would say something, anything. I am sick of my own voice" [10].

Gagarin would later write that after listening to his colleagues discussing the chamber he had entered it several times when it was unoccupied, in order to familiarize himself with the interior and its fittings. During his actual period of extended solitude he had imagined himself looking down at the Earth from space: "I thought about the future, not the past, as one usually does in such circumstances. I imagined myself in the cabin of Vostok. I closed my eyes and saw the continents and oceans passing below, the change of day and night, and far, far down, golden clusters of city lights. Although I was never abroad I pictured myself flying over Peking and London, Rome and Paris, and over my native Gzhatsh . . . All this helped me to sustain the strain of utter isolation" [11].

The only outside link was provided by a telegraph key and a radio line to the doctors. At certain times and at set intervals each cosmonaut had to report on his condition, and would have to press three keys coloured white, red and black in correct sequence in order to communicate. One typical report given by Gagarin was, "Chamber temperature, 27 degrees. Pressure . . . on the first hygrometer, 74 percent, on the second, 61 percent. My physical condition is normal; everything is proceeding well. I am lying down to sleep. I am signing off for now." But while the doctors sometimes responded, at other times they refrained, in order to gauge the subject's reaction to being ignored. At other times a few mathematical problems would also have to be completed, work done on some navigational tasks, a number of complicated instructions memorized, and physical exercises carried out.

As another part of the tests, the doctors would suddenly introduce classical music into the chamber and note the subject's response to this pleasant diversion. A favoured composer of this music was Ukrainian-born Isaak Dunayevsky. Most of the subjects would simply close their eyes and revel in the experience, although it was noted that some became mildly emotional, and the sight of tears was not uncommon.

Following his release from the isolation chamber, a report was prepared on the condition of Gagarin. It read: "At the end of the experiment, the subject was examined by a neuropathologist, and the physical condition of the subject was found to be good, as was his general condition. His appearance and behaviour were normal. There were no signs of emotional excitement or depression. He is quiet, and amiable."

When the examining physiologist asked Gagarin how he felt in the chamber, he is reported to have shrugged and said, "Not bad; like at home!" [11].

A TRAGIC ACCIDENT

On that fatal early morning of 23 March, Bondarenko's solitude had been broken by a signal light indicating that his exercise was finally at an end. He knew that the supervising technicians outside would be very slowly bringing the chamber back up to normal pressure. As Alexei Leonov would recall in the 2004 book, *Two Sides of the Moon*, the pressure in the chamber at that time had been lowered to simulate the Earth's atmosphere at a height of 5 kilometres. In order to compensate, the oxygen content had been increased to a partial pressure of 430 mm. This translates to around a 68% oxygen environment [12].

One fatal lapse in concentration

Bondarenko knew from experience that the depressurizing process would take about half an hour; a procedure necessary to prevent him from suffering the "bends" prior to opening the heavy door. Relieved that he had withstood the test and associated monotony, and in his haste to prepare to leave the chamber, the young man pulled the adhering medical sensors from his body as he sat at his desk and then cleaned the sticky paste residue from his skin with a small cottonwool pad soaked in alcohol.

Having finished this task, and probably displaying a not-unexpected lack of concentration after the ten days, Bondarenko unthinkingly tossed the pad away. According to some reports, it is said to have landed on top of the small electric hotplate the occupants used to warm their meals, which was switched on for making the first meal. But in an extensive interview conducted by Bert Vis with Pavel Popovich in Quebec, Canada, on 30 September 1996, the famed cosmonaut actually described the appliance in question as "a very simple, open heater with an open spiral—an open coil" that some of the cosmonauts used when it got too cold in the chamber. "I spent ten days in this pressure chamber myself," he recalled, "but I didn't switch on the heater. I sang all the time."

It must have been quite cold in the chamber that morning. "Well, it was chilly for Bondarenko," Popovich told Vis, "so he switched it on ... it was standing to his left on the floor. Without looking, he ... threw [the piece of cotton] directly onto the heated coil" [13].

The alcohol-soaked pad quickly burst into flames. Fuelled by pressurized oxygen the fire spread explosively over the interior fittings, finally taking hold on Bondarenko's woollen training suit, which had been literally soaking in oxygen for ten days. Unfortunately, the more he tried to beat out the tenacious flames, the more he was helping the fire to spread over his clothing and skin.

Outside the chamber a fire alarm began to sound. Duty doctor Mikhail Novikov immediately rushed over with the technical support team in an attempt to open the door into the chamber. Unfortunately the pressure differential between the interior and exterior of the chamber meant that, like submarine hatches, the door could not be opened until the pressure had been equalized. It was a task that usually took several minutes. Meanwhile Bondarenko's woollen training suit had melted onto his body, and he was screaming in agony and terror as the flames stripped away his skin. His eyes quickly puffed and closed with the intense heat and smoke, and his hair was on fire.

Eventually, the chamber decompressed sufficiently to allow the door to be opened, and the fire rapidly died. Bondarenko's smouldering body was dragged clear, but he was still conscious, crying over and over again, "It's my fault ... I'm so sorry ... no one is to blame!" [14].

Valentin Vasilievich Bondarenko perished in a horrifying training exercise.

The unknown "cosmonaut"

The Institute's medical staff members were quickly alerted to the situation; after a rapid assessment it was decided to rush Bondarenko to the nearest emergency facility. As the young man's body was placed on a stretcher and covered with a sheet and blanket, a military ambulance was quickly summoned and an urgent call placed to the nearby Botkin Hospital. Another call was made to the commander of the cosmonaut training centre.

Over the next eight hours the medical staff at the hospital tried desperately to stabilize the young cosmonaut, but it was all to no avail. Telephones at the hospital rang constantly as anxious officials tried to get any news on the patient's condition. At one time the chief surgeon of the Soviet army, General Alexander Vishnevsky, called the hospital and spoke to Golyakhovsky, saying that he had been under intense pressure from above to move the patient to a specialized burns unit, but he had argued that to move the patient would mean certain death from shock. Golyakhovsky took the opportunity to point out that his hospital lacked certain medicines vital to treating the victim. Vishnevsky promised to see what he could do in addition to sending a specialist named Schreiber and some assistants to help in the desperate fight. These men duly arrived and the more potent medications were administered, but for the most part all anyone could do was watch and wait for the inevitable to occur.

According to Golyakhovsky's memoirs, among those in attendance was an anxious young air force lieutenant wearing a white doctor's gown over his uniform, seated forlornly on a wooden stool in the hospital corridor. Sent to act as a liaison officer between the hospital and the military, he was taking all of the incoming calls and giving updated reports on the condition of the patient to his superiors.

The following morning, inevitably, the hapless victim died from his terrible burns and profound shock. Soon after, Golyakhovsky says that he noticed the young lieutenant was still in attendance, tired, pale and obviously upset, seated alone on a hospital sofa with his head bowed. When he saw the doctor approaching he jumped to his feet, but Golyakhovsky instructed him to sit down again, and then joined him on the sofa. He asked if the two men had been friends, and the young man confirmed that they had not only been very close, but were serving in the same unit at the time of the accident. After talking for a while the two men solemnly shook hands and the baby-faced lieutenant departed.

"He was very small in stature and his wrist was as thin as a child's," Golyakhovsky wrote of the experience in his book, "but he gave me a strong man's handshake. His face stuck in my memory and [just three weeks later] I saw his photograph in the newspapers: his name was Yuri Gagarin; he was the first man in space" [4].

While it's an amazing story, one would have to raise serious questions about the veracity of the good doctor's recollections. For instance, it is a known fact that Gagarin, Titov and Nelyubov had been flown to Tyuratum and the Baikonur launch site six days earlier on 17 March in order to prepare for and observe the launch of Korabl-Sputnik 5. This precursory flight would carry the dog Zvezdochka and a life-sized mannequin known as Ivan Ivanovich in an orbital test of the Vostok spacecraft, as well as its life support and parachute ejection systems. They were also there to check out the launch vehicle and to meticulously go over the spacecraft and procedures prior to the first manned flight. Two days after they arrived the Korabl-Sputnik 5 launch was delayed until the 24th or 25th due to technical problems, giving all three men more time to familiarize themselves with the Vostok systems and their spacesuits.

No corrections or additions

Word of the terrible death of Bondarenko would not reach Korolev and the cosmonauts until the evening of 23 March. One would have to conclude that with such a vital launch imminent, it is extremely unlikely that Gagarin could have been released to fly to Moscow and back. He was certainly at the Baikonur facility when Korabl-Sputnik 5 was launched into orbit early on the morning of 25 March.

The authors decided to check the details of that day with Dr. Golyakhovsky in 2007, but in his first response he stated that he did "not have any corrections or additions" to what he wrote in his 1984 memoirs. However, he finally agreed that

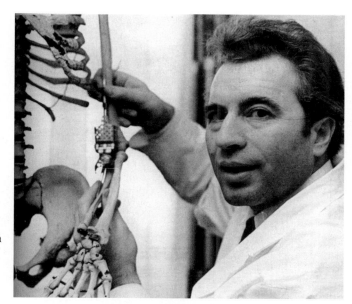

Dr. Golyachovsky in 1971, ten years after the tragic events of 23 March 1961 that took the life of Valentin Bondarenko. (Photo: Dr. Vladimir Golyachovsky)

"I certainly cannot assure you that the young lieutenant was Yuri Gagarin. He did not tell me his name, did not introduce himself, and he could not do this by the strict rules. I was extremely busy and tired and … just roughly paid attention to his appearance. It was my impression that I recognised him in a picture published in the papers a few days later. All those young pilots with short haircuts looked the same" [15].

It would therefore seem that if a cosmonaut was indeed in attendance at the hospital, it is quite improbable that he was Yuri Gagarin.

Loss of a friend and colleague

Valentin Bondarenko's funeral was held one cold spring morning a few days later in his Ukrainian birthplace of Kharkov, where his grieving parents and most of his relatives still lived and where he had spent his childhood. It was a quiet, sombre affair; Bondarenko was not buried as a cosmonaut, but as a regular air force senior lieutenant who had died in a tragic accident. Following the funeral his wife and their five-year-old son returned to Moscow, where it had been officially but quietly decreed they were to be housed and "provided with all the necessities, like the family of a cosmonaut." Bondarenko would be posthumously awarded the Order of the Red Star on 17 June 1960.

News of the young cosmonaut's death had been immediately suppressed, becoming a closely guarded State secret. Anyone involved in the episode in any way, including Golyakhovsky, was warned to say nothing. It would remain that way for the next 25 years, although in the early 1980s some intriguing and credible

Bondarenko's grave in Kharkov. (Courtesy: Bert Vis)

stories had begun to emerge regarding a cosmonaut named Boyko or Boychenko, said to have burned to death in a training accident just prior to Gagarin's flight. Although it contains several inconsistencies, such as giving the cosmonaut's name as Sergeyev, and the wrong month of the incident, Golyakhovsky's 1984 book had a definite ring of truth about it, which further encouraged Western space sleuths eager to learn the truth about this tragic episode. Soviet officials, however, continued to deny the story [16].

Then in 1986, on the eve of the 25th anniversary of Gagarin's flight, Bondarenko's identity would finally be revealed in a sanctioned article that appeared in the Soviet government's newspaper *Izvestia*. Soon after, changes were made to the inscription on his tombstone, which for a quarter-century had read: "With fond memories from your pilot friends." It now was inscribed: "With fond memories from your pilot and cosmonaut friends" [17].

Just weeks later, as recorded by Dr. Irina Ponomareva and discussed with researcher and author Rex Hall, the last two cosmonauts to undertake the same isolation training as Bondarenko would subsequently spend several days in the Chamber of Silence [18]. One can only imagine the thoughts that would have gone through the minds of Zaikin and then Filatyev during those long, quiet and lonely days.

Saving the Apollo 1 crew?

Many analysts now say that the secrecy surrounding the circumstances of Bondarenko's death may have indirectly claimed the lives of three fellow space explorers in a launch pad fire six years later. It has been widely speculated that had information about the perils associated with fire in an oxygen-rich pressure chamber been made available to NASA, it may have caused the space agency to think twice about subjecting the Apollo 1 crew to the same hazardous and potentially lethal environment. However John B. Charles, PhD., a long-time student of space-flight history, raised several interesting and concerning points in a 2007 article for *The Space Review*, asking if such a disclosure could actually have alerted NASA to the potential dangers.

Dr. Charles states that NASA should have taken warning from at least seven instances of oxygen-related fires in operational U.S. testing facilities, "four of which occurred between two years and nine months before the Apollo fire." Three of these, he stated, involved unmanned tests of Apollo life support systems, at least one of which involved the use of pure oxygen at a planned cabin pressure of 5 psi. "The remaining four fire events took place during manned US Air Force and US Navy chamber tests in the late 1950s and 1962. Three of those were tests of cabin atmospheres planned for Mercury and Gemini, and their crews escaped with injuries ranging from smoke inhalation to first and second degree burns." The fourth accident occurred in 1965 during tests involving pressure and gas combinations, not for spaceflight, but for deep-ocean operations. Two Navy divers would die in a fire in a chamber pressurized to 8.6 atmospheres.

"That NASA failed to grasp the lessons of these fires is regrettable," Dr. Charles concluded, "but it was not unusual. Only four days after the Apollo fire, the Air Force lost two veterinary technicians in a pure oxygen chamber fire. Clearly NASA's own object lesson was lost on the Air Force as well" [19].

Only 19 of the original 20 cosmonaut candidates would actually undergo isolation training and evaluation. Before his turn came around, Valentin Varlamov had suffered a serious injury that resulted in his removal from the cosmonaut team.

The isolation chamber still exists in the Institute of Biological Problems [18], and so does an enduring record the unfortunate Valentin Bondarenko holds to this day. He remains the youngest male candidate ever selected by any nation to train for a flight into space.

To help ease some of the tension, Nikolayev, Anikeyev and Rafikov engaged in games of chess.

Anonymous trips into Moscow were also popular with the trainee cosmonauts. Here Anikeyev, Nikolayev, Bykovsky and Zaikin are seen in Red Square.

A well-liked young man

"He was a very good-natured, merry fellow," Pavel Popovich once said of his former colleague Valentin Bondarenko. "He was nicknamed *Zvonochek* (Tinkerbell), but I cannot remember why." Popovich did, however, recall a vigorous young man who loved to joke around, but never took offence whenever the tables were turned on him; one who enjoyed a number of sporting activities, and like Popovich possessed an excellent singing voice. Valentin and Anya Bondarenko lived with their three-year-old son on the upper level of the apartment building they shared with other cosmonauts, and Popovich still remembers Bondarenko running up and down the stairs some mornings knocking on the doors of his colleagues, calling out and trying to entice them into a game of football.

Alexei Leonov also recalls Bondarenko's sporting prowess with fondness, saying he was "a good soccer player" and that no one could ever beat him at table tennis. He had quickly warmed to Bondarenko's shy humour, and the fact that he could laugh at himself whenever he fell victim to any jokes played on him. "And if someone's sense of humour extends to himself," Leonov added, "he is a good man, as a rule." As well, Bondarenko was never known to brag about anything he did, and these qualities endeared him to the older Leonov [2].

Georgi Shonin, who shared an apartment with Bondarenko for some time, said "Valentin could sometimes erupt, but without malice of taking offence. He would flare up literally for a moment, then blush and become embarrassed at his own lack of restraint. I admired his selflessness and determination."

Shonin also spoke of a time when Bondarenko courageously saved the life of a young boy, who had innocently and unnoticed clambered up and was standing on the open window sill of his fifth floor apartment. As he sat there looking down, someone noticed the child and yelled at him to get back inside. A few more people including Shonin began to gather as the drama unfolded, while the young child playfully looked down at them. The onlookers pleaded with him to go back inside, but it seemed the child, now becoming frightened, was unable to do this. Just at that moment Bondarenko walked by and saw what was happening. To everyone's astonishment he ran over to the wall and, displaying great athleticism, clambered hand over hand up an unsteady drainpipe towards the boy. "I still tremble even now when I recall Valentin climbing up that drainpipe," Shonin said. "Indeed each second he could fall down together with the creaking pipe." But everything ended satisfactorily: Bondarenko reached the young boy, picked him up and lowered him back safely into his apartment. He then climbed back down the drainpipe to the applause of the relieved crowd. Witnesses to the young pilot's feat greeted him as a hero, but he modestly dismissed the praise, saying it was just something he felt he had to do.

"Valentin loved his father very much," Shonin added. "He was proud of him, a former partisan scout. In the evenings, when we went out onto the balcony for a breath of fresh air, he would recount many interesting things about him, suddenly interrupting himself with the question: 'Did I ever tell you that dad's tall sheepskin hat is in the Museum of Partisan Glory?'" [2].

By way of tribute an 18-mile diameter lunar crater was named for Bondarenko, located on the far side of the Moon (selenographic coordinates 17.8°S, 136.3°E), northeast of the large Tsiolkovsky crater and south of Chauvenet crater [20].

THE FLIGHT IS IMMINENT

On the morning of 23 March the six core cosmonauts attended a briefing conducted by engineers at the cosmodrome on the best in-flight methods of correcting the Globus instrument that indicated the Vostok craft's position at any time over the Earth. They later met with Korolev, who wanted to ensure the trainees had everything they needed, and to discuss with them any concerns they might harbour. In the scheme of things it began as a relatively uneventful day, but things would change dramatically that evening, when they first heard of the horrifying death of Valentin Bondarenko at the end of his endurance test in the pressure chamber.

Korabl-Sputnik 5, and Ivan flies again

A sombre, introspective mood prevailed among the cosmonauts at the launch complex after they heard the shocking news about the man they had laughingly known as "Young Valentin". They were appalled at losing such a young and much-liked colleague in a tragic accident, and their hearts went out to his wife and young son. But the pressures associated with the task at hand demanded their undivided attention; they had to devote all their thoughts and energies to the upcoming and massively important orbital mission of Korabl-Sputnik 5, also known as Sputnik 10. With the smiling face of Bondarenko and the horrors he had endured undoubtedly fresh in their minds, they quickly knuckled down to their duties and training once again.

As before, there would be a female dog on board the Vostok spacecraft, with another batch of smaller animals and a raft of biological experiments to accompany her. The space-suited mannequin Ivan Ivanovich would once again ride a rocket into orbit as a passive substitute cosmonaut, albeit with a brand-new head. The previous one had been damaged during several parachute training exercises, and it was deemed unworthy of the occasion, so an urgent request had been sent to the manufacturers, the Moscow Prosthetic Appliances Works. They responded by shipping over a replacement that same day. Ivan was soon sporting a pristine new head, and ready to fly into space a second time.

Prior to the launch, some concerns were raised that anyone locating the spacesuit-clad mannequin in a forest or field after seeing it descend to the ground by parachute might mistake the silently inert figure for a corpse. With this in mind, and on the direct orders of Korolev, the word *Maket* (Dummy) was painted on the back of "Ivan's" orange spacesuit, and the same word was boldly written in black letters on a piece of white foam that was placed beneath the visor of the helmet. As well, in a touch of whimsy (but no doubt to also confuse Western listening stations), the ostensible cosmonaut had been wired for sound, ready to broadcast from space a tape recording of the Piatnitsky Choir in full voice, as well as a spoken recipe for cabbage soup [3].

A small street dog

While Ivan would fly again, this time a different canine passenger would be sent aloft. The animal selected and trained for the flight was another small orphan of the Moscow streets. A feel-good anecdote that seemed to gain momentum from this time concerned the first meeting the cosmonauts had with the friendly little dog. One of them is said to have asked the attending handlers for its name, to which the man casually responded, "Dymka [Smoky] or Tuchka [Cloudy]; something like that. I've forgotten." Gagarin (or Titov in some reports) is then said to have picked the playful animal up and stated that it needed a name better befitting the occasion. The cosmonauts are reported to have risen to the challenge, suggesting such names as Svetlaya, Kosmicheskaya and Laskovaya, but they couldn't reach an agreement. Then Gagarin spotted a "star of hero" badge on the jacket of one of the support team and said, "Let's call her Zvezdochka [Little Star]." And that, according to the oft-related story, is how little Zvezdochka got her name [3].

Dress rehearsal

In the early evening on the day prior to launch, with the rocket carrying the Vostok 3KA-2 spacecraft having been rolled out to the pad and fully erected a few hours earlier, Gagarin and Titov went through a trial run of the events leading up to their insertion into the confines of spacecraft. Each of them donned their orange pressure suits and both men were driven out to the launch pad by bus. Here they would practise riding up in the elevator to the hatch level. This dress rehearsal would not only give them a substantial feeling for the procedures and timing involved, but some added confidence that all was in readiness.

Initially, the launch of Korabl-Sputnik 5 was delayed for a short time due to a fault within the onboard communications apparatus. This delay was then compounded by a problem sensor on the third stage of the booster rocket. OKB-154 Chief Designer Semyon Kosberg ordered the faulty unit replaced, and the countdown resumed.

Meanwhile, in a blockhouse located just ten metres from the launch pad, Korolev and Kamanin were testing the spacecraft's communications system along with Pavel Popovich, who had been assigned the role of prime communicator for the test flight, a position very similar to that of the American astronauts' Capsule Communicator, or CapCom. Meanwhile, the other five cosmonauts had toured the launch pad together with General Leonid Goreglyad, Colonel Yevgeny Karpov and another Korolev aide named Azbiyevich, all from the cosmonaut training centre. They then retreated to view the launch from the IP-1 (*Izmeritelny Punkt-1*) tracking station, less than a mile from the pad, as the clock slowly wound down [21].

Unlike their American counterparts, the Soviets do not have a traditional vocal countdown to zero. At the correct time someone would simply press the launch button. As with the first Korabl launch, according to one of Korolev's designers, "We rejected a numerical countdown, fearing Western radio stations would monitor the human voice and raise a clamour throughout the world alleging that Russia has secretly put a man into orbit" [22].

Zvezdochka in orbit

At 8:54 AM Moscow time on 25 March, the booster carrying the latest Soviet satellite roared into lusty life, held fast for a while and then slowly hauled itself off the launch pad, rapidly accelerating upwards and penetrating the bright azure skies over Kazakhstan.

OKB-1 engineer Svyatoslav Gavrilov later recalled the excitement in IP-1 as the time approached to launch:

"I did not see Yuri's face as it was glued to the stereoscope viewer, but after the five-minute warning came he went up to the railings of the viewing platform and did not take his eyes for a single second off the rocket standing in the distance. Then I saw how his face drew in and his eyes darkened.

It occurred to me that he was imagining himself sitting in the spaceship's cabin. It turned out I was right, as he told me that evening that this was exactly what he was thinking.

'Launch!' The order came through to us over the loudspeakers. Yuri leant eagerly forward and pressed against the railings of the viewing platform; it even seemed to me that he stood up on tiptoe. There was still time before the rocket would lift, though. The seconds went by slowly and I saw incomprehension appear on Gagarin's face. He turned his head quickly towards me with a mute question in his eyes ... Then, as if afraid of missing a single moment of the lift-off, he locked his eyes on the launching pad again. 'Soon ...,' I said to him calmly and saw how he strongly expelled the air from his lungs.

'Ignition!' Yuri cast another look my way but at that moment a vague noise reached us, seeming rather to travel over the ground than to have come through the air. An instant later, a cloud of smoke mixed with steam and dust came pouring out of the exhaust evacuation duct and roaring thunder began shaking the earth and tearing the air.

'Lift-off!' Yuri froze still in his fear of missing the smallest detail of what was happening. His face was so open and candid that one could read everything in it. It then gradually relaxed as the rocket at last emerged almost unwillingly out of its cloud of fire and, literally tearing itself free from the earth's gravity, shot up with a victorious roar. Yuri was not able to hold himself back; a broad smile lit up on his face and he started to clap wildly. His comrades joined in. So did we. Then everybody was clapping" [23].

Once it had been established that the spacecraft was in orbit, immediate preparations were made for all the VVS officers and cosmonauts, 34 people in total, to return to Moscow aboard three Ilyushin IL-14 aircraft. Just prior to their aircraft taking off, word was received and relayed by the pilots to their passengers that transponder signals were indicating a successful completion to the orbital mission [21].

Because of the same freezing conditions encountered after the earlier flight, and snow piled up to around five feet thick in the spacecraft's touchdown area, it was not until the following day that the first members of the spacecraft retrieval squad were able to land in the village of Bolshaya Sosnovka, northeast of Izhvesk in the Perm district. Some local villagers were then cajoled into transporting the squad of officials to the landing area with the aid of a horse-drawn sled.

The spacecraft, still hot to the touch even after thumping down 24 hours earlier in freezing conditions, was found in a gully, nestled in a snowdrift. The mannequin and parachute were located some distance away. "The Earth seemed to have been expecting this to happen," commented Vladimir Efimov, who was a member of the rescue team. "Centenarian trees looked as if they had just parted to leave a small clearing, and in its middle, slightly reclining to one side, in deep snow, there was the orange-coloured hero."

However, when the villagers saw the figure resting against the snowdrift they immediately wanted to lend assistance, becoming suspicious and angry at the apparent lack of concern shown by the rescue team. At this time, the Russian population generally knew very little about cosmonauts or human space travel, but most were aware through a suitably-outraged Russian media that the previous May a Lockheed U-2 high-altitude spy plane flown by U.S. Air Force pilot Francis Gary Powers had been shot down over the Urals by a surface-to-air missile. Therefore, when local peasants saw the figure of an orange-clad figure lying apparently badly injured in the thick snow, they assumed it to be another American pilot who had been shot down, and were dismayed that no one seemed in any way concerned about helping the man. In order to assuage the mounting hostility, one of the team pointed out the sign inside the helmet, and then, as Efimov later wrote, "the incident was settled after the crowd of 'old believers' delegated their senior, who, showing an extreme dignity, walked unhurriedly ... toward the reclining figure and touched the rubbery, cold face of the

The recovery of "Ivan Ivanovich", showing the sign that reads "dummy" placed beneath the mannequin's visor.

dummy cosmonaut." He then confirmed to the others that it was nothing more than a lifeless mannequin [3].

Paving the way into space

Like the human-rated Korabl-Sputnik 4 flight earlier in the month, this second single-orbit flight was deemed to have been completely successful. The way had clearly been opened for the first manned spaceflight to take place just a couple of weeks later. As noted spaceflight researcher James Oberg would say of the mission: "Vostok 3KA-2 was the key in the door for Gagarin's flight."

However, one important question still remained in everyone's minds: Who would now be chosen for the first human spaceflight?

On 28 March, Academy of Sciences Vice-President Aleksandr Topchiyev held a press conference in Moscow, during which he outlined the significant results of the five Korabl-Sputnik flights to Soviet and foreign journalists. Innocently seated in the front row at the conference were several uniformed cosmonauts including Gagarin and Titov, but no one outside of a select few could have realized that very shortly one of these attentive but anonymous Air Force officers would make spaceflight history.

Gagarin had already begun displaying many of the qualities of leadership that not only impressed Korolev and Kamanin, but his fellow cosmonauts. Once, when the cosmonauts had been transported to a modified Ilyushin aircraft to carry out some onboard weightless training, they were told their instructor was running late. While most of the men were quite happy to simply sit back and relax until he arrived, Gagarin had suggested they use the time to discuss the Kepler curve, the principle behind the series of parabolic arcs that allowed them to be weightless on the aircraft for periods of up to 30 seconds. In effect, he assumed the role of instructor, and this enthusiasm and drive would be duly noted.

To help in the selection process, Korolev decided (like the NASA astronauts) to ask all of the cosmonauts to rate one another in a peer test and in a few words make a recommendation on who should be the first to fly. The results did not surprise him; a full 60% of the votes went to Gagarin. Some of the comments suggested that he was principled, had trained well and diligently, that he "never loses heart", was a "fit comrade", "bold and steadfast", "modest and simple" and "decisive". He would discuss the result of these peer ratings with Kamanin. The braggardly Mars Rafikov would further alienate himself by writing, "I should be sent, although I know that they will not send me. But my first name is 'cosmic', and this would sound good."

Elsewhere, pressure to conduct the flight on time was now being applied to Korolev, and even though he still had some massive problems to surmount he overcame them in turn in his usual brisk and uncompromising manner. Adding to the pressure being exerted on him was a cautionary message received from Marshal Vershinin that the Americans were preparing to launch their first astronaut into space on a suborbital mission on 28 April, and it was imperative the Soviet Union launch a man ahead of this time. Based on the success of the two precursory flights, the Central Committee of the Communist Party had issued Korolev with a top-secret decree: the first man would be launched into space somewhere between 10 and 20 April.

On 30 March Korolev sent a message to the State Commission, presided over by Chairman Konstantin Rudnev, certifying that all was in readiness for the launch of the first manned Vostok in the specified timeframe. The commission's members would in turn discuss and submit final recommendations in a memorandum to the Presidium of the Central Committee. The State Commission was weighty with names of eminent people at the very apex of space science and medicine, and the upper echelon of the VVS.

On 3 April the 11 eligible cosmonaut candidates who were not members of the Vanguard Six were informed that they would be undertaking a new screening process, set to begin that day. Khrunov and Komarov were the first to be interviewed, but soon after an urgent call came through to Kamanin for Gagarin, Titov and Nelyubov to attend a meeting at three o'clock that afternoon at which a final decision on proceeding with the first manned flight would be given. After an hour's debate Korolev was given the mandate he had sought from the Central Committee—the flight could go ahead as planned. The decree read:

1. The proposal of Ustinov, Rudnev, Kalmykov, Dementyev, Butoma, Moskalenko, Vershinin, Keldysh, Ivashutin, and Korolev on the launch of a satellite-ship "Vostok-3A" with a cosmonaut aboard is approved.
2. The plans for TASS to announce the launch of the space ship with a cosmonaut aboard an earth satellite is approved, and grants the Commission for Launch the right, if necessary, to introduce updates on the results of the launch and the Commission of the Presidium of the USSR Council of Ministers for Military-Industrial Issues, the right to publicise it [24].

Korolev immediately left for the cosmodrome at Tyuratam to make final preparations.

Test evaluations of the remaining 11 cosmonauts would also be concluded, and all were passed for flight status with the exception of Filatyev, Rafikov and Zaikin. They had passed the examinations, but due to their late inclusion in the cosmonaut team had not yet completed all of the required tests and training.

FIRST TO FLY

On Wednesday, 5 April, Nikolai Kamanin and Leonid Goreglyad departed for the airport, accompanied by the six prime cosmonauts, a group of VVS physicians and a film crew. Three twin-engine Ilyushin IL-14 aircraft (similar to the Douglas DC-3) were fuelled up and waiting for their arrival. As a necessary precaution, Gagarin and Nelyubov would fly down to Tyuratam with Kamanin, while first backup Titov would accompany General Goreglyad in the second aircraft. The third IL-14 was there to transport the VVS staff and film crew.

On reaching the cosmodrome, Korolev held a briefing for the upcoming flight, during which he revealed that the booster rollout was scheduled for 8 April, with the launch to take place as early as two days later. For those listening, and especially the three cosmonauts, the first manned flight had suddenly become an imminent reality.

A vital assessment

For Kamanin, there was still just one major question to be resolved. Who would fly first—Gagarin or Titov? Gagarin seemed to be just about everyone's choice, but he had recently and openly expressed concerns about the reliability of the 3KA-2's ejection system which demonstrated to some observers a slight but noticeable lack of confidence in the spacecraft. Titov, on the other hand, was showing great strength of character, reinforcing Kamanin's view that he would actually be better suited to the following, longer duration flight. He also felt that the very personable Gagarin might better handle the attendant storm of publicity and adulation that would follow the first mission, while Titov tended to be just a little uncomfortable and brittle in unfamiliar circumstances. Both men, however, were good-looking and very photogenic (as was Nelyubov), so this was not a selection issue.

The final decision on who would be the first pilot now rested with the State Commission, and this august body was heavily reliant on Kamanin's opinion as the head of cosmonaut training. His recommendations would no doubt have had considerable influence on their decision. A meeting to discuss and decide the issue was set for 8 April, the same day on which the R-7 rocket would be rolled out to the launch pad, with the shrouded Vostok spacecraft firmly mounted on its nose.

In his diary notes for 5 April, Kamanin acknowledged that he considered both cosmonauts entirely suitable. While he had placed Gagarin marginally in front after the January examinations identified him as the best of the three contenders, he had recently begun to lean towards Titov. "Both are excellent candidates," he wrote, "but in the last few days I hear more and more people speak out in favour of Titov and my personal confidence in him is growing too ... The only thing that keeps me from

The Vanguard Six, photographed early April outside a gazebo at the Air Force compound at Baikonur. From left, Nelyubov, Bykovsky, Gagarin, Nikolayev, Titov and Popovich.

picking him is the need to have the stronger person for the one day flight. The second flight, which will last sixteen orbits, will undoubtedly be more difficult than the first one-orbit flight. But the first flight and the name of the cosmonaut will never be forgotten by humanity, whereas the second and following ones will be as easily forgotten as new records." Kamanin then revealed some personal trepidation over making his selection, a choice which he knew would be quite influential in the final determination. "It's hard to decide which of them should be sent to die, and it's equally hard to decide which of these two decent men should be made world-famous" [25].

An historic decision

With all the recommendations in place, and the apparent approval of Premier Khrushchev to select either man without preference for either (although he is said to have liked Gagarin's boyish good looks and simple, peasant background), the State Commission, headed by Korolev, Keldysh and Kamanin, formally met at the cosmodrome on 8 April. No cosmonauts were in attendance. When it was time for him to place his recommendations on the record, Kamanin nominated Yuri Gagarin as the prime pilot, with Gherman Titov as his backup pilot, and Grigori Nelyubov as second backup. It was really all the Commission members needed to hear; the motion was passed and they soon moved on to other matters.

The following morning Kamanin invited Gagarin and Titov to his private office, and with very little formality other than a congratulatory handshake, informed Yuri Gagarin that he had been selected to fly the first Vostok manned mission, and in beating the Americans, would become the first man in history to fly into space. While Gagarin was elated, Titov was visibly shaken by the decision and did not even shake his friend's hand, leading Kamanin to note in his diary, "Titov's disappointment was quite obvious" [25].

Yevgeny Karpov, as commander of the cosmonaut training centre, would later fully endorse the selection, stating that Gagarin "possessed all the important qualifications; devoted patriotism, complete faith in the success of the flight, excellent health, inexhaustible optimism, a quick and enquiring mind, courage and resolution, self-control, orderliness, industriousness, simplicity, modesty, great human warmth and attentiveness to others" [6].

A pensive Gherman Titov.

Yuri Gagarin, selected as the first Soviet cosmonaut to fly into space.

According to Piers Bizony and Jamie Doran, the authors of *Starman: The Truth behind the Legend of Yuri Gagarin*, it was felt necessary to have some official footage recording the selection the day before, so the State Commission was reconvened with cinematographer Vladimir Suvorov invited along to film a re-enactment of the proceedings. This time, Gagarin, Titov and Nelyubov were prominently seated at a table, with Kamanin seated immediately to Gagarin's right. When Gagarin's name was announced he promptly stood up and recorded a message of thanks and commitment to the success of the flight. In more ways than one this was a farcical piece of deception, as problems with the film in Suvorov's camera caused Gagarin to have to repeat his speech, while a dejected-looking Titov squirmed uncomfortably to his left. Nelyubov, meanwhile, remained stony-faced [3].

"What did I feel at that session of the commission standing beside my friend and backup?" Gagarin later told Soviet space historian Evgeny Riabchikov. "Everything was clear and yet unclear—maybe even very complex. I was thinking of Gherman. He's a very good flier. He's an intelligent man and a wonderful friend. He should make the flight too. I felt rather awkward. Why me? Why not him? Of course the commission's decision explained everything. But it would have been better to make the flight together."

Kamanin told Riabchikov that while both men were fine specimens, the traits of "courage, sober judgement, a strong will, diligence, and dedication to his objectives" displayed by Gagarin had worked to his advantage. The dossier he kept on Gagarin provided further evidence that the right choice had been made.

"Throughout the period of training for flight, Yu. A. Gagarin displayed great accuracy in the performance of various experimental psychological tests. He manifested great equanimity when subjected to sudden and powerful stimuli.

This mock "announcement" of Gagarin's selection to the first Vostok mission was re-enacted especially for the cine cameras the day after it actually took place. Nikolai Kamanin is standing with Gagarin on his left. Titov and Nelyubov (partly obscured) are seated beside Gagarin. No cosmonauts were permitted at the actual selection by the State Commission on 9 April.

His reaction to 'novelties' (the state of weightlessness, prolonged sessions in the isolation chamber, parachute jumps, and other procedures) were always positive: he evinced the ability quickly to orient himself in new circumstances and skill in maintaining self-control in various unexpected situations.

Observations made during his confinement in the isolation chamber revealed a highly developed capacity to relax, during even the brief pauses provided for rest: to drop off to sleep quickly, and to awaken on his own at the scheduled time. Noteworthy among his character traits was his sense of humour—his good nature and fondness for joking.

His sessions in the trainer were characterised by a calm, self-confident performance, with clear, concise reports after the completion of each procedure. His self-confidence, presence of mind, curiosity, and cheerfulness made for a distinct originality in the elaboration of professional skills" [26].

Once the filming of the re-enactment was finalized, everyone trooped out and got down to the serious business of changing the history of the world forever.

They had just two days of intense work and pressure ahead of them if they were to succeed.

REFERENCES

[1] Nikolai Kamanin, "A minute's readiness has been announced" (English title), *Znamya* (monthly newspaper), No. 4 (April 1989), pp. 134–146.

[2] Yaroslav Golovanov, "Cosmonaut No. 1," serialized in *Izvestia*, 2–6 April 1986.

[3] Vladimir Suvorov and Alexander Sabelnikov, *The First Manned Spaceflight: Russia's Quest for Space*, Nova Science Publishers, New York, 1997.

[4] Vladimir Golyakhovsky, *Russian Doctor*, St. Martin's Press, New York, 1984.

[5] Mitchell R. Sharpe, *Yuri Gagarin: First Man in Space*, The Strode, Huntsville, AL, 1969

[6] Yevgeny Petrov, "Cosmonauts", *Men and Women of the Soviet Union*, pp. 277–292, Progress Publishing, Moscow, 1962.

[7] Gherman Titov and Martin Caidin, *I Am Eagle*, Bobbs-Merrill, New York, 1962.

[8] Gherman Titov, *700,000 Kilometres through Space*, Foreign Languages Publishing House, Moscow, 1962.

[9] Vladimir Lebedev and Yuri Gagarin, *Survival in Space*, Frederick A. Praeger, New York, 1969.

[10] Olga Apenchenko, *Truden Putdo Tebya, Nevo! (A Reporter's Account of Cosmonaut Training)*, Gosudarstvennoye Izdatelstvo Politicheskey Literatury, Moscow, 1961.

[11] Wilfred Burchett and Anthony Purdy, *Cosmonaut Yuri Gagarin: First Man in Space*, Panther Books, London, 1962.

[12] David Scott and Alexei Leonov, *Two Sides of the Moon*, Simon & Schuster, London, 2004.

[13] Bert Vis, interview with Pavel Popovich, Association of Space Explorers' Congress, Quebec, Canada, 30 September 1996.

[14] Jamie Doran and Piers Bizony, *Starman: The Truth behind the Legend of Yuri Gagarin*, Bloomsbury Publishing, London, 1998.

[15] Vladimir Golyachovsky, email messages to Colin Burgess, 21 January and 27 March 2007.

[16] James Oberg, *Uncovering Soviet Disasters*, Random House, New York, 1988.

[17] Colin Burgess, Kate Doolan and Bert Vis, *Fallen Astronauts: Heroes Who Died Reaching for the Moon*, University of Nebraska Press, Lincoln, NE, 2003.

[18] Rex Hall and Bert Vis, interview with Irina Ponomareva, Moscow, 17 June 2003.

[19] John B. Charles, "Could the CIA have prevented the Apollo 1 fire?" *The Space Review*, 29 January 2007. Online at *http://www.thespacereview.com/article/797/1*

[20] Wikipedia on-line dictionary ("Bondarenko Crater").

[21] Encyclopedia Astronautica, "Korabl-Sputnik 5" at *http://www.astronautix.com/details/korik565.htm*

[22] Sven Grahn, "The Flights of Korabl-Sputnik 4 and 5" at *http://www.svengrahn.pp.se/histind/sputnik910/sputnik910.htm*

[23] Yaroslav Golovanov, *Our Gagarin* (translated by David Sinclair-Loutit), Progress, Moscow, 1978.

[24] Asif Siddiqi, *Challenge to Apollo: The Soviet Union and the Space Race, 1945–1974*, NASA History Division publication, Washington, D.C., 2000.

[25] Jamie Doran and Piers Bizony, *Starman: The Truth behind the Legend of Yuri Gagarin*, Bloomsbury Publishing, London, 1998.

[26] Evgeny Riabchikov, *Russians in Space*, Doubleday & Company, New York, 1971.

6

"Poyekhali!": A man in space

"Why Gagarin and myself were chosen? It's difficult to say," Gherman Titov once mused during a translated interview with the CNN network. "The commanders chose us. But all the six of us were equally well-trained, and each could pilot the Vostok spacecraft. It was Gagarin's character that mattered most. You have to understand me correctly: the first man in space had to be a nice, attractive person. [We were] told [it would be Gagarin] on the 9th of April, and journalists say that I was so glad for Yuri that I almost went to kiss him. I was disappointed, because I also counted that I would be the first man in space. But as the decision had been made, what was there to do?" [1].

THE CHOSEN ONE

Titov would expand on his comments three years later, during one of his last interviews. "I was frustrated, of course, because up to that last minute I thought my chances were high enough that I could have been the commander of the Vostok capsule. We were all young and wanted to be first; not because we wanted to be heroes or get something but because it was new work for us. We had stopped being test pilots to go into space" [2].

Why Gagarin?

Indeed, as pointed out earlier, Titov had been the first choice of many involved in the decision-making process. One of the most influential was Nikolai Kamanin who had acknowledged in his diary notes that Titov was the more proficient of the two in certain areas of training, and displayed a greater self-assurance. Space journalist Yaroslav Golovanov, given unprecedented access to the cosmonaut corps, confessed he had no clear idea as to why Gagarin was selected, even suggesting that the man

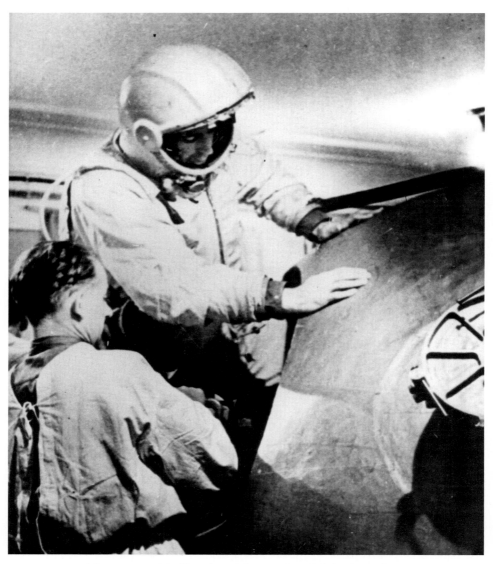

Titov engaged in Vostok training as Gagarin's backup pilot.

"was not shown to be an outstanding leader." However he adds that Gagarin's "main advantage was his intellect" [3].

In a reflective article published in 2001, *Pravda* stated that Gagarin's "humble roots saw him narrowly edge out Gherman Titov, whose intellectual family and penchant for spouting poetry seemed to make him too bourgeois for a Soviet Hero" [4].

Quite understandably the selection of Gagarin over Titov for the first flight seemed, for many historians, to be a reflection of the ideological symbolism then existing in the Soviet Union, and in particular the overt influence of Nikita Khrushchev on the space programme. It is true that Gagarin was extraordinarily personable and charming, and as the son of a poor tractor driver on a collective farm embodied and personified the unwritten preference of the Communist Party system for someone from a working-class background. But according to Golovanov in his biography of Sergei Korolev, the ultimate decision was not in fact one thrust upon space chiefs by Premier Khrushchev.

With a decision pending, biographies of the two men were evidently despatched as a courtesy to the head of the Defence Department of the Central Committee, Ivan Serbin, together with photographs of both candidates wearing civilian attire and in uniform. This information was then passed down to a member of the Communist Party Presidium and Secretary of the Central Committee, Frol Kozlov. Kozlov was an intriguing personality, once recognized as a top-level advisor to Khrushchev and even his likely successor. He was certainly regarded as Khrushchev's right-hand man in matters relating to the Soviet space effort; and a man once described by journalist Nicholas Daniloff as "the prime link between the party apparatus and the space scientists, dealing with their problems when Khrushchev was occupied with the other difficulties of running the Soviet Union" [5]. Kozlov in turn presented the folder to the Soviet premier. Having studied the profiles and photographs for a time, Khrushchev is reported by Golovanov to have then stated "Both pairs are excellent! Let them decide for themselves!" [6].

Years after the historic first spaceflight, Titov would be magnanimous in conceding that Gagarin had indeed been the right choice. "Yura turned out to be the man everyone loved," he admitted. "Me they couldn't love. They loved Yura. I'm telling you, they were right to choose Yura" [7].

The last days

On the morning of 10 April 1961, Gagarin, Titov and Nelyubov played a vigorous game of badminton prior to a midday meeting to discuss the upcoming flight, held in an ornate gazebo situated beside the Syr Darya River, on the edge of the air force compound. Korolev and Kamanin were there, together with Bykovsky, Nikolayev and Popovich and high-ranking members of the State Commission, including commission chairman Konstantin Rudnov, and the head of the Soviet strategic forces, Marshal Kirill Moskalenko. With most of the formalities over, Korolev made a short speech.

"It has been less than four years since the launch of the first satellite and we are ready for the first flight of a human into space. Six cosmonauts are present here and each of them is ready to make the first flight. It was decided that Gagarin would fly first, others will follow; as early as this year ten Vostok spacecraft will be ready. Next year we will have two- or three-seat Soyuz spacecraft. I think [the] cosmonauts who are present here wouldn't mind to accompany many of us

Yuri Gagarin would become the first Soviet cosmonaut.

into space orbits. We are confident, the [first] flight was prepared thoroughly and carefully and it will proceed successfully. All the success to you, Yuri Alexeievich!"

Kamanin and Moskalenko would also make speeches, offering their own congratulations and wishes for a successful mission. That evening, armed with the latest information from the launch team, controllers and medical staff, members of the State Commission sat down at a table to make a final determination on the launch date. Everything was in readiness, and the selection of Gagarin to fly the mission was ratified. The launch could proceed, and it would take place two days hence.

11 April 1961 would prove to be the very last day in our history in which no person had ever travelled into space. Most of the world's headlines were reporting that the trial of Nazi war criminal Adolf Eichmann would begin that day in Jerusalem, while in the United States a young mop-haired troubadour named Bob Dylan was making his singing debut that evening in New York City.

Meanwhile, at the Baikonur launch complex, a remote place once described by Titov as "a small inhabited island in the boundless ocean of the steppes," preparations were being carried out at a feverish pitch. These had begun at five o'clock

The pre-flight midday meeting held at the gazebo near the air force compound.

Moscow time that morning when a locomotive carriage supporting the horizontal R-7 core booster, earlier mated with its four strap-on boosters and the Vostok 3KA spacecraft, was trundling a slow and steady path down the rail tracks from the booster check-out building to the concrete launch pad. It was a journey of several miles across the flat desert terrain to the pad, and once the rumbling transporter–erector had rolled to a stop an anxious Korolev was on hand, checking to see that everything was still secure and no problems had occurred during the transfer. Once Korolev was satisfied that all was in order the 125.4-foot-long R-7 (officially designated Vostok 8K72K) was hydraulically raised to the upright position. The service towers were then raised, petal-like, to embrace the gleaming rocket, following which the task of fitting all the umbilical connections could begin. Fuelling would commence later that evening once the engineers had reported that final tests on the booster rocket and spacecraft had been successfully completed. At that time, hundreds of tons of fuel would begin to flow into the rocket's tanks.

The R-7 was a vehicle that could unleash considerable power, and it made for an impressive sight. Apart from the core rocket, each of its quad of identical strap-on boosters was fitted with four tapered RD-107 liquid-propellant rocket engines, with each pod capable of producing in excess of 222,000 pounds of thrust. Overall, the R-7 had 20 such engines, providing a lift-off thrust of 1,323,000 pounds. The vehicle's second stage, attached to the spacecraft, would in turn be powered by a single RD-119 engine capable of producing around 24,200 pounds of added thrust.

Final preparations

At 10:00 AM that morning of 11 April, following breakfast, Gagarin and Titov met with Konstantin Feoktistov and Boris Raushenbakh from Korolev's OKB-1 design bureau for a final but relaxed review of the flight plan. Feoktistov would tell Gagarin that the launch had been set for 9:07 the following morning. Raushenbakh clearly marvelled at the occasion he had helped to create. "I looked at [Gagarin] and in my mind I understood that tomorrow this kid was going to awaken the whole world," he would recall. "But at the same time I just could not make myself believe that tomorrow something would happen that the world had not yet seen—that this First Lieutenant sitting in front of us would tomorrow become the symbol of a new epoch" [8].

Prior to lunch, as a morale-boosting gesture to some of the lower ranked ground personnel and pad team, Gagarin was introduced to a number of excited soldiers, NCOs and officers, who all wished him a safe and enjoyable flight Following their visit to the launch pad, the cosmonauts accompanied Kamanin to a nearby small cottage, once the residence of Marshal Nedelin, where they would spend the night. There they had a late lunch consisting of the same food items Gagarin would be trialling in orbit the following morning. It comprised two types of pureed meat packed in squeezable tubes like toothpaste, followed by another tube containing chocolate sauce. "We had already been eating 'space food' for several days," Gagarin would later record in his memoirs, "pressing palatable and nourishing food out of tubes into our mouths" [9].

Then it was time for Gagarin to undergo a final physiological examination and to have a number of biosensors placed on his body, ready for the following morning. His body temperature was recorded as normal, his blood pressure also in the normal range at 115/75, and his pulse rate was a steady 64 beats per minute. Titov, there to replace Gagarin if he suddenly became ill, must have silently noted that his chances of flying were rapidly evaporating. Once all the readings had been taken by around 6:00 PM, the doctors instructed the men that it was now forbidden for there to be any more 'shop talk' about the upcoming missions. Instead they were to relax. Gagarin recalls soothing music and folk songs being played in the cottage as the two cosmonauts wound down, playing a short game of pool, and chatting about their childhoods. Titov even recited some poetry.

At around 9:30 PM the burly figure of Korolev arrived at the cottage to ensure the two men were ready. After a few words of encouragement he departed, after which Gagarin wrote that the physician (whom he named as "Yevgeny Anatolyevich") took final readings and told the men they could retire for the night. In fact, three of Yazdovsky's physicians were personally responsible for all the pre-flight medical preparations, namely Lev G. Golovkin, Ivan T. Akulinichev and Ada Kotovskaya. In giving the name of someone quite different in his heavily censored memoirs, Gagarin was likely acknowledging the presence that night of Yevgeny Anatolyevich Karpov, the director of the cosmonaut training centre (but certainly not a physician).

Before they retired to their room around 9:50 PM, Gagarin and Titov were asked if they would like to take a sleeping pill. Both men declined. "[Titov] went to sleep in the other bed in the same room," Gagarin would later recall. "We already had been living for several days together and according to the same schedule, and resembled each other in everything like twin brothers. To be sure, we *were* brothers. We were now closely linked by one great aim to which we devoted our lives" [9].

Within half an hour, when Karpov checked again, both men appeared to be sound asleep.

A DAY OF HISTORY

At 3:00 AM on 12 April 1961, it was a cool, grey spring morning over Tyuratam as pre-launch operations began at the launch pad. Excited controllers began taking their stations, knowing that their counterparts elsewhere in the Soviet Union were doing likewise. Pavel Popovich would be the chief spacecraft communicator, aided by Alexei Leonov, who had been dispatched from Moscow to a UHF radio station at Petropavlovsk-Kamchatskaya on the remote Kamchatka peninsula in the Soviet Far East to help monitor the flight and communicate with the spacecraft if deemed necessary. Leonov would later reveal that until the time his station picked up the first transmission from orbit, he had no idea which of his six fellow cosmonauts had been selected to fly the mission [11].

A restless night for all

An anxious Korolev had been unable to sleep, constantly thinking of ways in which problems might occur during the flight. After flipping aimlessly through a copy of *Moskva* magazine he had paid a brief visit to the small cottage shortly after two o'clock where Karpov, also sleepless, reassured him that the cosmonauts seemed to

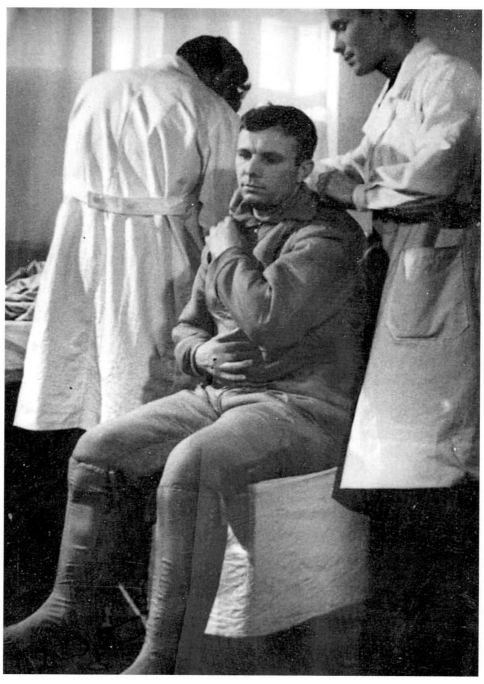

Gagarin suiting up on the big day.

be sleeping peacefully. "Sergei Pavlovich went off silently, indicating with gestures that everything was all right on his side," Karpov would later record. "I was later told that he went straight from us to the launching pad, where the final tests began at three o'clock in the morning" [12].

At 5:30 AM, as scheduled, Karpov crept into the small bedroom where the two men were still fast asleep. He paused at Gagarin's bedside and gently tapped him on the shoulder. The cosmonaut's eyes opened immediately, and he glanced at the clock by his bed before. "Am I to get up?" he asked simply. No answer was needed or given, as Titov was also being roused. A smiling Karpov had arrived bearing a small gift for each of them—a bouquet of early spring wild tulips presented to him earlier that morning by Klaudia Akimova, the woman who usually resided in the small cottage.

Some years later, according to Gagarin biographers Doran and Bizony, the cosmonaut revealed to Korolev what the attending physicians had long suspected—that he had not slept at all the night before his flight. Nor, he said, had Titov [7].

After some light exercises and a wash and shave the two men sat down to a very austere breakfast, once again packed into tubes and described by Gagarin as "chopped meat, blackberry jam and coffee" [9]. Jokingly, the men tried to pretend it was the most delicious meal they'd ever tasted, and Gagarin even laughingly said he'd have to take some home for Valentina to try. Yazdovsky and his physicians then conducted examinations of both men after which they were assisted into their Sokol spacesuits, with Titov going first so that Gagarin would not be in the hot suit for too long before they could plug into a power and oxygen source installed in their transfer bus. Their garments consisted of underwear and a sky-blue pressure suit followed by a protective orange overall. "All the instruments and apparatus with which the spacesuit is equipped were tested then and there," Gagarin reflected. "This procedure lasted rather a long time. On my head I put a white helmet with earphones and over that an airtight helmet on which CCCP (USSR) was inscribed in large letters" [9].

In fact, the lettering on Gagarin's helmet had been a last-minute addition. In the frenzy of pre-flight preparations, someone realized that on landing Gagarin might be mistaken for a foreign pilot, as the shooting down of Francis Gary Powers' U-2 spy plane the previous year was still fresh in the minds of the Russian people. "We thought there could be some misunderstanding," said Vitaly Svershchek, deputy director of the *Zvezda* spacesuit manufacturers. "Just before the flight we discovered this and a colleague of ours fetched a can of paint and painted [U.S.S.R.] on the helmet" [10].

With everything now in readiness, and with no evident problems, the flight was given official sanction to proceed.

Beginning of a new era

An hour after the two men had been woken, and as dawn began to streak the sky with pale colours, Korolev paid the cosmonauts a short visit, meeting them outside the cottage before they boarded the transfer bus that would carry them out to the pad. He wanted to make sure that the suiting-up procedure had gone well, and both men were ready to go. It was obvious to Gagarin that Korolev was exhausted and under considerable strain.

"It was the first time I had seen him looking care-worn and tired: he had clearly had a sleepless night. I wanted to give him a hug just as if he were my father. He added a few more recommendations and pieces of advice which I had not heard before and which might come in useful on the mission. It seemed to me that seeing the cosmonauts and talking with them cheered him up a bit.

The people helping me get into my space suit held out pieces of paper; one even held out his work pass, asking for an autograph. I couldn't refuse and signed several times. The specially equipped bus then drew up. I sat down in the "space" seat which was shaped like the seat in the module. The suit had its own ventilation apparatus, which required electricity and oxygen to run. The ventilation apparatus was plugged into the power and oxygen source that had been set up in the bus. Everything worked well" [12].

The two cosmonauts were ready for the ride out to the launch pad aboard the blue-and-white transfer bus, with the identically dressed Titov taking a seat directly behind Gagarin. Already onboard the bus were several excited people: cameramen, doctors and scientists, as well as fellow cosmonauts Nikolayev and Nelyubov. They all smiled and joked with each other and exchanged friendly salutes as the bus started up and then rumbled along a paved road towards where a silvery-grey rocket waited; one hundred and twenty-five feet of shining missile emitting wispy clouds of vapour into the crisp morning air, seemingly held upright by the clamps stretching out from the green work tower. "The closer we got to the launching pad, the larger the rocket grew," Gagarin recalled, "just as if it were changing size. It looked like a giant beacon and the first ray of the rising sun shone on its pointed peak" [9].

As popular legend has it, Gagarin was said to be a little quiet and edgy and unexpectedly asked that the bus be stopped before reaching the pad so that he could

A playful Nelyubov was one of those who accompanied Gagarin and Titov to the launch pad.

Gagarin strides towards the launch pad and the waiting rocket.

urinate. From that time on, tradition dictates that every male cosmonaut has to stop on the way to the launch pad and urinate on the left rear tyre, just as Gagarin is supposed to have done that day. However, the veracity of this story has often been disputed, particularly in light of a cine film showing the bus heading out to the launch pad, and it does not appear to make any stops along the way.

On arrival at the pad, Gagarin's suit was unplugged from the bus system and the hoses secured into a portable unit he would carry up to the spacecraft. He then stepped down from the bus at 6:50 AM and walked about 40 feet to where Chairman of the State Commission Konstantin Rudnev was waiting. He came to a halt before Rudnev and officially reported for duty, delivering a brief statement of readiness. This was followed by a number of warm hugs from officials and colleagues, including Andrian Nikolayev. "I was so nervous," Nikolayev later stated, "that I forgot he was wearing a helmet and tried to kiss him. I knocked my forehead against it so hard I even had a bump there. 'One for all and all for one, lads!' shouted Yura to us and set off for the launching pad" [12].

Gagarin then carefully made his way up the steps of the launch gantry to where the lift was waiting, after which he turned and waved to the crowd gathered below. The lift ride was a slow one, delivering the cosmonaut to the apex of the rocket nearly three minutes later, where his Vostok craft stood ready to be occupied.

The hatch is closed

With a film crew assembled to record the historic event, Gagarin was assisted into his seat by an engineer from the pad team headed by Oleg Ivanovsky, who personally ensured that all the connections and attachments between the cosmonaut and his craft were then carried out and secure. All of this was performed under the watchful eye of Alexei Ivanov, senior designer of the Vostok craft.

Once Ivanovsky and Ivanov were confident that all procedures had been properly carried out, Gagarin was able to conduct a suit pressurization test and communications check with primary CapCom Pavel Popovich, although the cosmonaut had to request that some piped patriotic music be switched off temporarily during the radio check. Popovich then asked Gagarin how he was doing, to which the waiting cosmonaut responded "Like they taught me."

A ludicrous anomaly was involved in this first human spaceflight. The flight would be under automatic or ground control from beginning to end, and Gagarin would be little more than a passenger throughout his journey. In fact, a comparison of the Vostok and Mercury instrument panels reveals that while the Mercury panel housed 56 switches and 76 indicators, the Vostok panel had only 4 switches and 35 indicators. However, there were only two manual control functions available to Gagarin in the event of an emergency—the ability to orientate the spacecraft into the correct attitude and the firing of the retrorocket to initiate descent. There were still lingering fears that weightlessness might adversely affect a spaceman's mental abilities, and it was even quietly feared that even the stable Gagarin could panic and try to do something unexpected, with potentially fatal consequences. One prime concern among the psychologists was that Gagarin might lose all reason and try to take over manual control of the Vostok craft. He was therefore "locked out" of assuming control. However, it was understood that should an emergency situation arise which actually necessitated the initiation of a manual re-entry, a three-digit code not revealed to Gagarin (now known to be 1-2-5) could be entered into a pad containing two rows of numbers that was attached to the manual control logic clock. Gagarin was informed that the code would be instantly communicated to him if a serious malfunction occurred.

Oleg Ivanovsky was acutely uncomfortable with the fact that Gagarin was not in immediate possession of the code that would allow him manual control of the spacecraft, especially in the added event of a communications failure, and so he decided to break some serious rules. He leaned into the spacecraft and rapped his knuckles against the cosmonaut's faceplate, signalling him to open it so he could communicate something without having to use the monitored radio link. Once Gagarin's faceplate had been raised, Ivanovsky leaned even closer and said in low tones, "Yura, the numbers are 1, 2, 5." At that Gagarin smiled, and told Ivanovsky that Nikolai Kamanin had already revealed the code digits to him [12].

An hour after his insertion into the Vostok capsule, the round hatch was closed and Gagarin was on his own. Outside, the technicians were busy securing the 30 bolts around the circumference of the hatch. With very few pre-launch duties to perform, and with the spacecraft's shrouds preventing him from looking outside, Gagarin settled back and asked Popovich if the piped music could be switched on once again. Soon after he became alarmed as the hatch was suddenly re-opened, but Korolev's voice came through his earphones, assuring him that there were indications one of the explosive hatch's contacts had not secured properly and this would need to be rectified. It was either a faulty contact or warping in the hatch. Either way, it would need to be recycled by Ivanovsky's ground crews. Gagarin, facing away from the hatch, sat calmly watching the operation through a small mirror attached to the sleeve of his spacesuit. Test engineer Lev Maryanin would later recall Gagarin sitting calmly and smiling on his couch, listening to music. As Maryanin would comment in a *Pravda* article on the first anniversary of Gagarin's flight: "Seeing the splendid mood of the cosmonaut the mechanics felt better too. They quickly inspected every-thing and found that a bracket which sealed the control contacts was a little bit loose. It took them literally one minute to put everything right. Again they wished the cosmonaut a good trip to the cosmos and locked the hatch" [13].

The final countdown

Fifteen minutes before launch Gagarin donned his flight gloves and ensured his helmet was fully sealed. Then, at 9:05 AM, with two minutes remaining until lift-off, the cosmonaut could feel a slight swaying of the R-7 on the launch pad. He

Lift-off on 12 April 1961, and the world's first manned spaceflight had begun.

knew that the service towers had been folded back and his rocket was ready to leap into the skies. "Roger, I'm in the mood," he announced soon after. "I feel fine, and I'm ready for the launch. I felt the working of the valves."

Then, at the appointed time of 9:07 AM, Gagarin could feel a growing rumble beneath him as the main engines ignited. A hold-down arm would restrain the rocket until the thrust had built up to the correct level, at which point the arm would disengage and the R-7 would be unleashed. He could hear the voice of Korolev in his earphones. "Preliminary stage ... intermediate ... main ... Lift off! We wish you a good flight. Everything's all right."

A delighted Gagarin, feeling the surge of power at lift-off radioed back: "*Poyekhali* [Off we go]! Goodbye until [we meet] soon, dear friends."

The very first human space mission had begun from the Baikonur Cosmodrome, and the name of Yuri Gagarin was about to be propelled into the history books.

THE WORLD'S FIRST SPACEMAN

On that historic April morning, Muscovites tuned to their radio sets would have been listening to the patriotic song, *How Spacious Is My Country*. Then, at 9:59 AM, the song ended abruptly and an excited newsreader broadcast an astonishing bulletin: "The world's first spaceship, Vostok [East], with a man on board, has been launched on April 12 in the Soviet Union on a round-the-world orbit." The announcer would later reveal that the Vostok craft was piloted by twenty-seven-year-old Senior Lieutenant Yuri Alexeievich Gagarin of the Soviet Air Force (he would be promoted in rank to major during his mission).

Regular bulletins would follow, sustaining the suspense for the Russian people and keeping them updated on the progress of the flight. Meanwhile, loudspeakers mounted in public squares throughout Moscow urged citizens to lay down their pens and tools and to join in celebrating this momentous occasion. As *Time* magazine would later report on the event: "From Leningrad to Petropavlovsk, the U.S.S.R. came to a halt. Streetcars and buses stopped so that passengers could listen to loudspeakers in public squares. Factory workers shut off their machines; shopgirls quit their counters. Schoolkids turned eagerly from the day's lessons. Somewhere above them, a Soviet citizen was arcing past the stars, whirling about the Earth at 18,000 miles an hour, soaring into history as the first man in space" [14].

Riding the rocket

During the ascent of the R-7 booster there was a temporary break in the communications line, causing immediate concern in the control centre. OKB designer Konstantin Feoktistov, who would later fly into space on the first three-man Voskhod flight, would wryly acknowledge that "such breaks considerably shorten a designer's life."

As his spacecraft settled into its orbital path, a delighted Gagarin reported over his high-frequency radio that the view he could see through his floor-mounted Vzor

Ground controllers were able to monitor Gagarin's movements through transmitted images of the cosmonaut.

periscope device was glorious. "I see the Earth!" he cried. "I see the clouds. It's beautiful; what beauty!" The ground communications system was limited by geography, having no stations outside of the Soviet Union, so Gagarin would eventually pass out of contact as he flew around the far side of the Earth. But while communications permitted he briefly reported on his condition, the effects of weight-lessness, and what he could see of the ground through his Vzor porthole. "The flight is normal. I am withstanding well the state of weightlessness," he reported over Africa. A highly stressed Korolev, who had experienced chest pains during the launch phase, was far more relaxed now and even excited, although his anxiety increased as the Vostok craft dashed inexorably out of radio contact. However, Gagarin was able to maintain a form of communication by tapping out messages on a telegraph key.

Gagarin may have only been a passive passenger throughout the flight, but television cameras were recording for the first time in history pictures of a person in space as the cosmonaut carried out minor duties, even playing gleefully with a weightless pencil on the end of a piece of string that quickly floated away and became lost. He also drank water through a tube attached to a polyethylene container, and squeezed pureed food into this mouth, proving that a human could eat and drink safely in space. At one time, while drinking, he sat fascinated as a stray globule of water slowly floated around his cabin.

Then, all too soon, what to that time had been a relatively flawless mission was coming to an end after a single orbit of the Earth as Gagarin prepared for the re-entry sequence to begin, 79 minutes into his mission. He checked and tightened the restraint straps holding him against the ejection seat and closed his helmet.

Meanwhile, American intelligence and national security operatives in overseas outposts had begun intercepting signals from the spacecraft as it passed over their stations, confirming reports that a man had been placed into orbit. Elint (electronic intelligence) station operators based in Shemya, Alaska, picked up live television transmissions beamed down from space just 20 minutes after the launch. News of the launch had been immediately wired through to the CIA in Washington, D.C., where a report was hurriedly prepared for President Kennedy.

When things go wrong

Over the following three decades, it was believed in the West that the retrofire sequence to end Gagarin's orbital flight had gone smoothly, eventually bringing the Vostok craft back safely through the atmosphere. But it is now known that Gagarin came perilously close to losing his life near the end of his flight. The story first came to light with a secret report filed by Gagarin after his flight, which was printed in 1991 in the Russian newspaper, *Rabochaya Tribuna*. Then, five years later, Colonel Yevgeny Karpov's notes from the time of the flight were auctioned at Sotheby's in New York, and they confirmed the earlier press reports.

According to Karpov, Gagarin's commander, the retrorocket situated in the Vostok's instrument module fired for precisely 40 seconds, slowing the craft before it plunged into the atmosphere. As the engine shut down, Gagarin felt a sudden jolt, and his descent module began spinning out of control. Under normal circumstances the instrument module would have been explosively jettisoned from the descent module some ten seconds after the engine shut down, and while separation did take place, the two modules were still tethered together by an umbilical cable that had failed to separate, causing the linked spacecraft to tumble as they arced back into the atmosphere. There was a real risk that the two components could also collide, causing a fatal breach in Gagarin's spacecraft [15].

Post-flight, Gagarin would prepare a comprehensive report on every stage and aspect of his flight, many details of which would remain top secret for more than three decades until published in *Rabochaya Tribuna* in 1991. This report would include—in the first cosmonaut's own words—how he felt during his hazardous re-entry.

> "When the light went out with the transmittal of the third command, I started watching the pressure in the RRS [retrorocket system] and the attitude-control system. It started to drop swiftly . . . I could feel the RRS kick in. You could hear a buzzing and a noise through the frame . . . The g-loads were growing . . . At that moment, the needles for the automatic attitude-control system and the RRS tank jumped to zero at the same time . . . The craft began to rotate . . . at a very high rate . . . I was just barely shutting the Sun out, so I could keep the light out of my eyes. I put my feet up to the viewport, but I couldn't close the shutters. I was wondering what was going on. I knew that, according to the calculations, it was supposed to happen 10–12 seconds after the cut-in of the RRS . . . But it felt like more time had

passed than that, and there was no separation. On the instrument panel, the 'Spusk 1' [Descent 1] light was still on, and the 'Prepare for ejection' was not on. The separation wasn't taking place ... I decided that something was wrong ... I figured I'd land one way or other ... somewhere before the Far East I'd land ... I transmitted an EN—everything normal ...

The spacecraft is beginning to turn slowly, about all three axes. It began to swing 90° to the left and right ... I could sense the oscillation of the craft and burning of the coating ... I could feel the high temperature ... Then the g-loads began to grow smoothly ... The g-load felt to be more than 10g. Things started turning a little grey. I squinted and strained again. That helped, and everything sort of went back in place" [16].

Meanwhile, Gagarin was acutely aware that if the problem persisted he would be incinerated in the unforgiving heat of re-entry, but he remained calm and kept reporting his situation over the radio. Finally, after ten perilous minutes, the umbilical sheared and burned through in the intense heat with an audible bang, and the instrument module fell away. The cosmonaut's descent module now assumed the correct orientation and continued to plunge through the atmosphere, its forward section quickly glowing white-hot from friction with the atmosphere.

"I'm waiting for ejection. At that time, at an altitude of about 7,000 meters, hatch No. 1 fired. There was a pop, and the hatch flew away. I'm sitting and thinking: it didn't eject me? So, I carefully looked up. At that moment, it fired, and I was ejected ... Went out with the seat. The cannon fired again, and the stabilizing parachute was deployed" [16].

In Pitsunda, where a visibly anxious Nikita Khrushchev was waiting for word of the end of the mission, he finally received a phone call from Korolev. It was a very poor connection, so Korolev spoke loudly right into the receiver. "The parachute has opened, and he's landing!" he told the Premier. "The spacecraft seems to be okay."

"Is he alive?" Khrushchev yelled into the phone. "Is he sending signals? Is he alive? Is he alive?"

Well before the facts of Gagarin's dangerous re-entry were known, another controversy had surrounded his flight, emanating from the Soviet stance that Gagarin had landed inside his Vostok spacecraft. Given the haste to launch a man into space, Korolev's designers had not been able to develop a satisfactory retro-rocket device that would fire just above the ground and brake the fall of the spacecraft to allow a reasonably cushioned landing. It was known instead that the Vostok would come down hard, perhaps hard enough to seriously injure the occupant, because there was no retrorocket device installed in any Vostok spacecraft. All of the cosmonauts involved in Vostok flights would be ejected from their craft at around 23,000 feet above the ground. The problem is that the Soviet space chiefs insisted post-flight that Gagarin had landed inside his Vostok craft so they could claim his flight as a new record. Under international aviation regulations, in order to establish a new aviation record, a pilot must take off and land inside his vehicle, and the Soviets wanted to be able to claim this distinction.

Touchdown

Shortly after being fired out of the Vostok spacecraft, Gagarin separated from his ejection seat and continued the descent by parachute. Below him he could see the Volga River, and he would actually land quite close to the pre-planned recovery site. There would be another moment of tension when his reserve parachute pack suddenly fell down and dangled beneath him. Fortunately the parachute did not deploy, as it could have become entangled with his main chute.

Gagarin touched down on soft soil on a collective farm named The Road of Lenin, 18 miles southwest of Engels near the village of Smelova, in the Saratov region of the Soviet Union, while his Vostok craft landed some two miles away. He was able to stay on his feet, and after removing the parachute harness opened his visor and breathed in the deep, rich smell of the land around him. "It was like a good novel," he would later reflect. "As I returned from outer space, I landed in the area [Saratov] where I had started flying" [12].

Both he and his spacecraft soon attracted the interest of puzzled locals who came to view the strange spherical craft and the man in the orange spacesuit and helmet. The first people to meet Gagarin after he had landed under his parachute were Anna Takhtarova and her granddaughter [17].

> "As I stepped down on firm ground, I saw a woman and a little girl and a spotted calf watching me curiously. I started walking towards them, and they moved towards me. But as they came closer, they slowed down. I was still wearing my bright orange space suit, which looked unusual and was making them feel uneasy. They hadn't seen anything like it before. I took off my helmet and cried out, feeling the chill of excitement.
> 'Don't be afraid, comrades. I'm no intruder.'
> The woman was the local forester's wife, and the girl was her granddaughter.
> 'Are you from space?' the woman asked uncertainly.
> 'You won't believe it,' I said.
> Then I saw a group of harvester operators running towards me and crying my name. Those were the first people I saw on Earth after my mission in space. We embraced and kissed one another" [18].

As he had landed fairly close to the target zone, helicopters were soon on the scene to pick up the cosmonaut and his heat-charred spacecraft. A post was erected on the spot where Gagarin had touched down, bearing a wooden sign reading "Do not remove. 12.04.1961, 10:55 Moscow time." It would later be replaced by a small stone obelisk. Nearby, a titanium obelisk now marks the landing site of the Vostok craft.

Meanwhile, people across Russia were nervously waiting for news of the cosmonaut's safe landing. In a broadcast at 11:10 AM they had been informed that Gagarin had completed a circuit of the Earth 45 minutes earlier, and the spacecraft's

At first a simple wooden marker was used to denote Gagarin's touchdown point.

braking rocket had been fired to bring Gagarin back to the ground. It would be another 75 minutes later, at 12:25 PM, when the news was broadcast that: "At 10:55 cosmonaut Gagarin safely returned to the sacred soil of our motherland."

A hero returns

Russia would explode in pride and exultation. As *Time* magazine reported: "Hats were heaved aloft. Russians cheered, hugged each other, and telephoned their friends. The celebration spread from factories to collective farms, from crowded city streets to clusters of huts on the lonely steppes. Newspapers blossomed with bright red headlines. Everywhere people paraded with banners hailing the Soviet leap into space. Not even for Sputnik 1 had the U.S.S.R. worked up such effervescent enthusiasm" [14].

Gagarin was first transported by helicopter to the military air base at Engels, where he received a congratulatory phone call from Premier Khrushchev and made a couple himself. After this brief stopover he transferred to an Iyyushin-14 which would fly him to the large town then known as Kuybyshev (now Samara), located on the Volga River some 560 miles southeast of Moscow. Here he would relax in an exclusive *dacha* while being thoroughly questioned about his flight and given a comprehensive medical examination.

Two days after his historic flight, having been given a medical clearance and recorded his immediate impressions of the flight, Gagarin was flown to Moscow aboard an Ilyushin-18 for a stupendous welcome. As reported in the *Los Angeles Times* on 15 April 1961 [19], "Weeping with emotion, Premier Khrushchev brought spaceman Yuri Gagarin home to Moscow Friday and the city's millions hailed the astronaut as the space age Columbus in a roaring welcome no Stalin or czar ever received ... Foreign observers presumed that never in the 805-year-old tumultuous history of Moscow has there been such an outpouring of public affection."

Life magazine would also report on the event: "Pictures of Gagarin suddenly appeared on postcards and in blow-ups stuck all over the city. The Soviet government which had jumped Gagarin from senior lieutenant to major just before the flight, now pinned on his chest its highest medal—Hero of the Soviet Union—and created a new title for him; pilot-cosmonaut" [20].

The massive celebration in Red Square (which some sections of the Soviet media wanted renamed in Gagarin's honour) had been followed by a celebratory reception held at the Great Kremlin Palace, and a news conference at the Scientists Club the following day, which was attended by over a thousand reporters.

President John F. Kennedy sent a grudging message of congratulation, which read: "The achievement by the USSR of orbiting a man and returning him to the ground is an outstanding technical accomplishment. We congratulate the Soviet scientists and engineers who made this feat possible. The exploration of our solar system is an ambition which we and all mankind share with the Soviet Union and this is an important step toward that goal. Our own Mercury man-in-space programme is directed toward that same end."

Gagarin soon after his
history-making
spaceflight had ended.

OFFICIAL!

RUSSIAN BACK FROM SPACE

LARGEST DAILY SALE IN N.S.W.

FINAL **Daily Telegraph**

And DAILY NEWS

The paper you can trust

SYDNEY, THURSDAY APRIL 13, 1961 (5d)

PHONE: B6666.   Newsagents: BM1044, BM1045
BOX 4088, G.P.O. 168 Castlereagh St.

WEATHER: CITY FORECAST: Cloudy, warm. Some rain and thunderstorms. Light winds. (Details page 39.) Forecast minimum temperature 77 degrees; Yesterday's maximum temperature, 79.4 degrees; minimum, 65.2.

TIDES—High: 6.42 a.m. (5ft. 5in.); 7.14 p.m. (5ft. 7in.). Low: 12.26 a.m. (1ft.); 1.85 p.m. (7in.). SUN: Rises, 6.16; sets, 5.35. MOON: Rises, 3.43 a.m.; sets, 5.31 p.m.

MOSCOW, Wed. — The world's first spaceman landed safely today after orbiting the earth in a Soviet rocket.

He stayed in space 108 minutes, completing slightly more than one circuit of the globe, before landing at a pre-determined spot in Russia.

The astronaut, Soviet pilot Major Yuri Alekseyevich Gagarin, 27, is married, with two daughters, aged two years and one month.

He made the flight in the spaceship Vostok, which means east.

Vostok weighs 11,200lb., or slightly more than five tons, excluding the weight of the final stage rocket-carrier.

It reached a minimum altitude of 109 miles and a maximum altitude of 187½ miles.

Vostok orbited the earth in 89min. 6 sec.

The first official Soviet announcement said briefly:

"Tovarish Gagarin is in very good condition."

American Associated Press said Soviet scientists watched the flight on television.

"Feel well"

The scientists released two messages from Major Gagarin:

OVER SOUTH AMERICA: "The flight is normal. I feel well."

OVER AFRICA: "I am withstanding the state of weightlessness well."

This is the official timetable of today's historic launching:

4.58 p.m. (Sydney time): Tass, the official Soviet news agency, announced that Russia had put the first man into space.

(Tass announced later that the spaceship had blasted off at 4.07 p.m. Sydney time.)

Moscow Radio interrupted its programmes to give the news.

5.41 p.m.: Tass reported the spaceman over South America.

6.12 p.m.: Moscow Radio reported that at 5.25 p.m. the spaceship's braking system was operating.

The Radio said Major Gagarin had begun his descent to a prearranged spot in the Soviet Union.

6.55 p.m.: Moscow Radio reported that Major Gagarin had landed safely.

Major Gagarin was

strapped into a custom-built couch, surrounded by a thin double wall of metal and devices designed to keep him alive.

The capsule separating him from the radiations, meteorites and temperatureless vacuum of space was specially designed for him.

But basically it was the

CONT. P. 3 UNDER THIS SYMBOL

Jew-killer's trial—P.10

SOVIET PILOT Major Yuri Gagarin, the first man into space, wearing test pilot rig. (U.P.I.-Radiophoto.) See "Spaceman is a moulder," Page 3

Newspapers around the world carried the astonishing news of Gagarin's mission. This is the front page of Sydney's *Daily Telegraph* the following day.

The view from space

At the massive Saturday press conference, Gagarin was asked to give his impressions as the first person ever to view our planet from Earth orbit.

"The earth ... can be seen very well. The appearance of the Earth's surface is roughly the same as we can see in flights at great heights in jet airplanes. Large mountain areas, large rivers and forests, coastlines and islands were easily distinguishable. During the spaceflight I was fully able to follow what we pilots call 'navigation' by geographical locality.

The clouds covering the surface of the earth could be seen very well, and the shadows of these clouds on the earth. The sky, however, was quite black. The stars in the sky look brighter and can be seen more clearly against this black sky. The earth had a most characteristic and very beautiful light blue halo. The halo becomes particularly distinct at the horizon where a gradual transition in colours takes place from the soft blue; it merges from light blue to dark blue to violet to black to a quite black sky. The transition is very smooth and beautiful.

On the surface itself, on the very horizon of the surface, one could see a bright orange colour, which then emerged into all the colours of the rainbow. The entry into earth's shadow occurs very quickly. Darkness comes at once, and nothing is visible. The stars are already visible. The stars are clearly visible."

One question everyone wanted to ask was how it felt to be weightless for such an extended period, and how Gagarin felt he coped with the sensation.

"Once I was in orbit, once I was separated from the carrier rocket, I experienced weightlessness. At first this sensation was somewhat unusual, even though I had experienced it before for short periods. I quickly became accustomed to this sensation of weightlessness, adapted myself to it, and continued to carry out the programme which was set for me during the flight ... In my view the condition of weightlessness does not affect the functioning of the organs, the performance of physiological functions ... I took food and drank water ... my capacity for work was fully maintained."

Gagarin's problems at re-entry were actually hinted at in an interview conducted for *Pravda* newspaper two months after his flight [21]. "The ship entered the dense layers of our atmosphere. Its external surfaces heated rapidly. Through the porthole I could see reflections of raging flames which encompassed my ship. In spite of the fact that I was inside a fireball, headed for Earth, the temperature inside the cabin was 20°C. Terrific *g*-forces pinned me down to my chair. The ship went into a spin. For a moment I was terrified. I could see the dear faces of my wife and children. The gravitation was much stronger than during the launching. The ship's rotation increased. I reported this situation to Earth. The spinning motion stopped suddenly and the descent proceeded in a normal manner."

While for most of the world's population Gagarin's flight around the world transcended deeply entrenched national and cultural posturing, it undoubtedly proved a massive propaganda coup for the Soviet Union. A centuries-old dream of space travel had finally become a reality, and Yuri Gagarin had instantly become the most famous and recognized person on Earth. As Trevor Rockwell of the University of Alabama would later observe: "To the Soviet state, Gagarin's flight was more than a technical and scientific achievement: in the context of the Cold War, it was an ideological victory which Gagarin's image was moulded to reflect" [22].

AN UNEASY FAME

This sudden global fame did not sit easily at first with Gagarin. For a person who would describe himself as "a simple man" the adulation was overwhelming. However, he toed the party line and while he spoke enthusiastically about his flight he was very guarded and disciplined in his responses.

The world wanted to see and greet the world's first space traveller, and the Soviet government was keen to show him to the best possible advantage, and Gagarin would soon embark on a highly publicised world tour, beginning at home with visits to cherished places associated with his childhood and military bases across the Soviet Union. From there he would travel around the world at the invitation of governments, kings and queens, presidents and various heads of state.

"I've made a lot of mistakes"

Privately, while realizing he was a travelling ambassador for communism and the people and ideals of the Soviet Union, it quickly became a role that Gagarin found exceedingly uncomfortable, and he felt embarrassed that all the attention was being focused on him. "I was far from alone in this achievement," he once recorded. "There were tens of thousands of scientists, specialists and workers who participated in preparing for this flight. I feel awkward because I am being made out to be some sort of super-ideal person. In fact, like everyone else I've made a lot of mistakes and have my weaknesses too. It's embarrassing to be made to seem like such a good, sweet little boy. It is enough to make one sick" [23].

It is certainly true that Gagarin had his weaknesses, and one of them was women. While on his national and world tour the young officer, short in stature but with an engaging smile, suddenly found that beautiful women were openly clamouring to share his bed, and he evidently found the lure to be irresistible. It was one aspect of the adulation that would soon be shared by his cosmonaut colleagues, and by many of their astronaut counterparts in the United States. All too soon, these instances of rampant womanizing and excess drinking by Gagarin and his fellow cosmonauts would lead to many profound problems, both for them and the Soviet space hierarchy, who found that the glories attached to their space triumphs would also have an unwelcome dark side.

One of Gagarin's first stops on his world tour was London, where a huge and exuberant welcome astonished British authorities, who had sent a minor official from the Ministry of Science to officially greet the cosmonaut. Britons, it seemed, loved the blue-eyed, smiling pilot. The conservative *Daily Express* newspaper welcomed Gagarin with a headline that read: "*Fantastichesky.*" He was even driven round in an open-top Rolls-Royce with the specially issued licence plate YG-1 and dined at Buckingham Palace.

Meanwhile thousands of people lined the streets wherever the first man in space went.

REFERENCES

[1] Interview with Gherman Titov, CNN *Cold War* series, conducted March 1996. Online at *http://edition.cnn.com/SPECIALS/cold.war/episodes/08/interviews/titov/*

[2] Adam Tanner, "Once adored cosmonaut looks back," Reuters, 28 April 1999.

[3] Yaroslav Golovanov, *Cosmonaut No. 1*, serialized in *Izvestia* newspaper, 2–6 April 1986. Extracts translated by Jonathan McDowell.

[4] "Gagarin smile still shines on Planet Earth," *Pravda*, Moscow, 12 April 2001.

[5] Nicholas Daniloff, *The Kremlin and the Cosmos*, Alfred A. Knopf, New York, 1972.

[6] Yaroslav Golovanov, *Korolev: fakty I mify*, Nauka, Moscow, 1994, pp. 634–635.

[7] Jamie Doran and Piers Bizony, *Starman: The Truth behind the Legend of Yuri Gagarin*, Bloomsbury, London, 1998.

[8] Asif Siddiqi, *Challenge to Apollo: The Soviet Union and the Space Race, 1945–1974*, NASA History Division publication, Washington, D.C., 2000.

[9] Yuri Gagarin, *Road to the Stars: Notes by Soviet Cosmonaut No. 1*, Foreign Languages Publishing House, Moscow, 1962.

[10] ESA News from Moscow, "40th Anniversary of Gagarin's flight: Gagarin's triumph inspires," No. 9, 16 April 2001.

[11] David Scott and Alexei Leonov, *Two Sides of the Moon*, Simon & Schuster, London, 2004.

[12] Yaroslav Golovanov, *Our Gagarin*, Progress, Moscow, 1978.

[13] Lev Mayanin, "The First Cosmic Flight a Year On," *Pravda*, 12 April 1962.

[14] "The cruise of the Vostok," *Time* magazine, Friday, 21 April 1961.

[15] William J. Broad, "Russian space mementoes show Gagarin's ride was a rough one," *New York Times*, 5 March 1996.

[16] "Manned mission highlights," Joint Publication Research Service, Report on Science and Technology—Central Eurasia: Space (JPRS-USP-92-004), 10 June 1992.

[17] "First spaceman," *Red Star* newspaper, Moscow, 31 May 1961.

[18] Anon., *Always Aim to be the First*, Novosti Publications, Moscow, 1991.

[19] *Los Angeles Times*, 15 April 1961.

[20] *Life* magazine, Vol. 50, No. 16, 21 April 1961, pp. 2–12.

[21] *Pravda*, Moscow, 18 June 1961

[22] Trevor Rockwell, "The molding of the rising generation: Soviet propaganda and the hero-myth of Iurii Gagarin," *Past Imperfect (Journal of the University of Alberta, Canada)*, No. 12, 2006.

[23] Rex Hall and David J. Shayler, *The Rocket Men: Vostok and Voskhod, The First Soviet Manned Spaceflights*, Springer/Praxis, Chichester, U.K., 2001.

7

Vostok flights continue

The flight of Yuri Gagarin had an immediate and sobering effect on NASA and its Mercury astronaut corps as the space agency was undergoing final preparations to launch an American into space on a suborbital flight. Not only had the U.S.S.R. comprehensively beaten the United States into space, but the manned orbital flight had already made the suborbital flight seem redundant by comparison. America, it seems, had paid the price for their open policy in announcing future space plans, and while the Soviet Union had run a secretive programme to launch a man into space, all the lead-up signs had also been there to provide significant clues as to their intentions. If dogs could survive orbital flights, then how far away was a manned mission?

THE IMPACT OF GAGARIN'S MISSION

Knowing that they were now considered behind in human spaceflight matters, NASA resolutely launched Alan Shepard into space just three weeks after Gagarin's orbital flight, and followed this just two months later by launching a similar suborbital Mercury flight with astronaut Virgil "Gus" Grissom aboard. Shortly after Shepard's successful mission, President John Kennedy committed America to landing a man on the Moon before the decade was out, and while NASA had originally envisaged several suborbital manned test flights, they decided to press on and commence orbital missions following Grissom's flight.

Second cosmonaut

With the suborbital flights of Shepard and Grissom successfully completed, the Americans must have felt they were at least some way towards bridging the gap with the Soviet Union in the space effort, but all too soon that gap would remorselessly open wide once again. Originally scheduled for late 1961, John Glenn would pilot

In May 1961 the cosmonauts and some other space dignitaries and trainers enjoyed a short break at Sochi, a resort on the Black Sea. In the top photo (back row, left to right), Filatyev, Anikeyev and Belyayev. Middle row: Leonov, Nikolayev, Rafikov, Zaikin, Volynov, Titov, Nelyubov, Bykovsky and Shonin. Front row (seated): Popovich, Gorbatko, Khrunov, Sergei Korolev, Gagarin, Karpov (head of training centre), Nikitin (head of parachute training) and medical chief Yevgeny Fedorov. In the bottom photograph, Khrunov has moved to the back and his place at the front has been taken by KGB General Vasily Titov, Secretary of the Central Committee of the CPSU.

The second Soviet man
in space, Gherman
Titov, would spend a
day in Earth orbit.

America's first manned Mercury orbital flight, and NASA openly announced details of the mission well in advance. As before, this open policy provided Soviet space chiefs with a preemptive timetable, and they would be quick to take advantage with yet another space spectacular.

On a day well in advance of Glenn's planned three-orbit flight, another Soviet cosmonaut was seated inside a Vostok capsule, ready to make history. As the Soviet news agency Tass would later announce, this latest feat in what had become known as the Space Race would begin in spectacular fashion:

"On August 6, 1961 at 09:00 [Moscow time] a new launching into orbit of an Earth satellite—the Spaceship Vostok 2—was made in the Soviet Union. The

Spaceship Vostok 2 is piloted by citizen of the Soviet Union, Pilot Cosmonaut Gherman Stepanovich Titov.

The aims of the flight are research into effects upon the human organism of a prolonged orbital flight and the consequent landing on the Earth's surface, and to study man's working capacity during a sustained state of weightlessness.

According to preliminary data, the spaceship has been put into an orbit close to the calculated one, with the following parameters: minimum distance from the surface of the Earth (at perigee) is 178 km [110.6 miles]; maximum distance (at apogee) is 257 km [160 miles]; the inclination of the orbit to the Equator is 64° 56'. The initial period of revolution of the spaceship is 88.6 min. The spaceship weighs 4,731 kg [10,430 pounds], excluding the weight of the final stage of the carrier rocket.

Two-way communication is being maintained with Cosmonaut Titov. The cosmonaut is transmitting on frequencies of 15.765 Mc/s, 20.006 Mc/s and 143.625 Mc/s. A signal transmitter operating on frequency of 19.995 Mc/s is also on board the spaceship. The equipment on board to sustain the cosmonaut's vital activity is functioning normally. The Cosmonaut Gherman Titov feels well. The flight of the Soviet spaceship, which is controlled by man, is proceeding successfully."

Initial statements declaring that Titov was feeling "well" were in fact far from the truth. Much to his chagrin and the puzzlement of medical specialists monitoring the flight, the Soviet Union's second cosmonaut was anything but well. In Titov's "auto-biography", penned by Pavel Barashev and Yuri Dokuchayev, the cosmonaut relates that he first began feeling unwell on the fifth orbit. "I felt nausea from time to time. In order to 'quiet' my 'naughty' vestibular system, I cautiously found the most comfortable position in the chair and—fainted. The nausea gradually went down, and things became much better" [1]. The doctors would eventually decide that some physiological abnormality must be to blame.

Despite Titov's illness, early progress reports indicated that the flight was proceeding well. His exuberance had even led him to cry out at one time, "I am Eagle!" During his third orbit, Titov transmitted greetings to the people of the Soviet Union and Europe, after which he gallantly consumed some food items from squeeze packs in accordance with his flight schedule.

Spending time in space

As the flight progressed into his fifth orbit, Titov reported the smooth functioning of all systems and equipment aboard his Vostok 2 craft. As reported in a Tass bulletin, an hour was spent as he "tested the spaceship's systems of manual control, after which he reported on the good controllability of the craft while manoeuvring with the aid of manual control." This had been an important facet of Titov's flight; after grasping the attitude control handle with his right hand he had fired the ship's thrusters, which rotated the Vostok around its axis. It was the first time a cosmonaut had physically manoeuvred his craft, but it fell far short of claims that Titov had

His flight aboard
Vostok 2 would prove
to be Titov's first and
only mission.

changed his orbital parameters. During the same orbit, Tass said that "During the
flight over Soviet territory the radio-television system transmitted pictures showing
the calm and smiling face of the Soviet cosmonaut. Through the multi-channel radio-
telemetric system extensive information of a scientific nature continued to come in as
well as detailed data about the functioning of the spaceship Vostok 2. While flying
over Kwangchow Major Titov sent greeting to the peoples of Asia, and while flying
over Melbourne he transmitted greetings to the people of Australia."

While on his sixth orbit, Titov exchanged greetings with Yuri Gagarin as he
passed overhead "in the neighbourhood of Canada". Gagarin was in Canada at that
time, visiting the successful businessman and philanthropist Cyrus Eaton, a recognized critic of the United States' foreign and military policies, who was then engaged
in a highly controversial bid to promote friendlier relations and even trade between
the U.S. and the U.S.S.R. In his message Gagarin jokingly said he would wave out of
a window. "However, without relying on a visual greeting, I sent out 'Thanks' on the
radio," Titov would later recall [1].

"In this orbit I used manual attitude controls again," Titov would tell a
COSPAR (Committee on Space Research) space science symposium in Washington,
D.C. several months after his flight. "After trying the excrement-removal arrangement, I reported my excellent spirits and started to prepare to sleep. Maybe I should
not have called my mood actually excellent, as I felt a certain discomfort similar to

the first symptoms of seasickness. It was mostly felt during an abrupt movement of the head, though this sensation did not affect my efficiency" [2]. As he passed over Moscow on the next orbit, Titov bade goodnight to the people below before settling down for a planned seven-hour sleep, with all radio transmissions suspended whilst he slept. It would be the first time any person had slept in space, so there was considerable interest in how he adjusted to this natural function while in weightlessness.

"I did not sleep badly, although I woke up several times," he would later record. "Weightlessness continued to play tricks with me and for a long time I could not cope with my arms. As soon as I began to doze, they would rise up and hang in the air. I was not used to sleeping in that position, and to curb my hands somewhat, I slipped them under my belt of the chair, and then, and only then, fell sound asleep. When I woke up, my arms were in the air again. What can you do?" [1].

Seventeen orbits and touchdown

Waking at 2:37 AM Moscow time, and having slept a little longer than planned, Titov was at first startled to hear transmissions coming from the ground, but quickly overcame his feeling of disorientation and realized where he was. Then he began to feel cold. "When I woke up, the temperature fell to six degrees Celsius," he would tell Radio Moscow International (later renamed the Voice of Moscow) two years after his flight. "I switched off the space suit's ventilator and pulled the helmet [visor] down leaving a narrow slit to let air in. Having warmed myself up, I decided to eat. My space food was stuffed into tubes—juices, liquid chocolate and soup grated like baby food" [3]. Tass then reported:

> "The cosmonaut's condition remains excellent. Having breakfasted well at 05:45 the cosmonaut is continuing the work envisaged by the programme of scientific research. The spaceship Vostok 2 carries much equipment, including radio-technical systems of trajectorial measurements, multi-channel telemetric systems ensuring objective observation of the condition of the cosmonaut and the control of the work of all the devices on board, short-wave and ultra-short-wave receiving and transmitting apparatus, and a tape recorder intended for recording the speech of the cosmonaut and the automatic accelerated reading of recordings upon command from Earth.
>
> The spaceship also carries scientific apparatus for obtaining supplementary data on the influence of cosmic radiation on living organisms. On board also are biological objects. During the flight, observation of the Earth and sky is carried out through three portholes. The cosmonaut may also, if he wishes, use an optical instrument of 3–5 magnification."

"By the thirteenth and fourteenth revolution I had become completely accustomed to weightlessness," Titov would later tell biographers Barashev and Dokuchayev, "and decided in my next few free rest minutes to take pictures of the earth with the Konvas moving [motion] picture camera. It is very handsome, our Earth! At every turn, new colours, new spectra, irreproducible panoramas. The outlines of continents, mountains and forests were clearly to be seen; the ribbons of the big rivers, the seas and the lakes gleamed."

As his flight neared its end, Titov checked the ejection seat straps, secured items, checked his survival equipment and lowered his helmet visor, waiting for the ground-controlled braking procedure to begin. "The terrestrial day was drawing to an end," he recalled. "It was time to prepare to land." He would also record his description of the re-entry phase.

"And here came the command. The ship changed its direction unhurriedly. It was quiet. When the braking apparatus was put on, the objects that had been 'floating' in the air lay silently down at my feet, as if trained; the movie camera, the logbook, the pencil, the exposure meter. They too felt that soon they would land and 'soaring' there without reason is prohibited. The frame heated up with the friction of the atmosphere, there was a bright flame surrounding me, and I could not help remembering the Chief Designer [Korolev] and his unshakable faith in the technical coatings. The braking apparatus worked precisely, and when there were a few kilometres left to Earth, I turned on the landing system" [1].

At a press conference convened in Moscow on 11 August, four days after his return from space, Titov surprisingly disclosed that he had not landed inside his Vostok craft, instead ejecting from the descent module and parachuting to the ground. "I felt very fit and decided to try the second landing system," he stated, as if it was a decision he could make arbitrarily. "At a low altitude the seat was ejected and my further descent was by parachute. The ship landed successfully nearby." This statement only deepened prevailing suspicions that Gagarin had also ejected and landed by parachute, but this would not be officially confirmed for several years to come.

As he descended to the ground, Titov reflected, "I took in vast fields spreading below, stacks of hay; I saw a railway and a train that appeared as though stuck on the rails, a river and two big cities in the distance. The ground comes nearer and nearer. It seems that I am going to land close to the railway. I touched the ground safely two hundred metres from it, on the field. The flight was accomplished." He had landed at 10:18 AM, Moscow time, near the settlement of Krasny Kut in the Saratov Region after completing just over 17 orbits of the Earth, in a mission lasting 25 hours and 18 minutes.

Once again, however, there is a marked disparity between what was contemporarily published as fact and the real truth behind these early missions. Many years after his flight, Titov would admit that his landing was anything but easy. As he recounted: "I passed about fifty metres from a railway line and I thought I was going to be hit by the train that was passing. Then, about five metres from the ground, a gust of wind turned me round so I was moving backwards when I landed and rolled over three times. The wind was brisk and it caught my parachute so I was dragged along the ground. When I opened my helmet, the rim of the face plate was scooping up soil."

An excited Titov chats with Premier Khrushchev soon after landing.

The train Titov had observed screeched to a halt, as the passengers and rail workers had seen him coming down. Everyone rushed over to the orange-clad figure, but he wasn't in the best of moods. "What are you staring at?" he cried out. "Help me take off my space suit. I'm very tired."

To further annoy Titov, as he recalled, "There was supposed to be a clean lightweight overall for me to get into, but as usual somebody forgot to pack it in my emergency kit" [4].

There was more drama to come at the landing site, when a local woman driving by became intrigued by what was going on and veered off the road to investigate. She lost control when her car hit a pothole and the woman banged her head badly on the steering wheel. Titov heard all the commotion, and was soon told that the woman was bleeding badly. The cosmonaut quickly retrieved his medical kit and assisted in applying bandages to the woman's head. It was hardly the sort of welcome home he had been expecting.

In May the following year Titov would tell the COSPAR meeting in Washington D.C. that he had withstood the g-forces of lift-off and re-entry well, but that more research would need to be done on the human condition in weightlessness. "At the present time it would be highly ill-advised to assume that the problem of weightlessness has been understood," he told the crowded assembly. "Thus, beginning with the fourth orbit, and particularly in the sixth–seventh orbits, my state underwent certain changes which, generally speaking, did not affect my efficiency. I felt changes in mood during abrupt movements of my head which entailed unpleasant sensations resembling seasickness. I felt giddy and nauseous. This was accompanied by [a] deterioration of the appetite. Falling asleep was a bit difficult. Later sleep became deep and restful. By the end of the flight this unpleasant sensation abated. We may assume that this was caused by unusual conditions for inner ear otolith organ functioning in weightlessness" [2].

The truth is that the hapless cosmonaut had become the first victim of what is now known as space adaptation sickness (SAS), a reaction to spaceflight that could

Titov chatting with his backup pilot, Andrian Nikolayev, who would be the next Soviet cosmonaut into space.

seemingly strike at random on any flight, no matter how healthy or well-prepared the subject. Forty years on, SAS is still of concern to space travellers, although a greater recognition and understanding of the illness has now led to chemical and physical ways to minimize its debilitating effects.

"In orbit Titov suffered some vestibular problems, which he made no attempt to conceal from doctors," reported Academician Andrei Grigoriev two years after the flight of Vostok 2. "Some blamed it on lack of character. As a rule, cosmonauts don't complain of their problems. They all want to be supermen. But Titov, aware of how important such information is for doctors seeking to minimise the negative effects of weightlessness on [the] human body, was absolutely frank. He felt giddy, nauseous and had the impression that everything around him was upside-down" [3].

Unfortunately for Titov his honesty would cost him dearly. So little was known about his unexplained illness that after he had supplied doctors with much-needed data on his condition, serious concerns would be expressed about his well-being on any future spaceflights—a bitter blow for the cocky young cosmonaut to endure.

COSMONAUTS BEHAVING BADLY

The timing of Gherman Titov's Vostok 2 mission was a noteworthy example of the deceptive way in which Soviet propaganda functioned to suppress news of less praiseworthy events. A week after Titov's spaceflight, Berlin was abruptly and cruelly divided when the East Germans erected a formidable wall effectively separating West Berlin from what was then the German Democratic Republic.

Vodka, women and song

The despised Berlin Wall would become an iconic symbol of the Cold War until it was torn down some 28 years later. There was a palpable tension in the air as Soviet and American troops maintained watchful posts within sight and sound of each other at the notorious "Checkpoint Charlie" and the nearby Brandenburg Gate.

An evident nervousness existed among Warsaw Pact nations at this time, but a supremely confident Khrushchev was able to assuage any high-level concerns by

Wherever they toured, massive crowds such as this one in Berlin greeted Gagarin and Titov.

Titov giving a few words to a huge crowd of adoring Berliners, standing beside a fanciful representation of his space capsule.

pointing out the technological superiority of the U.S.S.R. over the United States, as evinced by Titov's record-breaking mission, and the huge amount of worldwide interest and positive publicity garnered by this latest space feat. Following on from the tremendous public interest in Gagarin's flight, and his subsequently popular global tour, it was decided to send the popular and photogenic Titov on a similar tour to raise awareness and admiration for his—and his nation's—most recent achievement in space.

Post-flight, at his debriefing, Titov had impressed medical doctors with his willingness to discuss the inexplicable nausea and disorientation he'd suffered while in space. What did not impress them at all was the fact that having just spent a day orbiting the Earth and suffering a mystery malady, he had nevertheless consumed a bottle of beer prior to his debriefing. It smacked to them of arrogance.

If anything, Titov became even more cocky and self-assured after his flight aboard Vostok 2. Whenever his good friend Gagarin was in Moscow in between propaganda tours of the world or the U.S.S.R. the two cosmonauts would take advantage of the more lavish and sometimes carnal rewards offered to them by grateful citizens—parties, alcohol, and certainly as many willing women as they could handle. The drinking and womanizing finally became so openly rampant that late in 1961 the two men were ordered to stand before a closed-door meeting convened by the Communist Party echelon, at which they were warned in no uncertain terms about the grave consequences of negative publicity. Although it did not seem to slow the men's activities down at all, towards the end of his life Titov confessed that even though he had a constant stream of woman eager to share his bed, "I had to limit myself. It would be one thing for a simple pilot, but not for a cosmonaut because of the reputation."

The Foron incident

Early in October 1961, Gagarin suffered a serious injury as a result of a philandering episode while vacationing with his family at the Crimean sanatorium of Foron, by the Black Sea. Titov was there as well, together with several high-ranking space officials including Nikolai Kamanin. As usual, Gagarin and Titov were relieving some of their touring pressure taking advantage of the resort facilities to play and drink hard, much to the displeasure of Valentina Gagarin. According to Kamanin's diary notes, a drunken Gagarin had tried to take advantage of one of the resort's nurses and forced his way into her second-floor room. Soon after, a concerned Valentina began looking for her missing husband, and soon her suspicions led her to begin knocking on the nurse's door and calling out to her husband [5].

Gagarin could hear his wife on the other side of the door and knew that her discovery of his amorous adventure was imminent. Still quite drunk, he decided to escape by leaping from the room's balcony to the ground six feet below. Unfortunately, as he jumped, one of his feet caught on some ornamental grape vines growing on the side of the sanatorium and he fell face-down on the cement edge of an asphalt pathway.

"The story was kept under tight wraps," observed Dr. Vladimir Golyakhovsky. "Gagarin was taken to the exclusive Hospital #6 set up by the Third Department of the Ministry of Public Health for the cosmonauts, and there his wound was treated.

The gash over Gagarin's left eye would leave a deep, permanent scar to remind him and others of his folly.

The outer table of the frontal bone was found to be fractured and bent inward; the wound was contaminated. The surgeons feared that the infection had extended to the front sinus."

However, Golyakhovsky acknowledges that the cosmonaut's youth and fitness had helped his body to cope with the trauma, but it would result in a deep and permanent scar. "Gagarin complained to me that the surgeons had been unable to make his face scar invisible. I examined and felt the scar. To be sure, it was the epitome of crude surgery. The skin was mottled and the scar soft, but it crossed the eyebrow in an ugly line, extending downward in a stepwise fashion. The forehead dent was totally exposed" [6].

The injured cosmonaut would remain in hospital for three weeks. As researcher Bart Hendrickx deduced from Nikolai Kamanin's translated diary notes of that time: "The incident, which happened on 3 October, could not have come at a worse time, because Gagarin and Titov were supposed to be the big stars at the 22nd Communist Party Congress opening in Moscow on the 17th of that month. Gagarin did eventually make a brief appearance at the Congress on the 24th. Answering a reporter's question about the scar, he said he had stumbled while playing with his baby daughter and had landed with his face on a stone. Several weeks later Gagarin told Kamanin that he did not know there was a women in the room and that he had merely been playing hide-and-seek with his wife."

Gagarin was depressed about the scarring, but when Golyakhovsky suggested that he should consider having plastic surgery in the West the cosmonaut apparently confided that the Politburo would not allow that, and had already slapped him with a Party reprimand for the indiscretion that had led to his injury. "I was told to keep quiet about that incident so as to avoid unfavourable publicity in the West," Gagarin added. "So there's no way I could go for treatment there" [6].

A cavalier attitude

Meanwhile, as well as drinking heavily, Gherman Titov also had a reckless penchant for driving fast cars, and would often combine the two activities, sometimes with disastrous results. Post-flight he would be involved in a string of drink-driving and other incidents that enraged his superiors, and he was often rebuked for this cavalier attitude to his fame. But his popular status as a nation's hero ensured that he could not be dismissed or shunted out of the spotlight.

In February 1962 Titov drove his Volga car into a bus while intoxicated—his third driving accident in a year. Three years after his flight, on 26 June 1964, he would be involved in a far more serious traffic accident in which a woman passenger in his car not only died, but a drunken Titov fled the scene and later lied to police. That, however, was in the future.

Instant dismissal

As if Gagarin and Titov were not enough to contend with, cosmonaut trainee Mars Rafikov was causing no end of trouble for Nikolai Kamanin with what would be

Mars Rafikov in happier times
with his son, and playing tunes
on a piano.

described as a variety of offences, including overt womanizing and "gallivanting".
While holidaying in Sochi with other cosmonauts the previous May, Rafikov had
openly courted a number of women, which is said to have led to a particularly heated
confrontation with this wife. He had already been given a stern warning about his
heavy drinking habits in Moscow hotels, and the head of the training centre could

only control his patience so far. On the evening of 12 March 1962, Rafikov and Anikeyev had imprudently decided to have a night on the town. They managed to leave the air base without permission and spent several hours drinking and carousing in the old Moskva Hotel, just off Red Square.

On 24 March, after reports of this drunken escapade in Moscow and further complaints from his distressed wife, Rafikov was called before Kamanin together with Anikeyev and asked to explain their behaviour. Rafikov took it upon himself to use the behaviour of Gagarin and Titov as a vague form of excuse for disobeying orders and getting drunk. Although he had earlier denied marital problems and said that his wife and five-year-old son wanted to stay with him, a divorce was going through, and this was another reason for him to fall into disfavour with his superiors. In his unpublished memoirs he blamed his pending divorce as the principle reason for his unpopularity with his superiors at that time. However, Vershinin and Kamanin knew that any future publicity would definitely not favour a divorced cosmonaut.

While Rafikov's pathetic excuses and pleas fell on deaf ears, Anikeyev was lucky; he would escape with a severe reprimand. Three days later the decision on Rafikov was handed down. He was dismissed without recourse from the cosmonaut corps "for violation of military discipline", according to Gherman Titov. There was no leniency given; the unflown cosmonauts would potentially become ambassadors of the Soviet Union to the world after their space missions, and the nation could ill-afford such a renowned braggard and troublemaker.

Prior to his fall from grace, Rafikov was photographed with Popovich and an unknown person at the cosmonaut training centre.

"He loved himself very much, Rafikov," observed fellow trainee Dmitri Zaikin in a 1993 interview in Star City. "He thought that all women were fond of him ... loved him. So he began to have some troubles with his wife, and his wife went to his commander. He didn't come home at nights to sleep, and when the commander heard about it he said that he wasn't living up to the regime and that he did so because he didn't really want to be a cosmonaut. And so, they dismissed him" [7].

In his memoirs, Rafikov also writes that, after his dismissal, Kamanin had called all the cosmonauts together and explained the situation. According to Rafikov, Kamanin said that the errant cosmonaut candidate would be allowed to rejoin the cosmonaut corps after two years.

On 27 March 1962, Kamanin backed Rafikov's statement to a certain extent when he wrote in his diary that the cosmonaut candidate had admitted his mistakes and that they had not only damaged him, but also the team and its leaders. Rafikov, he said, had asked for forgiveness and that he be allowed to return after two years. "We promised we would help him but that much would depend on Rafikov himself," Kamanin wrote. He added that he had spoken with Rafikov's wife who foolishly said that she still loved him and their five-year-old son very much, and would follow him everywhere.

As a footnote to the affair, Kamanin suggested that it had primarily been Titov and Anikeyev who had gotten Rafikov "on the wrong path" [5].

TITOV ON TOUR

Once things had settled down in the wake of his history-making flight, Gherman Titov embarked on an extensive goodwill trip that officials hoped would emulate the success enjoyed by Gagarin on his own whirlwind tour earlier that year. They would soon come to realize that the second cosmonaut was much more of a rogue and less of a diplomat than his predecessor. In Romania, for instance, he astonished officials by leaving the seat of his official limousine during a motorcade through city streets and borrowing a motorcycle belonging to one of his police escort to continue the parade. Citing youthful exuberance, Titov may have seen this as a bit of spontaneous fun, but it certainly did not sit well at all with his Soviet minders. "They were not happy," he later reflected.

Astronaut meets cosmonaut

In late April 1962, as part of his arranged goodwill tour, Titov arrived in the United States on a two-week visit with his wife Tamara and General Kamanin. Happily, Titov spoke reasonable English, so communication was not a problem. There was a huge amount of interest in his arrival, not only because he was the first cosmonaut to pay an official visit, but the popular John Glenn had recently completed America's first orbital spaceflight aboard *Friendship 7*. A meeting between the rival spaceflight pioneers—two of only three men who had flown into orbit—was therefore a widely anticipated media event. It could have proved a major publicity coup for the Soviet

Union, but Titov's brittle lack of enthusiasm for anything American quickly began to grate with local officials.

In its issue of 11 May 1962, *Time* magazine commented negatively on the "uncommunicative attitude" displayed by the visiting cosmonaut. "Sent to the [United States] to share his hard-won knowledge of travel in space with Glenn and COSPAR (Committee on Space Research), Titov seemed under orders from home to do nothing of the sort," the article pointed out. "In press conferences and TV interviews, he was always guarded and reluctant in his replies, though often breezy when it came to enjoying the crowds" [8].

In fact, rather than becoming a public relations favourite, the cosmonaut's graceless responses to anything shown to him caused some unflattering criticism of his efforts in the press. On a rushed tour of New York City, for example, he displayed very little interest in a visit to the Stock Exchange and was even booed by floor traders for his lack of enthusiasm, and for saying he would not like to work there. He also spoke disparagingly about the city's famous skyscrapers, dismissing them as something that hid the Sun from the people, and said he found the proliferation of neon signs repulsive. On a tour of the Museum of Modern Art he continually outraged his hosts by ridiculing many of the exhibits on display.

Some bad press

In Washington, D.C., where the cosmonaut would make a speech before the COSPAR delegation, John and Annie Glenn took Titov and Tamara on a heavily

Titov at the COSPAR convention alongside U.S. astronaut John Glenn.

escorted whistle-stop tour of the capital's better known sites, including a trip to the Smithsonian Institution to see Alan Shepard's *Freedom 7* spacecraft. In a recent interview, NASA Public Affairs spokesman Paul Haney said Titov seemed to be going out of his way to appear unimpressed with anything he was shown, and would always add that they had something bigger and much better back home. "It didn't matter what we showed him," Haney recalled. He said they took Titov to the biggest rolling steel mill in the United States, and the cosmonaut shrugged it off by saying that there was one in Novorossiysk that annually produced three times as much steel as the American plant. The Washington trip ended with a brief visit to the White House for a photo-shoot meeting with President John F. Kennedy [9].

Western media representatives following the tour were not slow in pointing out an evident difference between the taciturn, tactless Titov and his more open and gracious astronaut counterpart. For instance, when Titov was asked if he would like to fly into space aboard an American aircraft, he blurted out "No. Quality not good enough!" That response certainly set a few teeth on edge. Also, when asked for his opinion on perils associated with the arms race and the threat of nuclear war between the two nations, he took the opportunity to loudly rattle his country's sabres with a truly undiplomatic statement. "We do not want war; we do not threaten," he emphasized. "But our scientists can make a bomb equal in power to 100 million tons of conventional explosives. And they can make a rocket big enough to carry that bomb anywhere on Earth" [10].

There was further disappointment in store for those who attended the main reason for Titov's visit: his Washington address at the COSPAR symposium. Whereas Glenn had earlier given a beautifully structured and articulate post-flight address before Congress, in which he had discussed the emotions involved in making his nation's first orbital flight, Titov's descriptions were far more general in nature and technically bereft. He blandly imparted very little of interest to the assembly. As the *Time* article dryly observed, "If [American] scientists want to know how a space traveller feels after more than three orbits of the earth, they will have to wait until they have sent one of their own astronauts on the trip" [8].

Despite some bad press regarding the tour, there were some light-hearted moments. If anything, the one thing that the Titovs really did enjoy for many reasons was a hurriedly arranged private barbecue on their second evening in Washington, held in the carport of John Glenn's home in nearby Arlington. It was so impromptu that Annie Glenn had to send a couple of motorcycle security police to get some frozen peas and other supplies, while she canvassed her neighbours in order to get enough steaks for everyone.

While waiting for the Russian contingent to arrive, Glenn lit the barbecue in his car port and set up a couple of electric fans nearby to speed up the charcoal-heating process. "I threw the steaks onto the grill over the charcoal fire," Glenn recalled, "and in a few seconds the fat started dripping and making some flame. I went in the house to get water to control the fire, and when I came back there were flames about three feet high." He had to drag the barbecue out into the carport; thick smoke was billowing everywhere, and some steaks had fallen into the fire. Glenn then threw water onto the barbecue, which only aided in producing more thick smoke.

"Everything was pretty much in turmoil, and just at that moment the big black limousines pull up and the Russian visitors step out and start up the driveway! So I met them on the driveway and told them ... if they expected to get anything to eat they'd have to pitch in and help. Titov took his coat off and helped me with the barbecue. The Russian general with them, Nikolai Petrovich Kamanin, went in the house with Titov's wife, Tamara. The next time I saw Tamara, she had her shoes off and was helping Annie and Louise [Shepard] with the salad. As it turned out we had a very enjoyable evening" [11].

Fatal accident

Nikolai Kamanin would do his utmost to retain strict disciplinary control over the cosmonauts under his command, but it was a difficult and ongoing task. His continued demands for unquestioning obedience certainly did not sit well with his charges. According to Belgian space analyst Bart Hendrickx, "While it was relatively easy for Kamanin to keep a tight reign on the new trainees, keeping discipline among the flown cosmonauts was a different matter altogether. Unlike the rookie cosmonauts, they were known to the outside world, making it virtually impossible to ban them from the cosmonaut team without giving some kind of official explanation" [5].

Kamanin, as observed by Soviet journalist Yaroslav Golovanov, was probably jealous of the attention lavished on the cosmonauts, and while he said that they did not deserve the attention heaped on them, it was a spotlight he was covertly only to eager to share [12]. Kamanin wrote of the problem of discipline in a diary entry of 1 December 1962, curiously referring to himself in the third person: "Our country's highest leaders are parading [the cosmonauts] like a child with a new toy and are showering them with tributes, titles and invitations as if from the horn of plenty, while Kamanin has to keep them in check and has to be prepared to take responsibility for the fact that they get drunk at government receptions and say and do stupid things when they are in their cups ..." [13]. Kamanin's mounting concerns about the reckless behaviour displayed by flown cosmonauts, his nation's beloved human national assets, would be fully realized late on the evening of 26 June 1964, when a female passenger in a car driven by Gherman Titov was tragically killed in an accident fuelled by alcohol.

Earlier that year, in January 1964, Titov's undisciplined behaviour and perceived lack of responsibility had led to him being severely censured by Commander of the Soviet Air Force Konstantin Vershinin, but it appears to have had very little impact on the headstrong young man. On the night of the fatal crash, Titov was said to have been driving home, and would later admit under cross-examination that he was in a state of intoxication. As the story goes, he stopped to pick up a female hitchhiker named Fomenkov, but as he drove on he was recklessly speeding along a poorly illuminated road and failed to see what he later described as "a pile of old asphalt and stones". In the ensuing crash the woman passenger was injured, while Titov would somehow sustain only minor cuts and bruises. This is where the events become a little murky, as Titov would later say that Fomenkov did not appear to be seriously injured, and had crawled out of the wreckage by herself. It would seem an ambulance was then summoned, but inexplicably Titov caught a taxi home, where he went to bed. However, Fomenkov was actually critically injured internally, and died on the way to hospital.

The following morning Titov was woken by police investigating the accident, and gave the first of several false statements. Later he would also be accused of bribing the taxi driver to tell lies in order to protect his version of events.

An official dressing-down

Five days after the accident, on 1 July, and in the presence of Titov's 1960 cosmonaut colleagues, Nikolai Kamanin unleashed a stern reprimand on the disgraced cosmonaut, later recorded in his diaries. "There is every reason to exclude you from the party and deprive you of all your titles," Kamanin wrote in recalling his stinging dressing-down of Titov. "Deputy, Hero [of the Soviet Union], pilot-cosmonaut, lieutenant-colonel ..." Much as he would like to have taken further action against Titov, Kamanin was all too aware of inherent difficulties in recommending such a move, as he recorded in his diaries. "Titov is not only known by the Soviet people; the whole world knows him. The disgrace of Titov will be the disgrace of all cosmonauts, the disgrace of our people. We cannot allow this and will make one final attempt to keep Titov in the party and in the cosmonaut team."

On 17 July, following an official investigation into the accident, the case against Titov was heard before the military procurator. Apparently ignoring the blatant fact that Titov had been far from truthful in his version of the events and had made several false statements, he was essentially cleared, and found not culpable of any wrongdoing in the death of Fomenkov. His only penalties were all temporary admonishments: a driving ban, being forbidden from attending important meetings, travelling overseas and going to the cosmodrome—leading Kamanin to record in his summing-up of the hearing: "It has become absolutely clear to me that if an ordinary officer had been in Titov's place, he would have been tried."

It later emerged that a lot of deals had been struck to keep Titov's name and the cosmonaut team out of the headlines. Fomenkov's only close relatives were an eight-year-old son and her 62-year-old stepfather, so Titov was made to pay for the funeral, and on his behalf the stepfather was promised that the boy, when old enough, would be admitted to one of eight prestigious Suvorov Academies. Founded during the Second World War, these military academies were originally intended as shelters for wartime orphans, but would evolve into key formative elements for the Russian armed forces.

Spiral, and some redemption

Titov's earlier demeanour, when combined with his dishonourable actions during and subsequent to the accident, undoubtedly led to his being removed later that year from the position he had been awarded as deputy head of the cosmonaut detachment, replaced by Valery Bykovsky. Bykovsky would remain in this position until 28 April

An early version of the Spiral spaceplane.

1966, when he was made to stand down after being involved in a fight with his brother-in-law, and was also temporarily excluded from training with the Soyuz team. He was already under notice for breaking his leg in August 1964, having fallen down some stairs while intoxicated. With Bykovsky's removal as deputy head of the cosmonaut team, Kamanin once again looked to Titov to fill the position, as his behaviour was said to have markedly improved, but when it was offered to him he declined.

Instead, Titov undertook instruction at the Chkalov school, qualifying in 1967 as a test pilot. He then began working on the reusable Spiral spaceplane project. The Spiral, also known as the Mikoyan-Gurevich MiG-105, was an innovative project endowed with the aim of creating a Soviet orbital spaceplane and was developed in the design bureau headed by Artem Mikoyan. The head designer appointed to this project was Dr. Gleb Lozino-Lozinsky. Originally conceived as the Soviet answer to the USAF's X-20 Dyna-Soar military spaceplane (which was subsequently cancelled), it also drew on the work being done in the United States on lifting body technology. The eight-metre, single-seat Mikoyan Spiral spaceplane also picked up the humorous nickname of *Lapot* or "Shoe", due to the pronounced ski-like shape of its nose.

Work had begun on the project in 1965, after early design work by Sergei Korolev and Soviet aircraft designer and the head of OKB-23 in Fili near Moscow, Vladimir Myashishev. The plan was to launch the spaceplane from a specially modified Soyuz booster or from a hypersonic launch air-breathing aircraft then in the planning stages. Gherman Titov was appointed the head of a cosmonaut group who would receive training on the spaceplane, which initially included Georgi Dobrovolsky, Anatoli Filipchenko, Anatoli Kuklin and Aleksandr Matinchenko. For his part, Titov would spend a full year training hard on the MiG-21 and Sukhoi Su-7 and Su-9 aircraft in preparation for the Spiral project, and much to his pleasure Nikolai Kamanin noted that it had lent the rebellious cosmonaut a new maturity as he devoted his efforts to the training team and to the success of this space programme.

Titov in conversation with
Admiral Nikolai Amelko, then
Commander of the Soviet
Pacific Fleet, and later a
Deputy Minister of Defence.

However, the Spiral programme was a long-drawn-out and under-funded programme, and Titov soon lost faith, becoming disillusioned with its seeming lack of progress. He would quit the training group in 1970.

Seven years later, in 1977, the first of eight eventual air drop launches from a Tu-95K aircraft took place in order to test the Spiral's subsonic aerodynamic characteristics and air-breathing systems. Guided down from 15,000 feet by test pilot Aviard Festovets, it would land using skids on an old dirt runway near Moscow. Eight of these drop tests would be carried out before the prototype hit the runway too heavily the following year and was damaged so extensively it could not be repaired.

The Spiral project was eventually cancelled when a decision was made to scrap the programme and concentrate instead on the winged Buran space shuttle, which relied heavily on the technology and experience of the Spiral.

TANDEM ORBITS

By 11 August 1962, a little over a year had slipped by since the successful day-long orbital flight of Gherman Titov aboard Vostok 2. Puzzled by the lack of a follow-up mission, Western scientists and observers had been openly wondering what the ominously silent Russians would get up to next, and what space spectacular they might have in store. In the interim the United States had gained some ground and prestige in the Space Race by launching John Glenn and Scott Carpenter on successful solo three-orbit flights. Suddenly, on that day, things were about to change dramatically.

Nikolayev and Popovich fly

At 11:30 AM Moscow time, Vostok 3 thundered into the skies from the Tyuratam launch complex, carrying 32-year-old Major Andrian Nikolayev into an initial elliptical orbit that varied between 113.6 miles and 155.9 miles above the Earth, very close to the desired apogee and perigee. Using the call sign *Sokol* ("Falcon") Russia's third cosmonaut to travel into space would circumnavigate the globe every 88.5 minutes, quickly establishing communications to reassure ground controllers and scientists that he was in good health and spirits. "I feel well," he reported. "Everything is normal on board. The Earth is clearly visible through the porthole."

On his first day in orbit, Nikolayev was kept busy. On the fourth orbit he transmitted a radio greeting directly to Nikita Khrushchev, reiterating that the flight was proceeding smoothly. "I feel fine," he told the exultant Premier. "All systems of the ship are functioning perfectly."

"I am glad that you feel well," Khrushchev responded. "I am proud of the courage you have displayed in making this flight."

To cap off an interesting day, Nikolayev also appeared in the first ever (although delayed) televised pictures of a person beamed from space, and twice operated his spacecraft manually. Unlike Titov, and much to the relief of medical doctors, he had displayed no early symptoms of illness and his appetite seemed quite healthy. He proved this by consuming some plastic-wrapped roast veal, pastries, chicken fillets, fruit pieces and bite-size sandwiches. These consumables were designed to be eaten during the first part of the flight; beyond that there were the more familiar but obviously less appealing tubes of strained space food. By the time Nikolayev settled down for his first night in space, at 10:00 PM Moscow time, Vostok 3 had already completed seven orbits, covering some 186,000 miles.

As the lone cosmonaut slept, there was intense speculation around the planet and in the press, most of it centring on the purpose and duration of Nikolayev's flight. Fuelling some rumours was a somewhat non-committal announcement by Radio Moscow that Vostok 3 was in orbit to "obtain more information on the effects of weightlessness and other space flight conditions." Further reports suggested that the flight would also test improvements in the Vostok spacecraft. It was widely felt, therefore, that Nikolayev would return to Earth with a new spaceflight duration record against his name. Very few people were aware of an intense flurry of activity at the same blackened launch pad Nikolayev had left only hours beforehand. Or that a second Soviet cosmonaut was making ready to fly into space. The launch team was tense; anxiously waiting for Vostok 3 to begin its 17th orbit, by which time it would be in the same orbital alignment as when it was shot into space, meaning that a second craft could be launched on a similar "catch-up" trajectory.

Showing due caution, no announcement had been made of Nikolayev's launch aboard Vostok 3 until 86 minutes after he had successfully left the launch pad and entered orbit. Similarly, just under 24 hours later, life carried on as usual in the Soviet Union as another R-7 rocket carrying Vostok 4 ripped a path into the heavens from the remote Tyuratam launch complex, thrusting Pavel Popovich into orbit. He would be using the codename *Berkut* ("Golden Eagle"). The launch came at 11:02 AM and Popovich's spacecraft eventually slid into orbit with an initial apogee given as 157.73 miles, and a perigee of 11.78 miles.

Once a successful orbit had been achieved, and Popovich had reported that all was well with him and his ship, permission was finally given to Radio Moscow to

Soviet cosmonauts Nos. 3 and 4: Andrian Nikolayev and Pavel Popovich.

announce the amazing news that two cosmonauts, later described in the Soviet media as the "Heavenly Twins", were in space at the same time in separate spacecraft on a new "joint" mission. It was an outstanding feat that in retrospect should have been regarded as unachievable at that time: launching two men less than 24 hours apart from the same firing pad. It took a gargantuan effort, but it resulted in triumph and a massive propaganda coup. The West was truly rocked back on its heels.

The announcement from Radio Moscow reported in deliberately couched terms the safe orbital insertion of Vostok 4, carrying Popovich, and said that "the task of setting two spaceships in orbits close to each other is to obtain experimental data on the possibility of establishing contact between the two ships, coordinating the actions of the pilot-cosmonauts, and to check the influence of identical conditions of space flight on human beings." All of which was quite true, but the media instantly cottoned onto the words "contact between the two ships" and the world's headlines speculated on a link-up in space between the two craft now flying in tandem orbits. It was a heady time for the Soviet propagandists.

Even those involved in launching Vostok 3 and 4 could barely believe what they had pulled off. Despite a number of non-controllable variables such as atmospheric density and high-altitude winds, Vostok 4 had been launched with unmatched precision, slipping into an orbit which allowed Nikolayev to report soon after that he was able to see Popovich's craft for a short time through his porthole. However, different orbital parameters, the immutable laws of physics, and the fact that their spacecraft could not be manually manoeuvred, meant that the closest they ever came was around three miles, and they would inexorably drift apart. Nevertheless, it was a stunning achievement at such an early stage of the Space Race. Somewhat half-heartedly, one would assume, President John F. Kennedy issued a statement on 12 August, saying, "I congratulate the Soviet Union on this exceptional technical feat and salute the courage of her two new cosmonauts. The American people, I know, wish them a safe return."

The following day, Kennedy was a little more circumspect during a televised economic address, admitting that many Americans were interested in, and concerned about, this latest Soviet achievement. "I have said from the beginning that this country started late in the 1950s," he proclaimed with obvious regret. "We are behind and we will be behind for a period in the future, but we are making an effort now, and this country will be heard from in space ... in the coming months and years."

The non-Russian cosmonauts

Ironically, while the world's press lauded this latest feat of launching two Russian cosmonauts into space, neither man was in fact a true Russian. Nikolayev was a Chuvash national, coming from a race of Tartarized Bulgaric people who spoke in a distinctive Turkic language, while Popovich was born a Ukrainian. Not that the world's media would have been interested; all they knew or cared about was that the two cosmonauts had been launched aboard rockets lifting off from the Soviet Union.

Both spacecraft were actually slightly modified versions of those that had carried Gagarin and Titov into space. An upgraded radio communications system meant that the two cosmonauts could now be in direct radio contact with each other, as well as having the ability to transmit television pictures from their respective craft to the ground. Modifications had also been made to the Vostok's interior and personal gear to improve living conditions for the occupants. There was also new biomedical equipment loaded to measure the cosmonauts' physical condition, brain-waves, skin electrical conductance and eye movements, and to monitor cosmic radiation. This data was relayed by telemetry to ground stations throughout the joint flight.

A rapt Soviet nation stayed tuned in as regular radio and television bulletins kept them informed about the condition and well-being of the two cosmonauts, punctuated by patriotic music and constant reminders of the superior Soviet technology that had enabled this extraordinary and historic mission to take place. Elsewhere in the world there was a lot of hand-wringing and head-shaking going on. Eminent radio astronomer Sir Bernard Lovell from the Jodrell Bank Observatory must have sighed deeply as he told media representatives that he found himself astonished at the Soviet virtuosity. "I think," he confessed, "the Russians are so far ahead in the technique of rocketry that the possibility of America catching up in the next decade is remote."

If Wernher von Braun was at all impressed, he refused to allow it to show. He quickly dismissed the latest Soviet effort by saying he saw no significant advance in the tandem mission, and added, perhaps a little contemptuously, "It does not look like the Russians used any new equipment."

Several times during the tandem flight both cosmonauts were given permission to unbuckle their restraining straps and float free from their contoured couches, which they greatly enjoyed. In fact, the two men were in high spirits and in constant contact, discussing their experiences and even singing to each other. Nikolayev sang about his native Volga River. "In reply I sang my favourite Ukrainian song," Popovich later told reporters. They were both quite at ease with the way in which their joint flight was proceeding, although on the fourth day in orbit Nikolayev became a little testy with a Soviet tracking station that had given him the incorrect time. "You were wrong by five minutes," the obviously tired cosmonaut barked back. "Please give me a new time recording now. Can't you hear what I say? Start the timing, for heaven's sake!"

On that morning of 15 August, Nikolayev and Popovich were also informed that, as scheduled, their mission programme had been successfully completed, and they were to prepare for re-entry. Right on 9:24 AM Moscow time, Vostok 3's retrorockets fired into life, and the spacecraft started its descent. Six minutes later, Vostok 4's retrorockets also fired, and it too began to dip into the upper reaches of the atmosphere. Both spacecraft then safely passed through the volatile passage of ferociously high temperatures and crushing g-loads of re-entry. As their charred spacecraft finally plunged toward the ground, the two cosmonauts were automatically ejected at the desired altitude and gently drifted down beneath their parachutes to the steppes of Kazakhstan.

At 9:52 AM, after a flight lasting 94 hours and 25 minutes, and having completed 64 revolutions of the Earth while travelling a total of 1,649,750 miles, Nikolayev touched down near the city of Karkaralinsk in the Karagandinskaya region, 1,500 miles southeast of Moscow. One of the first officials to arrive on the scene was Vitaly Volovich, a renowned physician also trained in parachute jumping, whose expertise in both fields went back many years.

Volovich to the rescue

Recognized today as an accomplished polar explorer and survival specialist for the Soviet and Russian air and space forces, but originally assigned to aerospace work in Moscow, Vitaly Georgievich Volovich was one of a group of doctors selected in late 1960 to serve as recovery doctors for the Vostok flights. Earlier, on 9 May 1949, he and Andrei Medvedev had become the first parachutists to successfully carry out a landing on the North Pole.

In his book *Experiment: Risk*, Volovich recalls being summoned before Nikolai Kamanin after submitting a comprehensive report stating that Air Force physicians should be on hand at the landing sites to examine the cosmonauts immediately after touchdown and recovery.

"We know," he told Kamanin, "that certain aspects of space flight, particularly increased pressure during lift-off and weightlessness, can have detrimental effects on a cosmonaut, which may alter his physiological functions. These changes may be either short-term or long-term. It is important to examine the cosmonaut as soon as possible after he lands in order to determine these effects. This can be done by a physician-parachutist. He should be on board the search plane, parachute down to the cosmonaut and conduct a medical examination, or, if need be, render necessary medical assistance" [14].

Kamanin was in full agreement with Volovich's proposal, and they agreed that the ideal candidates should be Air Force physicians already familiar with medical practices in the space programme, who could be given supplementary training at the N.V. Sklifosovsky Institute in Moscow. They would, however, have to be trained as parachutists. Volovich consulted with Oleg Gazenko from the Ministry of Health's Institute of Biomedical Problems (IMBP) and Vladimir Yazdovsky, in charge of the spaceflight medical programme, to set up a medical aviation commission to screen suitable applicants. Unfortunately, as Volovich relates, many of his colleagues assumed this would be a relatively easy assignment, and consequently failed to pass their physical examinations before the commission. In fact, only four men would eventually make the grade—Volovich, Ivan Kosolov, Viktor Artamoshin and Boris Yegorov. As part of their training, the four men had to undergo rigorous parachute training under instructor Vasily Sarayev, which is believed to be a possible source of a photograph showing Yuri Gagarin talking with Yegorov, who is dressed in a soft helmet and goggles.

Volovich had been the first physician to examine Gagarin after his flight, but on that occasion there had been no necessity to parachute down. The aircraft carrying him, an Ilyushin-14 (a twin-engine prop-jet similar to a DC-3), had landed at an airfield where the world's first cosmonaut had already been delivered prior to his transfer flight to Kuibyshev, a town in the Volga region, where the State Commission and Korolev were waiting to greet him. Volovich would carry out his medical examination during this transfer flight.

He would also miss out on greeting Russia's second cosmonaut after Titov landed by parachute following his day-long mission aboard Vostok 2. Flying over the spacecraft's landing site Volovich had leapt out of the Ilyushin-14 aircraft into a strong cross-wind. Due to this swirling wind his landing in a ploughed field three hundred metres from the landing site was so wildly unstable that he hit his head and was knocked unconscious. By the time he had fully regained his senses the cosmonaut had already been picked up and driven away in a car, although he did later manage to run medical tests on Titov in Kuibyshev, in which he says everything checked out fine. "And the minor dizziness and nausea which Titov experienced for a certain time while in space did not affect his vestibular function after landing."

Once again, Volovich's physician-parachute team was on hand on 15 August, following Andrian Nikolayev's landing in the desert steppes south of Karaganda.

End of a mission

At 9:55 AM the navigator aboard the Ilyushin-14 signalled the medical team that Nikolayev and his Vostok 3 spacecraft had successfully touched down in the designated area. At his signal those making the jump, including cameraman Mikhail Besschetnov, attached the rip cords of their parachutes to a rope stretched along the body of the plane, and made their way down to the tail. The flight mechanic then opened the door, letting in a sudden burst of light and cold air. As the IL-14 swept low over the landing area Volovich could easily make out a figure dressed in a bright orange flight suit, waving his hand in greeting. The aircraft again gained altitude, circled and began to approach the target area. A beeping siren sounded, and a signal light flashed above the door. The order came to jump, and Volovich pushed off with his leg, falling into nothingness.

"Like a feather, my body was caught up in a powerful air current," he recalled in his book. "One second, two seconds, three ... A jerk and my chute opened up over me. I checked the canopy with a quick glance; it was fine."

After landing, Volovich quickly deflated his parachute canopy, scrambled out of his harness, and ran over to Nikolayev.

"Calm and unhurried as always, he got up to greet me, smiling behind a four-day growth of beard. 'How is Pavel?' was his first question. I didn't know if Popovich had landed or not, but I was sure he was alright. Viktor Artamoshin, in whom I had full confidence, was to meet him where he landed."

In fact, Popovich had touched down by parachute some 120 miles away, just six minutes after Nikolayev. His flight had lasted 71 hours and three minutes, during which time he had completed 48 orbits, and travelled 1,242,500 miles around the Earth. Volovich continued:

"Nikolayev had already unpacked the portable emergency supply kit each cosmonaut carried with him in case he landed in an uninhabited area, set up the radio transmitter and changed out of his bulky space suit into dark-blue sweatpants and a sky-blue knit shirt with a white collar and stripe across the middle.

'How are things, Andrian?' I was too excited to think of anything more pertinent to ask.

'Everything's fine. Just great. You fellows did a good job getting here so fast. I hadn't even changed my clothes when I saw the plane flying over. Everything worked like in training. Only when the parachute opened I hit my chin against something and bit my lip. But that's nothing. When I was coming down I opened my visor right away. There was a nice wind. I looked down and saw a small stream in the distance and a rather even field. Of course, it only looked like that from above. You see how many rocks there are around. But everything went fine."

Nikolayev would later say of his landing that "A strange feeling of complete, almost contentment suddenly overcame me ... The weather was foul, but I smelled

Earth, unspeakably sweet and intoxicating. And wind. How utterly delightful; wind after long days in space."

Soon after, the other parachutists came running up to where the two men were talking, including the cameraman, Besschetnov. It was time for Volovich to give Nikolayev his preliminary medical check-up.

"The emergency supply box, from which a silver antenna wire was poking out, served as a table," Volovich recalled. "Andrian held his arm out, and I began to write down my observations: skin colour—normal; visible mucous membranes—pink; pulse—96, but rhythmic and strong; blood pressure 120 over 90. I looked at the first page where I had noted Nikolayev's pulse and pressure before his flight; there was virtually no difference. His pulse was more rapid now, but that was not surprising. This was the usual case after even an ordinary parachute jump.

"Then I conducted a few special neurological tests. They too indicated that Nikolayev had tolerated four days in space perfectly well. That was it. I held out my book to Nikolayev. The first page already showed the autographs of Yuri Gagarin and Gherman Titov. Andrian carefully wrote: '15/8/62. 25 minutes after landing. A. Nikolayev'."

Shortly after, they heard the rumble of an engine and saw a tractor appearing over the crest of a small hill. Two excited young peasant men clambered down and shook the cosmonaut's hands with great gusto, shouting excited praise for his courage. Not long after a Mi-4 search helicopter swept in and landed. Volovich's group then helped Nikolayev to load his belongings onto the craft, after which they all boarded the helicopter for the flight to a reception centre in Karaganda, where a far more extensive medical check and debriefing of both cosmonauts would take place [14].

Celebrations in Moscow

Following the news of the safe landings, hordes of exultant Muscovites had begun converging on Red Square, cheering, singing and waving banner photographs of the nation's two latest space heroes as they listened to patriotic marches and the latest news reports blaring from a series of loudspeakers.

By 18 August, Nikolayev and Popovich had been thoroughly examined and questioned about their flight and reactions to weightlessness at the Karaganda centre. It was time to celebrate, so they boarded an Ilyushin-14 aircraft for the flight to Moscow's Vnukovo Airport, some 20 miles from the capital, and a rapturous welcome by an adoring public numbering in their hundreds of thousands, who had already congregated in Red Square in anticipation of the cosmonauts' arrival back in the capital.

Leaden skies hung over Moscow that day, but the crowded Moscow square was gaily festooned with colourful flowers, and the milling throng was in an effervescent mood. They cheered in near-hysterical anticipation as the Ilyushin turbo-prop carrying the cosmonauts swept low over the city, escorted by seven MiG fighter aircraft, and dipped its wings before descending into the airport. Nikolayev and Popovich then deplaned onto a long red carpet that stretched out before them across the

The Soviet Union's newest heroes: Nikolayev and Popovich, who had become known as "The Heavenly Twins".

tarmac, at the far end of which stood a beaming Premier Nikita Khrushchev, Deputy Premier Anastas Mikoyan, and other members of the Soviet Presidium. After numerous kisses on the cheeks and mighty hugs they received further hugs and kisses from Gagarin and Titov, and then they were swept up into the arms of their families. A band from the Moscow garrison began playing the national anthem and a 21-gun salute rang out across the airport.

A convoy of official vehicles stood at the ready, and the two cosmonauts were ushered into an open, flower-lined Zil limousine for the triumphant motorcade journey into Red Square, along a route lined with cheering spectators who eagerly waved flags, flowers and photographs of the two men, dubbed "The Heavenly Twins" by the Russian news agency Tass.

On arrival in Red Square, Nikolayev and Popovich were quickly escorted to a stand high above the crowd on the Lenin Mausoleum, where they were joined once again by a beaming Khrushchev. The shouting and applause was almost deafening, only falling quiet when they each gave a brief speech of thanks. Then the cheering intensified into what was almost a solid wall of clamour when they were joined on the stand by the immensely popular Gagarin and Titov. Afterwards, at a reception held in the Kremlin's St. George's Hall, the nation's two latest cosmonauts were awarded the title of Hero of the Soviet Union and received the Order of Lenin, as well as being named Pilot Cosmonaut of the Soviet Union 3 and 4.

Vitaly Volovich would later assist at the landing of Valery Bykovsky after his Vostok 5 flight. Among his numerous awards he was presented with the Order of Lenin, the Red Banner Order, the Labor Red Banner Order, the Order of Great Patriotic War, three Orders of the Red Star and 20 other medals. In all, he made 175 parachute jumps. In 1999, Volovich celebrated the 50th anniversary of his historic parachute jump over the North Pole at a function held to commemorate the event. The venue for this celebration was the North Pole.

Describing the tandem flight

On 17 August, Nikolayev and Popovich attended a press conference for Soviet reporters and writers, but the two cosmonauts were under strict instructions not to volunteer any information or discuss potentially controversial issues arising from their tandem flight. Among the press-friendly titbits of information Nikolayev did give was the fact that he had unstrapped himself from his couch four times for a total of $3\frac{1}{2}$ hours and discovered to his satisfaction that he could work and eat easily while floating in the cabin in a state of weightlessness. Popovich then stated that he had also floated unrestrained for about three hours aboard Vostok 4 without experiencing any unpleasant sensations.

A second press conference held before 500 media representatives at Moscow University on 21 August was a more open affair, with several Western reporters in attendance. Supervising the proceedings was the President of the Soviet Academy of Sciences, Mstislav Keldysh. However, the openness did not extend to answering questions without notice—all questions had to have been previously submitted in writing and vetted. Anything else would be ignored.

At the interview, Nikolayev confirmed that neither he nor Popovich had suffered the mysterious malady that had affected Titov on the previous space mission. When asked why Vostok 3 and Vostok 4 had not landed in the same area as the two previous manned spacecraft, Popovich would only volunteer that: "The landing was made strictly in accordance with the programme." He was also asked why the weight of the two spacecraft had not been made public. "I am glad to supply the information," Popovich replied, and the journalists' pens were quickly poised in anticipation. But again the response was harmlessly vague: "The weight of each spaceship, Vostok 3 and Vostok 4, was about five tons."

Popovich also commented on mysterious particles outside of the spacecraft that had been seen by previous space explorers, particularly John Glenn, saying, "We feel we understand these; they are merely the exhaust of the rocket motors." As well, in describing weightlessness, he mentioned having a bottle half-filled with water, and that "in the weightless state the water gathered around the edge of the bottle, and the air collected in the middle in a little sphere."

"Weightlessness," Nikolayev would later recall, "is an amazingly pleasant state of both body and soul, not to be compared to anything else. You don't weigh anything, you aren't supported by anything, and yet you can do everything. Your mind is clear, your thoughts precise. All your movements are coordinated. Both

The first four Soviet cosmonauts: Gagarin, Nikolayev, Popovich and Titov.

vision and hearing are perfect. You see everything and hear everything transmitted from the ground."

REFERENCES

[1] Pavel Barashev and Yuri Dokuchayev, *Gherman Titov: First Man to Spend a Day in Space*, Crosscurrents Press, New York, 1962. Translated from the original Novosti Press Agency (Moscow) publication.
[2] *Flight International* magazine, "Day in Orbit," 17 May 1962, p. 802.
[3] Voice of Moscow, "Seventeen space sunrises for Gherman Titov, 2001." From website *http://www.vor.ru/Space_now/Cosmonauts/Cosmonauts_1.html*
[4] Alex Simpson, *Gherman Titov: The Man Who Came Second*, Capella (Cambridge Astronomical Association) Newsletter No. 94, January/February 2002, pp. 6–9.
[5] Bart Hendrickx, "Translation of the Kamanin Diaries 1960–1963," *Journal of the British Interplanetary Society*, **50**(1), 33–40, January 1997.
[6] Vladimir Golyakhovsky, *Russian Doctor* (translated from the Russian by Michael Sylwester and Eugene Ostrovsky), St. Martin's/Marek, New York, 1984.
[7] *Time* magazine, Friday, 11 May 1962. Online at *http://www.time.com/time/magazine/article/0,9171,939388,00.html*
[8] Dmitri Zaikin, interview with Bert Vis, Star City, Moscow, 13 August 1993.
[9] *The Times*, London, Friday, 22 September 2000
[10] Francis French and Colin Burgess, *Into that Silent Sea: Trailblazers of the Space Era 1961–1965*, University of Nebraska Press, Lincoln, NE, 2007.
[11] John H. Glenn, Jr., Oral History interview for the John F. Kennedy Library and Museum, recorded 12 June 1964 in Seabrook, TX. Interviewer Walter D. Sohier.
[12] Yaroslav Golovanov, *Korolev: fakty I mify*, Nauka, Moscow, 1994.
[13] Nikolai Kamanin, *Skrytyi kosmos: kniga vtoraya (1964–1966)*, Infortekst, Moscow, 1997.
[14] Vitaly Volovich, *Experiment Risk*, Progress Publishers, Moscow, 1986.

8

The "missing" cosmonauts: Rumour and reality

It was very late on the mild spring evening of Wednesday, 27 March 1963. Grigori Nelyubov, along with fellow cosmonaut trainees Ivan Anikeyev and Valentin Filatyev, was returning to the cosmonaut training centre after a long night of drinking and relaxation. The three men were dressed in their air force uniforms, impudently ignoring a regulation stressing that these were not to be worn off-base.

THE DISMISSAL

It was likely that Nelyubov was ready to let off a bit of steam. Even though he was a senior cosmonaut, once in the top three after Gagarin and Titov, he had been overlooked for the flights of Vostok 3 and Vostok 4 because of a decision to fly nationalities other than Russian: Nikolayev was a Chuvash and Popovich a Ukrainian. Nelyubov then felt he should have been in line for either Vostok 5 or Vostok 6 in the middle of that year, but first Valentina Tereshkova had been suddenly thrust into the cosmonaut team ahead of him to fly on Vostok 5 with minimal training, and then Valery Bykovsky was given the nod ahead of him for Vostok 6. It would be fair to say that he was one annoyed, impatient cosmonaut.

That evening the trio had ended a night out by drinking heavily in a small restaurant on the platform of Chkalovsky railway station, which serviced the training centre and nearby military airfield of the same name. According to Alexei Leonov, the restaurant was meant to be strictly off-limits to the cosmonauts [1]. Soon the three men had brought attention to themselves by their boisterous behaviour as they engaged in a ribald arm-wrestling competition. Twice, laughing and joking out loud, they had knocked glasses filled with beer onto the floor of the restaurant, by which time the bar owner decided it was time to call in the authorities and have the men asked to move along. At this, the cosmonauts decided it was time to leave and made their way out onto the platform, ready to call it a night.

Nelyubov undergoing acceleration and *g*-force training. All too soon his cosmonaut career would come crashing down through his inappropriate arrogance.

When arrogance and anger meet

As they meandered along the platform in high spirits the cosmonauts had a two-mile walk ahead of them back to the training centre, but their boisterous behaviour in the bar had resulted in a military security patrol being called out, and the two groups met on the platform. It was a fairly amicable confrontation at first, with the patrol leader suggesting they should simply quieten down, produce their identification allowing them to be in a restricted area and move on. When none of the trio could produce the necessary credentials for passing through the security checkpoint things began to get a little heated, and Nelyubov started to react loudly to what he regarded as unnecessarily rigid officialdom.

While Anikeyev and Filatyev quickly realized the gravity of the situation, Nelyubov stepped forward, arrogantly deciding that their position as Soviet cosmonauts should merit a little more respect. Already four of their colleagues had flown into space and become national heroes, and he was well in line to be one of the next to fly. It was a huge mistake on his part. Partially fuelled by alcohol, he began to argue with the patrol, resulting in all three men being temporarily placed under guard in the duty office while their identities were being confirmed with a phone call to the training centre. Here the patrol commander spoke with Karpov's political officer, Colonel

Photographed in happier times
with the towering KGB
headquarters in the
background are Nelyubov and
TsPK trainer G. Titarev.

Nikolai Nikiryasov, who was angry at the men's behaviour, but felt it was not all that serious.

Displaying a brash egotism that had been noted during his training, Nelyubov still refused to moderate his openly defiant stance. The duty patrol commander was becoming increasingly tired of the drunken airman's belligerent attitude, but out of respect for their uniform offered Nelyubov and his companions a chance to have the incident pass with little more than a reprimand on the spot. He suggested that if they were to apologize for their behaviour all charges would be dropped. Anikeyev and Filatyev readily agreed, but Nelyubov's increasing arrogance would not allow him to follow suit. He boasted that he had friends in high places, and they would not be happy with the guard company if an incident report was filed. It was the last straw for the duty commander, who had heard enough. Initially prepared to forgive a relatively harmless drunken episode, he now insisted that a full apology was in order from

Nelyubov. Before releasing the three men the officer said that a suitable apology would have to be made before 2:00 PM the following afternoon, or the matter would go much further.

At that time Gherman Titov was acting as head of cosmonaut training while Gagarin was fulfilling a tour obligation, and he would reluctantly become involved in the aftermath of the incident. The following morning, after hearing of the late night call to confirm the identities of the three men, Titov rang the duty commander to discuss what had happened. The officer related the details and stressed that Nelyubov, as the main offender, had until that afternoon to extend an apology, freely and without being ordered to do so, or he would have to take the matter further. Titov, certainly no saint himself, would undoubtedly have cursed Nelyubov's obstinacy after hanging up.

"At 2:00 PM ... the commander called and said no one came," Titov recalled in a 1999 interview with journalist Adam Tanner. "So he sent a report to the chief of staff on disciplinary violations at the base." That damning report quickly found its way onto the desk of Nikolai Kamanin. "I asked Grisha [Nelyubov] 'Why didn't you go?'" Titov related to Tanner. "He said 'He (the commander) can go to hell'" [2].

The axe falls

In order to stress the need for strict discipline within the ranks, and to act as a potent warning to any future transgressors, Kamanin decided to leave the punishment up to

the other cosmonauts. A meeting to be attended by all those present at the centre was called for that evening. "It was a short meeting," Leonov later recalled in his memoirs. After much discussion, Leonov said they very reluctantly but unanimously agreed that the penalty had to be severe. "We took a vote; all those who believed the four [three] should resign were asked to put up their hands. The decision was unanimous ... it was a sad day."

"In effect, they disrupted their and others' training programmes," Leonov said of the decision. "All of us had agreed on certain rules of behaviour and the penalty for violating them was expulsion from the group. So a vote was taken and they were expelled" [1]. The decision was passed to Kamanin, and the expulsion was later confirmed by the commander-in-chief of the Soviet Air Force, Air Marshal Konstantin Vershinin.

On 17 April the official documents of

Nelyubov was dismissed from the cosmonaut corps for his unseemly behaviour.

expulsion were signed, with immediate effect. Shocked and still disbelieving, Anikeyev and Filatyev left the training centre that day, but for some unknown reason Nelyubov stayed on until 4 May, when he too walked out and closed the door on his all-too-brief career as a cosmonaut. Curiously enough, his wife Zinaida recently said that initially he had felt no remorse at the time, despite being kicked out of the cosmonaut corps, the Zhukovsky Academy, and their comfortable three-room apartment.

Life without a dream

The cosmonauts undoubtedly felt that they might have treated Anikeyev and Filatyev a little too harshly, but their two now-former colleagues were judged guilty by

Too much partying would also result in Anikeyev's dismissal from the cosmonaut team.

association, or "burnt down from company" as one put it. Anikeyev, it has to be emphasized, was already under a probationary cloud resulting from the incident the previous year while undertaking studies at the Zhukovsky Academy, when he and Mars Rafikov had been caught off-base without permission, drinking in a restaurant in the large Moskva Hotel [3]. Obviously, he hadn't learned his lesson.

"I feel sorry for Anikeyev," Leonov recently lamented. "He had a strong personality and he was an excellent pilot. The same goes for Filatyev." However, he added that while the two men were contrite and willing to apologize, they had to convince Nelyubov to do likewise, and this had not happened [1].

A shattered man, Nelyubov could never fully come to terms with the fact that he had been sacked and would never fly into space, and that he had caused his friends to be likewise dismissed from the programme. He subsequently moved with his family to the village of Kremovo, north of Vladivostok in the Mikhailovski District of the Primorski Region. Here, in the Far East Military District, he returned to flying as a military pilot, first class, assigned to an interceptor squadron based near Vladivostok.

According to space historian Asif Siddiqi in his book, *Challenge to Apollo: The Soviet Union and the Space Race, 1945–1974*, the loss of Nelyubov in particular, one of the brightest and most qualified in their group, was felt quite deeply by the other cosmonauts:

"There was some discussion among Kamanin and the cosmonauts in later months on bringing Nelyubov back into cosmonaut training, based on Nelyubov's performance at his new assignment in an Air Force unit in the Soviet Far East. This never happened. It seems that Nelyubov suffered from a psychological crisis through the following years, as cosmonaut members junior to him started flying their space missions" [4].

The fact that Nelyubov was given another chance is borne out by an interview given by his widow Zinaida in the 2007 Roscosmos documentary *On Mog Byt Pervym* ("He Could Have Been the First"). She told the interviewer that some time later in 1963 Popovich had been in the Far East armed with a last chance for Nelyubov. If he personally apologized to Nikolai Kamanin there was a chance he would be readmitted to the cosmonaut corps. With the permission of his Far East commander, Nelyubov made his way to Moscow, but much to his chagrin Kamanin—then busy with the 1963 cosmonaut selection—would not see him. Returning to the base near Vladivostok he then applied to be a test pilot, but this was refused. According to Zinaida Nelyubova, his regiment colleagues told her he was very angry at these let-downs [5].

Two years later, a chastened Nelyubov decided to make one more try, and sent a personal appeal to Sergei Korolev, who had always liked the young pilot. However, just as things were looking up, Korolev died on 14 January 1966 following a botched operation in hospital.

A sad and lonely death

A despondent Nelyubov now began drinking heavily. As his former colleagues and then newer generation cosmonauts continued to fly into space, he would brag to anyone who cared to listen that he had once been a member of the first cosmonaut team—even a backup pilot to Yuri Gagarin. Not surprisingly, no one had heard of him, so his stories were mostly scoffed at behind his back; dismissed as the wishful dreams of a pathetic loner. A deep depression had now taken root and the former cosmonaut had nowhere to go and seemingly nobody to whom he could turn.

The spectre of suicide has hung over the death of Nelyubov for many years now, so in September 2007 Alexei Leonov was asked by Bert Vis for this book to give his impression of what happened to the once promising young cosmonaut.

On the evening of 17 February 1966, it is known that 31-year-old Nelyubov returned home from a night flight operating out of Chernigovka aerodrome and decided, apparently not out of the ordinary, to have a few drinks before retiring. "He had something to drink before he went to bed," Leonov recalled. "His wife closed the door . . . locked the door, and went to work in the morning. He got up and left the house. Actually, he tied the sheet to the first floor window and climbed out. He went along the railway to the station where they had a bar and he could get something to eat." This meant crossing the rail bridge at Ippolitovka station, northwest of Vladivostok.

"There was a snowstorm," Leonov stated, "and he raised the high collar of his bomber jacket [to protect him from the snow]. At the edge of the railway line a train was approaching and there were some wooden planks lying on the open platform." With his collar turned up, Nelyubov apparently did not hear the train approaching. The end of one of the heavy planks was jutting out into the path of the train. Leonov told Vis that the train hit this plank, which became airborne, striking Nelyubov as he walked by [6]. Despite rumours to the contrary, Leonov has always been quite adamant that the train itself did not strike Nelyubov, nor did he throw himself under the train or into its path.

It had been widely assumed that the disgraced cosmonaut decided to end his life in a moment of drunken despair, but officially the incident has been determined as nothing other than a tragic accident. The report into his death only states that following "a serious spiritual crisis . . . he was killed in a state of intoxication by a passing train."

However, despite Leonov's insistence that it was an unfortunate, drunken accident, Zinaida Nelyubov knows differently. She harbours no doubts at all that he took his own life in deep despair, having found notes he left for her that night, saying that she must carry on and live her own life [5].

When asked if the cosmonauts were told about the incident soon after Nelyubov's death, Leonov shook his head. "No, we learned about it quite a long time after it happened, when I visited the place. I was in the area recruiting new cosmonauts. I recruited three: Onufriyenko, Gidzenko and Padalka [in the latter part of the 1980s] . . . it was at that point that I visited the grave. Each person is the creator of his own destiny; there was no point in being angry. Lots of journalists now state that he was an outstanding person, badly treated by us, but this is not the case. It's a very mundane ... a very banal story" [6].

Nelyubov was buried in Kremovo, where, and in recognition of his flying achievements within the Soviet Air Force, segments from a crashed aircraft were used to form part of his grave. In later years, however, vandals would smash his grave and steal the aluminium to sell.

It was not for some 25 years after Yuri Gagarin's pioneering flight that Grigori Nelyubov was officially identified as a previously unknown member of the first group of cosmonauts. Once his status as a cosmonaut was revealed in 1986, the Mikhailovski district newspaper and council began campaigning for a far more suitable gravestone to recognize his achievements. They managed to raise a total of 37,000 roubles, and a new black marble headstone was commissioned. His last resting place is now a revered monument to his short life and career. As well, the Mikhailovski museum created a special Hall of Nelyubov within the museum. Nelyubov's widow, who had subsequently moved to Shchelkovo, a few miles north-east of Moscow (where she resides to this day), was very supportive of this tribute to her late husband and donated to the museum various personal effects, documents and photographs.

"In the general opinion of almost all cosmonauts," wrote spaceflight journalist Yaroslav Golovanov in his brief official history of the first cosmonaut team for *Izvestia*, "Nelyubov could, with time, have become one of the top five Soviet cosmonauts," but added that his dismissal highlighted the "exactingness with which the centre approached training for selection as cosmonauts."

But had Nelyubov's arrogance and a failed desire to be awarded the first flight contributed to his eventual downfall? "Perhaps," Golovanov observed. "It was precisely his unconcealed craving for leadership that prevented him from being a leader" [7].

Guilty by association

On 30 September 1996, Dutch spaceflight researcher Bert Vis was presented with an opportunity to ask Pavel Popovich if he had maintained any sort of contact with Nelyubov following his departure from the cosmonaut team. Popovich suggested that the two families had stayed in touch for a while.

"I even met him after his dismissal, and the meeting took place in the [Soviet] far east," Popovich told Vis. "He experienced his dismissal as very painful—a great amount of pain. It resulted in a very rapid drop in his spirit. I talked to him on several occasions and sometimes he even accused me of being responsible for his dismissal, because at the time I had been a commander of the detachment ... of the team. I [tried] to prove to him that I wasn't involved and had nothing to do with his dismissal, but he didn't believe me."

When asked if Nelyubov's sacking was the man's own fault, Popovich paused momentarily before agreeing that it was. "For instance, if I was rude with you and if I am an honest, decent man, all I have to do is have the courage to apologise. I would say, 'Please excuse me, I just exceeded my powers; I was over-excited, please excuse

Filatyev's grave in Orel.
(Courtesy: Bert Vis)

me.' Nelyubov didn't do that; that's why he suffered. It was the result of his own fault."

Popovich was a little more forthright when asked if any of the cosmonauts had attended Nelyubov's funeral, and it was obvious from his carefully chosen words that their former colleague's death did not cause any excessive grief in the training centre. "I was probably the first one to hear about his death," Popovich recalled, "but I decided not to fly down to his funeral. And even if all the cosmonauts had known about his death, nobody would have gone. Usually, we do not forgive the betrayal. And he betrayed us" [8].

Sadly enough, all three of the expelled cosmonauts would die at a comparatively young age, but Anikeyev and Filatyev lived long enough to know they had finally been recognized as former members of the first group of cosmonauts. At least it was something.

After being discharged from the cosmonaut team, Filatyev returned to the Air Defence Forces before retiring from the Air Force with the rank of major on 29 November 1969. He was then employed in the State-run industrial design institute

Gipropribor in the Russian administrative city of Orel until 1977, when he became a teacher of civil defence courses, also in Orel. He retired from teaching in 1987 and became a pensioner. Three years later, on 15 September 1989, he passed away in Orel from natural causes, aged 59.

Anikeyev also returned to active flying in the Air Force and was posted to the Soviet Union's northern territory as a military pilot, second class. He remained active until 1965, and then served another ten years with the Air Force as a ground controller/navigator, rising in rank to captain [9].

In his co-authored book, *Two Sides of the Moon*, Alexei Leonov tells an interesting story from Anikeyev's post-cosmonaut life, although he does not reveal when this incident is alleged to have happened. "Anikeyev was at a party one night when someone stole his car keys from his pocket. They took his car and, in their haste to get away, ran someone over, killing them. The thief then put the keys back in Anikeyev's pocket. Anikeyev was arrested and sent to prison for a year until it was discovered he was innocent. But after he was released he never flew again" [1].

After transferring to reserve status, Anikeyev moved to the town of Bezhetsk in Tver Oblast, Russia, where he died of cancer on 20 August 1992, aged 57.

LOST AND PHANTOM COSMONAUTS ABOUND

"Little is known of the Russian astronauts, or even how many are undergoing training. Names that have been released include Mikhailov, Grachov, Begoloniev, Kachur, Zovodosky and Vladimir Ilyushin, the son of the well-known aircraft designer" [10].

So read an article in the December 1960 issue of *RAF Flying Review* magazine, although some of the names were unintentionally misspelled. Beneath a photograph said to be that of "highly trained Russian astronaut Ghennadi Vassilievich Zavadoski" (which also differed in spelling from the name given in the text), the article decried the lack of authoritative data on the Soviet space programme, suggesting that it was being conducted with "a degree of secrecy normally reserved for vital military operations" [10].

Lack of denial

In an odd twist of fate, it was such articles that provided names to many of those who would come to be incorrectly known as the "lost" or "phantom" cosmonauts, said to have been killed in top-secret spaceflights before and around the time of Gagarin's history-making orbit of the Earth in April 1961. However, most of those named as cosmonauts were simply aerospace engineers and military subjects engaged in test work and systems evaluation, whose participation in the actual Soviet space programme was to varying degrees just doing what they were paid to do.

It has to be said that if Russia had wanted the Western world to harbour any doubts that Yuri Gagarin was the first person to fly into space and return, they did so

Test subject Gennady
Vassilievich Zavodovski.

in a most effective manner. Contradictions and discrepancies assumed the place of
facts, and rumours flourished wildly as Soviet space chiefs remained tight-lipped and
stoically refused to comment on speculation about the deaths of Soviet cosmonauts in
undisclosed spaceflights. To some researchers, a lack of denial was every bit as good
as a wink to them. They all wanted to be the first to seize upon some irrefutable proof
of a State-ordained cover-up and reveal this conspiracy to the world.

Even the eminent academician Anatoli Blagonravov of the Soviet Academy of
Sciences fell under suspicion for a statement he made on 22 November 1959, and he
was once again forced to issue a statement denying the existence of any covert man-
in-space programme. According to a report of a committee on aeronautical and space
sciences issued by the United States Senate on 1 December 1959, "Blagonravov
termed reports of a training programme for Russian astronauts 'ungrounded and
stemming mainly from journalistic imagination'. He said Russia has no man-in-space
programme as such—just a programme on flight safety ... Blagonravov's denial of a
specific man-in-space programme at the ARS [American Rocket Society] meeting
clashes with earlier statements by Professor Andrei Kuznetsov, head of the Soviet

aerospace medical programme. Kuznetsov told delegates to the 52nd General Conference of the Fédération Aéronautique Internationale [FAI] in Moscow last summer that the Soviets have selected four astronauts for their first manned space capsule programme" [11].

Brigadier General Don Flickinger, a former US Air Force flight surgeon deeply involved in space medicine research was one of those attending the FAI conference in Moscow. He was not at all satisfied with evasive responses to his questions about a Soviet manned space programme, and later reported that after badgering Soviet officials for answers he had been privately told that up to five decorated Soviet pilots were undergoing specialized training.

The *New York Times*, intrigued by Blagonravov's denial of an active manned space programme, offered two conclusions of its own. "One is that the Soviet scientists were playing dumb ... either because of restrictions of Soviet secrecy or because of a deliberate attempt to lull the United States into false security. The alternative is the one now generally held by space administration officials. It is that the civilian Soviet scientists are not involved in the man-in-space programme, which is being managed by the military. In fact, it is believed the civilians have no great enthusiasm for putting a man into space at this point" [12].

Myths and legends of the Space Age

According to a wide-spread but spurious story seized upon by those seeking out the so-called "phantom cosmonauts", an outstanding Soviet parachutist by the name of Pyotr Dolgov was reported to have been launched into space aboard a Vostok spacecraft on 11 October 1960, some six months before Yuri Gagarin. At the time, Premier Nikita Khrushchev was visiting the United Nations building in New York City and he seemed inordinately keen to extend his visit.

Meanwhile reports were coming in that American tracking stations had been alerted to the possibility of an imminent Soviet space feat aimed at boosting Khrushchev's prestige. Following several days of mounting speculation, listening posts in England, Italy, Japan and Turkey were reported to have homed in on what could have been a manned Soviet launch. Tape recorders began preserving the signals and sounds, among which was said to be the unmistakable, regular beating of a human heart. Approaching the first staging point the heartbeat was said to have pulsed faster, indicating a normal human reaction at this critical phase. Then, just as the second stage should have ignited, all contact was lost and the signals abruptly ceased. A cosmonaut was subsequently assumed to have been incinerated in a fiery explosion.

If one was to place credence in all the stories then doing the rounds, the year in which Yuri Gagarin became the first man to fly into space was a particularly bad one for Soviet cosmonauts. Three men specifically named as Belokonev, Grachev and Kachur were said to have died in precursory Vostok flights during 1961, with the sounds of their heartbeats supposedly picked up by ground monitoring stations. In fact, these and other names had been circulating for some years. Although their names were misspelt, "Alexis Belokonev, Alexis Grachev and Ivan Kachur" were

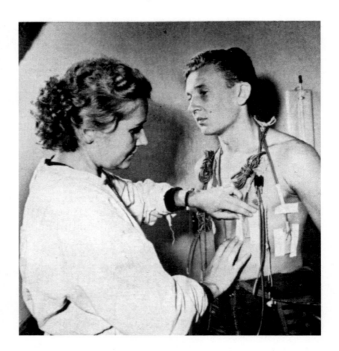

Test subject Alexei Grachev
has sensors taped to his body
by space physiologist Dr.
Natalya Arabova.

said to have been physically involved in bioastronautics research (later seized upon as a possible euphemism for space training) in the 22 October 1959 issue of the weekly illustrated Soviet magazine *Ogonyok*—quite some coincidence.

Then, just one day before Gagarin was launched on his epic voyage, the British Communist Party newspaper *Daily Worker* carried a dramatic article by Moscow-based correspondent Dennis Ogden, reporting on a flight mishap involving "cosmonaut" Vladimir Ilyushin, a well-known test pilot and son of legendary aviation and spacecraft designer Sergei Ilyushin. Ogden, who resided in the same building as Ilyushin, had been investigating steadily strengthening rumours about an impending Soviet spaceflight. He was aware that Ilyushin had been hospitalized following an unexplained accident, so he created a story relating how a Russian pilot had been launched into space in December 1960, orbiting the Earth three times before being brought back down. As part of Ogden's speculations, he wrote that Ilyushin, whom he called "the world's first cosmonaut", was critically injured during a flawed ejection landing. On learning of this, according to Ogden, the Soviet government had decided they could not present him to the world as a national hero, and news of his spectacular feat was immediately quashed.

Although Russian space administrators were acutely secretive about their nation's early cosmonauts, it is known without any shadow of doubt that Ilyushin was not one of the first cosmonaut team. It is also known that he had been seriously injured on 8 June 1960: not in a Vostok spacecraft, but in an automobile accident [13].

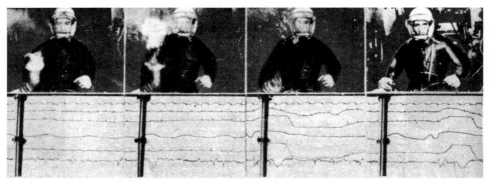

This strip of film purports to show the effect of a meteorite hitting a hermetically sealed space cabin. Identified in the caption as "future Russian cosmonaut Alexei Belokonev," the subject sits calmly while a glass of water bubbles, boils, then disintegrates as air rushes from the cabin. A graph below the film shows heart and lung changes.

Belokonev inside the pressure chamber.

Naming the "lost" cosmonauts

On 17 May 1961, a little over four weeks after Gagarin's triumphant return from space, another two cosmonauts are said to have died in space, uttering unanswered, desperate calls for help as they orbited without any hope of returning alive from space. These two lost cosmonauts were said to be a man and a woman.

A monstrous solar flare that took place on 14 October 1961 was then rumoured to have been responsible for dramatically altering the course of a Soviet spacecraft carrying a crew of three, which disappeared forever into a stellar oblivion with its hapless occupants. It then seems that Alexei Belokonev died a second time after he

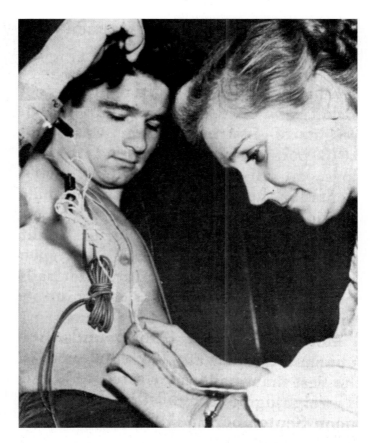

Mikhailov has electrodes taped to his body by Dr. Arabova prior to entering the high-altitude chamber to test a Soviet spacesuit.

was launched on yet another failed spaceflight on 8 November 1962. Following this, an unnamed female cosmonaut was reported lost during her flight into space just 11 days later on 19 November 1963, five months after the well-publicized flight of Valentina Tereshkova.

An article in *Newsweek* magazine late in 1961 discussed prevailing rumours of dead cosmonauts and determined after investigation that there had been no humans involved in fatal space accidents prior to the flight of Gagarin. "For their part," the article concluded, "U.S. monitors, who have eavesdropped on Soviet rocket shots since 1956 with all the tools of electronic intelligence, are inclined to believe the Russian declarations that Gagarin was the first cosmonaut to be despatched into space" [14].

In May 1962, while on an official tour of the United States, Gherman Titov was questioned about persistent rumours concerning spaceflight fatalities during a television interview in New York. When specifically asked if the Soviet Union had lost any cosmonauts before the successful flight of Yuri Gagarin, Titov dismissively responded: "I have heard about that. According to your newspapers, the Soviet

Union has lost not one, but five cosmonauts. They even named them. We had *no* cosmonauts in our country before the first cosmonaut, Yuri Gagarin."

But these insidious rumours would simply not go away. A cosmonaut named Alexei Ledovsky was said to have perished as early as 1957 during a suborbital flight launched from the Kapustin Yar cosmodrome in the Astrakhan region of southern Russia, when "transmission was abruptly halted" at a height of 200 miles. An explosion at stratospheric altitude in an experimental "space aircraft" was also reported to have taken the life of a female cosmonaut named Mirija Gromova. In 1960 an unknown cosmonaut was yet another said to have died when his capsule flew into orbit, but somehow headed off in the wrong direction and sent him on a lifeless journey into eternity.

In August 1961, just four months after Gagarin's mission, a book titled *Cosmonaut Yuri Gagarin: First Man in Space* was published in the United Kingdom. Written by former Fleet Street journalists Wilfred Burchett and Anthony Purdy, who seemed to have gained unprecedented access to many higher-ups in the Soviet space programme, the book's introduction stated of the two authors that "they interviewed scores of scientists and space-workers, including Gagarin himself." At the time, it seemed to be an authoritative and credible account of Gagarin's life and mission, as distinct from the propaganda-fuelled Soviet "biographies" that been churned out to commemorate Gagarin's flight. In their book, Burchett and Purdy mentioned a trio of onlookers at the historic launch as three of his fellow cosmonauts, namely "Vladimir Mikhailov, 28, Alexei Grachov, 27, Vasilievich Zavodoski, 27." This would cause many Western observers to put two and two together when they saw photographs and footage of these three men actively engaged in spacesuit and pressure chamber exercises. However, while they were in fact real people involved in testing equipment and procedures for supersonic aircraft and spaceflights, they were not cosmonauts [15].

Was Dolgov a cosmonaut?

Early in 1963, the Canadian Press Agency reported that at least two cosmonauts had perished in space. They substantiated this claim by stating that a Russian-speaking correspondent of the *Vancouver Sun*, Arthur Karday, had located a reference to these fatalities in the 31 December 1962 issue of the Soviet government newspaper *Izvestia*, within an article on the recently completed tandem flight involving cosmonauts Nikolayev and Popovich. According to Karday, the article read: "One of the truly magnificent and awe-inspiring monuments to the human spirit is the self-sacrifice of two other heavenly heroes—Andreyev and Dolgov. One of them, Dolgov, was destined not to return from the stratosphere. He, like the Gorkian eagle, sacrificed his life in order to save hundreds and perhaps thousands of other spacemen through his heroism."

No explanation was given in the *Izvestia* article, although it was later revealed that Dolgov had actually been killed while testing an ejector seat system with recovery applications for cosmonauts. His companion on this occasion was Yevgeny

Pyotr Dolgov.

Andreyev, who survived the test. For the record, however, this occurred late in 1962, and not two years earlier.

Colonel Pyotr Ivanovich Dolgov of the Soviet Air Force became heavily involved in testing parachute systems after the Second World War, both as an experienced test jumper and as a designer of aircraft escape systems from supersonic aircraft. He would soon turn his attention to testing capsule ejection and parachute landing systems for future Vostok flights and Soviet cosmonauts. Meanwhile, in June 1960, he set a new world record for the highest parachute opening at 48,671 feet. Initially, these drop tests from high-altitude balloons were carried out using mannequins, and were scheduled to be completed around May 1961. After two successful drop tests had been successfully completed, there were worrying signs that America was making serious progress in its attempts to launch an astronaut on a suborbital mission, and pressure was subsequently exerted from high levels on Dolgov's commander, Voronin, to have all tests completed by the first day of March.

These tests required Dolgov and other volunteers, wearing pressure suits and helmets, to be strapped into ejection seats within the confines of hermetically sealed spherical capsules, which would then be transported to about 32,000 feet aboard a heavy military aircraft. The capsule would then be dropped from the aircraft and allowed to freely plunge nearly ten thousand feet. By this time, and 23,000 feet above the ground, it would be travelling at roughly the same vertical speed as a Vostok craft returning to Earth. Having reached this altitude the hatch would be automatically and explosively discarded. After a further pause of one second the ejection seat containing the parachutist would then be catapulted out through the open hatch.

One of the stratospheric gondolas produced for the Volga programme is shown on display at the Central Air Force Museum at Monino, Moscow. (Courtesy: Bert Vis)

A small drogue chute would deploy, followed by a larger stabilizing parachute. Two-and-a-half miles above the ground the main parachute would finally blossom out, the ejection seat drop away, and the test subject would touch down in a conventional parachute landing.

Major (later Colonel) Yevgeny Nikolayevich Andreyev was an experienced test-parachutist involved in these exercises. He and Dolgov had been attached to the parachute rescue division of the Soviet Air Force's Flight Test Centre in the 1950s and 1960s, and he had also been working closely with the early cosmonauts in their training and the testing of Vostok space hardware and systems. High-altitude balloon test flights were then being conducted under what was known as the Volga programme, with Andreyev and Dolgov set to trial a range of spacesuits and ejection systems using a new, non-explosive type of Vostok ejection seat. Both men had been preparing for the flight, which would take place on the Volga Steppes, for about six months. As Andreyev later recounted, "Our pre-flight training had been tough and thorough, and had included instruction in steering the 100-metre-high balloon, pressure-chamber testing of the high-altitude equipment, and contingency practice."

Two experienced Volga programme parachutists would serve as backups for Andreyev and Dolgov. Andreyev later named one as Ivan Kamyshev, while the other participant was 34-year-old Vasili Lazarev, selected as a cosmonaut two years later.

On 1 November 1962, a few weeks after the tandem flight of Nikolayev and Popovich, the balloon carrying Andreyev and Dolgov lifted off from the city of Volsk at 7:44 AM Moscow time, slowly ascending to an altitude of 83,523 feet. Dolgov was wearing a full Sokol spacesuit as part of his test. At 10:13 AM, after an ascent lasting just under two-and-a-half hours, Andreyev leapt out of the balloon. Dolgov would remain in the balloon until he too jumped, this time from 93,970 feet. Andreyev takes up the story:

"Having depressurised my section of the cabin, I waved goodbye to Dolgov, who was getting ready for the jump behind a transparent airtight partition, and ejected into the void of the stratosphere. I didn't feel the usual swirl of the air stream. To prevent the front glass of my pressure helmet from freezing over, I turned on my back and saw the black velvet of the night sky studded with the countless diamonds of bluish-silver stars. Plummeting to about 12 kilometres at some 900 km/h I began to feel the growing resistance of the atmosphere. I turned again to face the earth, and after free-falling spread-eagled for a while, I pulled the ripcord.

After a successful soft landing—I even managed to remain on my feet—my first thought was for Dolgov. I looked up and saw the canopy of his chute—everything was OK. And it simply couldn't be otherwise; an ace parachutist, Pyotr had tested quite a few sophisticated ejection seats and capsules, had ejected from planes at supersonic speeds, and had taken part in many record-breaking high-altitude and delayed-opening jumps. However the pilot of a support party plane monitoring the descent reported to the control post that Dolgov appeared limp in the harness. I felt a knot in my throat.

An ambulance was at the landing spot even before Dolgov's body touched the ground. A small hole in his pressure helmet told the story. It seemed that when leaving the nacelle Pyotr had hit some metal part with his helmet and had been instantly killed by depressurisation. Later, a special timer had automatically opened his parachute" [16].

It had been Dolgov's 1,409th parachute jump. Today in the Tushino district of Moscow there is a street named in his honour, and one of the district's schools, No. 829, maintains a small museum dedicated to Dolgov's life and achievements. He was made a posthumous recipient of Hero of the Soviet Union, and is buried in a military cemetery at Chkalovsky Air Base.

Following the fatal accident, chief designer Sergei Korolev would demand several crucial modifications to the ejection system; one of which was to make the exit hatch much larger. He also increased to two seconds the interval between the hatch being explosively discarded and the automatic operation of the ejector mechanism.

For his part, and despite the fatality, Andreyev had claimed a new world freefall record, plunging 79,560 feet before opening his parachute, breaking the previous record created by another Soviet parachutist, Nikolai Nikitin, whose record for freefall had previously stood at 46,965 feet.

Falling on deaf ears

Sensational rumours of dead cosmonauts were constantly being rehashed in the Western media during the 1960s, leading at one stage to a rather prickly response from the editor-in-chief of *Izvestia*, Alexei Adzhubei, who also happened to be the son-in-law of Premier Nikita Khrushchev (and who also lost his cushy job along with Khrushchev the following year). In May 1963, coinciding with a series of U.S. Senate committee hearings on Soviet space failures, the *New York Journal-American* had run a feature article on the death of "phantom cosmonauts", which was picked up by the *Baltimore Sun* and the Washington, D.C. newspaper *Space Business Daily*. The *New York Journal-American* mentioned two cosmonauts named Belokonev and Mikhailov, who had graduated from the Space Research Centre of the U.S.S.R. but disappeared from public life after what was referred to as a failed expedition to the Moon by the two men on 17 May 1961. Fed up with constant accusations of cosmonaut deaths, Adzhubei decided to refute these time-worn allegations by hauling the two "deceased cosmonauts", Alexei Belokonev and Gennady Mikhailov, into the newspaper's office on 27 May 1963 and having them pose for a photograph. An accompanying article that appeared the following day above Adzhubei's name stated in part that the newspaper "denied last night, May 27, a report by the *New York Journal-American* that they died after launching into space" [17].

In following years other cosmonauts were said to have perished in a number of ground and spaceflight accidents. Because of recurring stories in the popular media, many people became convinced that rumours of deceased cosmonauts simply had to be credible. Soviet space officials, meanwhile, remained steadfastly reluctant to discuss anything but successful space missions, contemptuously dismissing any

Alexei Belokonev (left)
and Gennady
Mikhailov
photographed at the
office of the government
newspaper *Izvestia* on
27 May 1963 to refute
persistent claims in the
Western press that both
men were cosmonauts
who had been killed in
unreported spaceflight
accidents.

such rumours. However, they would issue no real clarifications when pressed for information. This only served to exacerbate the matter, and fuel further unsupported speculation of a massive State-ordained cover-up.

The inherent secrecy surrounding the real Russian space programme, when combined with traditional Soviet practices of gross exaggeration, untruths, re-inventing events and the commonplace retouching of official photos, led space experts worldwide to believe that accidents and fatalities may indeed have occurred and been hushed up. Soon, photographs of men who were quite obviously cosmonauts in training began turning up in translated articles and books originating in Russia, but official versions of these same photographs would have some men clumsily airbrushed out of the image, replaced by scenery or blank walls. It was becoming abundantly clear that a cover-up involving the Soviet cosmonauts was taking place.

Well-known spaceflight writer, historian and investigator James Oberg was at the forefront of those seeking to resolve the true story of the early Soviet space programme. His ground-breaking book *Red Star in Orbit* not only revealed many previously unknown Soviet failures in space, but exposed forever Russian attempts to conceal the identities of several cosmonauts who had either died or been dismissed from the space team. He raised numerous questions in his book, and most would be answered over the next few years.

Airbrushed out of history

James Oberg had just one question after comparing two group photographs: just who was the missing man? He had come into possession of two seemingly identical photographs—one in a 1971 book written by a Soviet journalist but published in New York. This group shot, taken in the resort town of Sochi in May 1961, showed Sergei Korolev sitting in the midst of several men from the training centre clearly

The retouched photo (above) from which the mystery cosmonaut named Grigori has been crudely airbrushed. Below is the photograph as originally taken, with Grigori Nelyubov included in the group shot.

identified in the caption as "Soviet cosmonauts". To his immediate left sat the centre's training director Yevgeny Karpov, and to Karpov's left in turn was parachute instructor Nikolai Nikitin. On Korolev's right were the well-known figures of Andrian Nikolayev and Yuri Gagarin. But it was the back row that Oberg found most intriguing. It comprised Pavel Popovich, Gherman Titov and Valery Bykovsky, but in between Popovich and Titov was another smiling, young, but unidentified man. Amongst some official photographs Oberg had received from another source was another in which the tall young man had been completely obliterated, airbrushed out with a substitute background clumsily painted in by a graphic artist.

Through diligent research, Oberg was able to establish that the person in the book photograph was a cosmonaut named Grigori. He had also heard informed speculation about another young cosmonaut who had died in a training fire but did not know his name, apart from saying "It might have been Ivan or Vasily or Valentin. Nobody outside of Russia knows." He was very definitely on the right track [18].

Russian space journalist Golovanov would acknowledge the detective work of Oberg in expertly dismissing rumours of Soviet cosmonauts killed in phantom space flights, and this is most likely what led to Golovanov being given official sanction in 1986, on the 25th anniversary of the flight of Gagarin, to tell the story of the first cosmonauts in a series of articles in *Izvestia*. Through his words, hidden truths and names emerged from exile as Golovanov not only confirmed the identities of Russia's first cosmonauts, but for the first time told how two of those 20 men had perished in tragic circumstances.

Decades of denial

Today, through the mighty efforts and dedication of scores of researchers, the chronology of Soviet spaceflight has been substantially revealed, although much of this information has been hard-won through incredible persistence. These researchers know that there was a natural sequence of missions and events leading up to Gagarin's epic flight, in fact remarkably similar to those in the U.S. space programme. This included precursory launches of Vostok spacecraft, carrying a number of dogs and other smaller animals as test subjects, and the dual-flight mannequin named Ivan Ivanovich, in order to study orbital and ejection processes. While Ivan could never be truly considered a "lost" cosmonaut, in April 1994 he did suffer the appalling ignominy of being sold to the highest bidder at a Sotheby's auction house sale in New York City. Sergei Korolev would have turned in his grave!

After decades of denial, documented injuries and even deaths attributable to spaceflight training were finally revealed. Without any shadow of doubt the first group of 20 cosmonauts has also been fully identified, and none of them has mysteriously vanished. Leading experts on the history and chronology of the Soviet space programme have assiduously put together lists of every R-7 rocket and every spacecraft manufactured by the Soviet Union, and each has been accounted for by these researchers. No matter what stories have been told, and how much alleged 'evidence' exists of dead or phantom cosmonauts, no one has ever explained how several R-7 launch vehicles could have been manufactured without appearing in official Soviet records. Nor could they have been launched, particularly with cosmonauts onboard, without this fact being recorded by the Soviets or at least noted by Western intelligence agencies.

Despite this, a bizarre cult of fabrication still exists today. In 1998 the now-defunct American spaceflight magazine *Final Frontier* ran a sensational article in which a writer claimed he had "uncovered" archival material in a Moscow vault, detailing the deaths of unnamed cosmonauts in the early 1960s. When asked about these fresh "revelations", Oberg and other Western analysts of the Soviet space programme could only grimace and shake their heads, reiterating yet again their painstaking and meticulous research over several decades.

The Rudenko rumours

Has the passage of more than four decades brought fresh enlightenment to such rumours? One would think so, but sadly that is not the case, and it does not take much to once again fan the fires of speculation based on long-disproved stories. It would happen again on 12 April 2001, which should have been respected as the 40th anniversary of Yuri Gagarin's flight. On that day, Reuters news agency published an Interfax news agency report relating how three Soviet pilots had perished in secret test launches well ahead of Gagarin's flight in 1961.

The source of these rumours was reported to be Mikhail Rudenko, a "senior engineer" known to have worked "in one of the main Soviet space centres" [later identified as Valentin Glushko's OKB-456, in Khimki]. Rudenko claimed that three manned suborbital flights had taken place between 1957 and 1959. He claimed that all three parabolic trajectory flights, which briefly touched outer space at their highest point, had ended in failure, resulting in the death of all three sole cosmonauts.

The names Rudenko attached to these failed flights were Ledovskikh, Shaborin and Mitkov. "Obviously, after such a series of tragic launches, the project managers decided to cardinally change the programme and approach the training of cosmonauts much more seriously in order to create a cosmonaut detachment," Rudenko added. At the time Sergei Gorbunov was the spokesperson for Russia's space and aviation agency *Rosaviakosmos*, and when he was asked for an opinion on this report, he quickly dismissed it as: "Foolishness, just plain foolishness. I don't think this even deserves comment. It is just an old legend associated with Gagarin's flight" [19].

The names of the three cosmonauts allegedly killed in accidents at Kapustin Yar came unsurprisingly close to those revealed in James Oberg's 1988 book, *Uncovering Soviet Disasters*, although the spelling of two names (apart from Mitkov) is slightly different. Rudenko's "Ledovskikh" is Ledovsky in Oberg's book, while "Shaborin" is Shiborin. While Reuters tried to make out that the report was some sort of latter-day revelation, weary Western researchers had known for some years of stories concerning two alleged cosmonauts named Serenty Shiborin and Andrei Mitkov and their allegedly fatal suborbital flights. They could only add these latest rumours to their "lost cosmonaut" files. Oberg, for his part, would dismiss speculation surrounding these phantom flights and names as completely unsubstantiated.

In fact, James Oberg is quite adamant in his summing-up of these rumours, both old and new. "After considering their sources and their details in the hindsight of subsequent space activities, I concluded that all such stories dealing with alleged flight fatalities were baseless" [20].

REFERENCES

[1] David Scott and Alexei Leonov, *Two Sides of the Moon*, Simon & Schuster, New York, 2004.
[2] Adam Tanner, "Once adored cosmonaut looks back," 28 April 1999. Online at *http://www.cdi.org/russia/johnson/3264.html*
[3] Encyclopedia Astronautica (courtesy Mark Wade), "Gagarin." Online at *http://www.astronautix.com/astros/gagarin.htm*
[4] Asif Siddiqi, *Challenge to Apollo: The Soviet Union and the Space Race, 1945–1974*, NASA History Division, Washington, D.C., 2000.
[5] Zinaida Nelyubov, from 2007 documentary *On Mog Byt Pervym* (He Could Have Been First), Roscosmos Russian Space Agency.
[6] Alexei Leonov, interview with Bert Vis at Association of Space Explorers Congress, Edinburgh, Scotland, 17 September 2007.
[7] Yaroslav Golovanov, "Cosmonaut No.1", *Izvestia*, 10 April 1986.
[8] Pavel Popovich, interview conducted by Bert Vis, Ottawa, Canada, 30 September 1996. Quoted with permission of Bert Vis.
[9] Michael Cassutt, *Who's Who in Space: The International Space Year Edition*, Macmillan, New York, 1993.
[10] Maurice Allward, "Is It Russia's Moon?" *RAF Flying Review*, December 1960 (Vol. XVI, No. 3).
[11] *Project Mercury: Man-in-Space Program of the National Aeronautics and Space Administration*, Report of the Committee on Aeronautical and Space Sciences, United States Senate, United States Government Printing Office, Washington, D.C., 1 December 1959.
[12] *New York Times*, "Space chief reveals space plans", 3 December 1959.
[13] James Oberg, *Uncovering Soviet Disasters*, Random House, New York, 1988.
[14] *Newsweek* magazine, Vol. LVII, No. 17, pp. 23–29, 24 April 1961.
[15] Wilfred Burchett and Anthony Purdy, *Cosmonaut Yuri Gagarin: First Man in Space*, Panther Books, London, 1961.
[16] "Hero of the Soviet Union: Pyotr Dolgov." Online biography at *http://www.warheroes.ru/main.asp?I=3*
[17] Alexei Adzhubei, editorial in *Izvestia*, 28 May 1963.
[18] James Oberg, "Alexei Leonov: Space walk and space handshake," *Space World*, June 1975, pp. 10–28.
[19] *Pravda*, "Gagarin was not the first cosmonaut," 12 April 2001. Online at *http://english.pravda.ru/main/2001/04/12/3502.htm*
[20] James Oberg, "Phantoms of space," *Space World*, January 1975. Online at *http://www.jamesoberg.com/russian/phantoms.html*

9

First woman of space

Just as with NASA's space programme administrators, Soviet space chiefs soon came to the realization that they would need to recruit new pilots for future spaceflights—in their case, beyond Vostok. Among those who supported this further and ongoing recruitment was Nikolai Kamanin, in his role as director of cosmonaut training. But it had also been noted in American magazines and newspapers that a number of women pilots were currently undergoing extensive (albeit unofficial) training as possible NASA astronaut candidates, and that did not sit easily with the influential Kamanin. "We cannot allow that the first woman in space will be American," he wrote in his diary. "This would be an insult to the patriotic feelings of Soviet women" [1].

SEEKING OUT SUITABLE CANDIDATES

General Kamanin subsequently began pressing for the inclusion of several women—around five—in the next intake of cosmonauts. On 31 December the Central Committee of the Communist Party approved the selection of a new group of candidates to satisfy the demand for pilots beyond the Vostok programme. Heeding Kamanin's concern about American women involved in space training the committee further decreed that five of this number would be females, to begin training well in advance of their male counterparts, who would not be selected until 1963. Premier Khrushchev was personally delighted with this innovation; it would be another chance for the Soviet Union to post an impressive new and attention-grabbing space "first" ahead of the Americans.

A dream for Valentina

The DOSAAF [Freewill Society for the Army, Aviation and Navy Support] was given initial responsibility for seeking out suitable females, and its members covertly scouted out any candidates involved in military, aerobatic or sport flying, sky-diving, or those with advanced parachuting qualifications. Because of the dimensions and capabilities of the Vostok spacecraft, there were certain restrictions in regard to age, height and weight. They would eventually submit 400 names, but the vast majority proved to be unsuitable under the guidelines. Among the names that survived was that of young textile mill worker Valentina Vladimirovna Tereshkova, who had taken private parachuting lessons. Swept up in a moment of national triumph following the flight of Yuri Gagarin, and like so many other proud Russians, she had written a passionate letter to the authorities volunteering her services should a woman cosmonaut ever be required.

Plans for the initial screening of the remaining female candidates reached a stage on 15 January 1962 when the files of 58 women landed on the desk of Nikolai Kamanin. After a review of the files he expressed extreme disappointment with what he perceived as a lacklustre choice, and decided to personally reduce that number yet again by almost two-thirds. On 19 January he wrote in his diary: "Yesterday I considered the files of fifty-eight female candidates. Generally disappointed and dissatisfied. The majority are not suitable for our requirements and have been rejected. Only twenty-three will be brought to Moscow for medical tests because DOSAAF did not examine their credentials correctly. I told them I needed girls who were young, brave, physically strong and with experience of aviation, who we can prepare for spaceflight in no more than six months. The central objective of this accelerated preparation is to ensure that the Americans do not beat us to place the first woman in space" [1].

The field had now been narrowed to just 23 candidates, and of these 5 did not pass preliminary medical tests, bringing the number down to just 18 candidates. Their names were Borzenhova (first name not known), Valentina Daricheva, Svetlana Ivleva, Galina Korchuganova, Galina Korolkova, Tatyana Kuznetsova, Vera Kvasova, Natalya Maslova, Tatiana Morozitcheva, Valentina Ponomareva, Marina Popovich, Marina Sokolova, Ludmila Solovyeva, Irina Solovyova, Valentina Tereshkova, Yefremova (first name not known), Zhanna Yorkina and Rosalia Zanozina [2].

The women were split into two equal groups of nine: Group 1 and Group 2. Each group was then subjected to far more searching interviews and extensive examination by the medical commission. Seven out of the nine from Group 1 would go on to the next phase of selection—namely, Kuznetsova, Kvasova, Sokolova, Solovyeva, Solovyova, Tereshkova and Yefremova. Only four from Group 2 passed this screening: Borzenkova, Ponomareva, Popovich and Yorkina. It is suggested that one of the candidates actually dropped out of the running before these further tests.

Although he was mostly unimpressed with the qualifications of the female finalists, Kamanin was not looking for an extraordinarily high level of proficiency. He knew that the simple design of the Vostok and its systems, and the fact that once in orbit most of its functions would be controlled from the ground, meant that someone suitably self-demanding who began training in April should be sufficiently prepared by the originally targeted launch date of August that year.

Final five

By 3 April a final selection of five women had been made and approved, and as they were all civilians involved in a military endeavour they found themselves enrolled with the lowly rank of Private in the Soviet Air Force; a ranking that would rise to Junior Lieutenant upon completion of their basic training that November. The five women were

— Tatyana Kuznetsova, 20, a qualified parachutist with a number of world records to her credit.
— Valentina Ponomareva, 28, married, and the mother of a three-year-old son, Aleksandr, a graduate of the Moscow Aviation Institute, where she had learned to fly and become a parachutist.
— Irina Solovyova, 24, who had set a number of world records while completing 2,200 parachute jumps.
— Valentina Tereshkova, 24, a textile mill worker and amateur parachutist who had completed over 100 jumps.
— Zhanna Yorkina, 22, also an amateur hobby parachutist.

It is a known fact that the incumbent cosmonauts were not at all pleased when told they would be training alongside five women. They were not only highly trained and brash young Air Force officers with prodigious flight records, but they also felt that women had no place in space or space training at that time. Further souring their lack of appreciation for this news was the fact that one of the women would move in and take a coveted Vostok flight seat they felt was rightfully theirs. Nelyubov in particular was extremely annoyed, as the recognized third cosmonaut in line behind Gagarin and Titov. He had not only lost an early assignment when it was decided to fly Nikolayev (a Chuvash) and Popovich (a Ukrainian) on the "non-Russian" tandem flights of Vostok 3 and Vostok 4, but it looked as if he might miss out on the following flight as well, with that assignment being virtually presented to the successful woman candidate.

Having already flown, Gagarin was fairly philosophical about the whole process, but did express concerns about Ponomareva in particular. "Cosmonautics is an unknown and unsafe world," he openly declared. "Can we risk the life of a mother?" As leader of the cosmonaut detachment, however, it was Gagarin's responsibility to pave the way for the arrival of the five women, and to act as their guide. At a meeting of the cosmonauts he told them, "Some lady cosmonauts are on their way to join us. Let's be considerate and helpful toward them. There must be no teasing or anything else offensive to them" [3].

Tereshkova was actually the first of the five women to arrive in Moscow, on 2 March 1962. Here she was welcomed by General Kamanin, who then introduced her to several future cosmonauts, including Nikolayev, Popovich and Bykovsky. Together with Kuznetsova and Solovyova they would report for training on 12 March. The other two women, Yorkina and Ponomareva, were delayed and would not arrive until the following month. On their arrival, Gagarin welcomed the women

Yuri Gagarin assisting in the training of Valentina Tereshkova.

to the cosmonaut training centre (then just a collection of office buildings). One of the first things he organized was to show them around the newly organized museum, pointing out models, photographs and drawings illustrating the brief history of Soviet cosmonautics. The women, who would be quartered in a rehabilitation centre, were totally awestruck on meeting the cosmonauts, and to have the god-like Gagarin acting as their tour guide.

"Let me tell you, when I entered the Star City of cosmonauts, my heart was about to stop," Tereshkova would later state in her biographical *Stars Are Calling*. "How will they meet me here? I didn't do anything great in my life; in fact, I didn't see much of a real life and these were real pilots. And two of them, Gagarin and Titov, were heroes whose names were known all over the world!" She was not to know it then, but the following year Tereshkova would become almost as famous and well-known as her dashing new mentor [4].

A hard-won acceptance

Despite Gagarin's earlier admonishment, the male cosmonauts were still openly derisive when it came to the women, and kept them under close scrutiny. Titov was especially sceptical, insisting that women did not belong anywhere near an aircraft, and openly contending that only men could cope with the rigours and tasks associated with space travel. "The men thought that we were completely redundant," Ponomareva reflected, "that there was no place for a woman on a spacecraft" [5].

Looking decidedly uncomfortable, Tereshkova would undertake dual flight instruction.

However, the women would slowly gain the reluctant admiration of the stoic cosmonauts as they began to spend time in the centrifuge, the thermo-chamber and the decompression chamber. Tereshkova stood up particularly well to the unsettling loneliness of the isolation chamber over several days, emerging smiling and unruffled by the experience.

The two latecomers joined them and soon all five were in serious training. With a need for them to become junior officers in the Air Force, each received basic cockpit training under the supervision of Colonel Vladimir Seryogin (who would later be killed in an air crash along with Gagarin) to qualify them as flight-trained passengers in Ilyushin-14 turboprop aircraft and MiG-15 dual-seat trainers. While they would be allowed to take the controls to get a feel for flying, they were never allowed to fly solo. With this very minimal flight experience to their credit, the women would each receive the commissioned rank of junior lieutenant.

By this time, the other cosmonauts had noticed to their amusement that Andrian Nikolayev always appeared to be at hand whenever Tereshkova was undergoing any training exercises, assisting the attractive young women and offering advice when anything became too complicated. Romance, it seemed, was very definitely in the air. In August 1962 Tereshkova enjoyed a break in her training when she was flown down to the Baikonur cosmodrome to witness Nikolayev being launched in his Vostok 3 spacecraft, and Popovich the following day in Vostok 4.

As their training intensified, so a couple of the women began to feel the strain. Zhanna Yorkina apparently performed poorly during a three-day simulated Vostok mission, and is even said to have fainted, while Tatyana Kuznetsova was stood down from training in its latter stages, "because of illness" This left just three women eligible for the upcoming Vostok 6 mission—Ponomareva, Solovyova and Tereshkova. Meanwhile the pilot's seat on Vostok 5 was being hotly contested between Valery Bykovsky and Boris Volynov, and the five flight candidates would often undertake training together until the final selections were made. By now, Nikolayev and Tereshkova were regularly seen together, many times deep in whispered conversations in the library.

THE NEXT INTAKE

Around the time that Tereshkova was reporting for training in March 1962, serious consideration was being given to recruiting another cosmonaut group. On this occasion the qualification criteria were relaxed to allow not only pilots, but engineers and navigators, while the age limit was shifted upwards to 40 years. However, in another reflection of future directions for the space programme, all of the applicants were required to be graduates of a military academy or civilian university.

Eventually a total of 15 male candidates were selected, which included eight pilots. These future pilot-cosmonauts were

> Major Georgi Dobrovolsky
> Major Anatoli Filipchenko
> Major Alexei Gubarev
> Major Anatoli Kuklin
> Engineer-Captain Aleksandr Matinchenko
> Major Vladimir Shatalov
> Major Lev Vorobyov
> Captain Anatoli Voronov.

The seven engineers were

> Engineer-Captain Yuri Artyukhin
> Senior Engineer-Lieutenant Eduard Buinovsky
> Engineer-Lieutenant Colonel Lev Demin
> Senior Engineer-Lieutenant Vladislav Gulyayev
> Engineer-Captain Pyotr Kolodin
> Senior Engineer-Lieutenant Eduard Kugno
> Senior Engineer-Lieutenant Vitali Zholobov.

A year later another name was added to the list of cosmonauts when Colonel Georgi Beregovoi, a highly decorated pilot who had arrived at the training centre to act as a flight instructor, was encouraged to become a cosmonaut by Kamanin, Korolev and Kuznetsov (who would succeed Yevgeny Karpov as director of the training centre in

November 1963). Beregovoi would commence his training in February 1964, linking up with the group selected 13 months earlier.

Interestingly, another Air Force flight instructor would also begin cosmonaut training at the same time. Lieutenant Colonel Vasili Lazarev was a pilot and physician who had earlier taken part in Volga-class high-altitude balloon experiments and equipment tests, and with the encouragement of Kamanin he began simulation work with the other cosmonaut trainees. For some unknown reason, however, he had earned the personal dislike of Sergei Korolev, who prevented him from being enrolled as a cosmonaut candidate. In late January 1966, three weeks after Korolev's sudden death, there were no longer any lingering impediments and Lazarev was officially accepted into the cosmonaut team [6].

By the beginning of February 1963 there were 34 cosmonauts in training, basically divided into six training groups.

- Group 1: four women candidates—Ponomareva, Solovyova, Tereshkova and Yorkina—in training for a tandem Vostok mission.
- Group 2: three cosmonauts—Bykovsky, Komarov and Volynov—training for possible individual flights in late 1963.
- Group 3: four flight-experienced cosmonauts—Gagarin, Nikolayev, Popovich and Titov—undergoing academic training and public relations duties.
- Group 4: eight cosmonauts from the first group—Belyayev, Filatyev, Gorbatko,

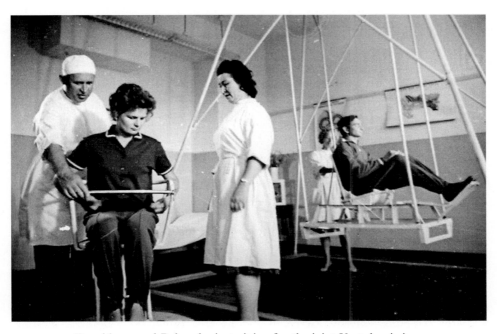

Tereshkova and Bykovsky in training for the joint Vostok mission.

Khrunov, Leonov, Nelyubov, Shonin and Zaikin—in general training for future
Vostok and Soyuz flights.
- Group 5: seven newly selected pilot-cosmonauts undergoing initial training.
- Group 6: eight newly selected engineer-cosmonauts also undergoing initial training.

"Gagarin in a skirt"

It soon fell to Nikolai Kamanin to assess the women candidates in line for the flight of Vostok 6, and after consideration his choice fell to Tereshkova. She had rated highly in all her tests, and according to Kamanin, "We must send Tereshkova into space first, and her double [backup] will be Solovyova. Tereshkova," he added, "she is Gagarin in a skirt." Premier Khrushchev fully endorsed the selection, as Tereshkova had an enticing propaganda profile—attractive, single, hard-working, introspective and the daughter of a collective farm worker who had been killed in 1940 fighting in the Soviet–Finnish War.

The women candidates were gathered together on 21 May 1963, before they left for the Baikonur cosmodrome, at which time Sergei Korolev revealed to them that Tereshkova had been selected for the flight, with Irina Solovyova to act as her backup. In an endeavour to placate those who had missed out, Korolev said that eventually all of them would have a chance to fly into space. But Ponomareva was upset with the decision, and made her displeasure known to Yevgeny Karpov. "Trying to console me, Karpov said that political considerations prompted the sending of a person 'with particular roots', and I had the misfortune of being a

A group photo taken prior to the dismissal of Nelyubov and Anikeyev shows (back row, from left) TsPG trainer G. Titarev, Shonin, Belyayev, Titov, Nikolayev, Nelyubov, Khrunov, Komarov, Gagarin, Volynov, Gorbatko and Leonov. Front row: Anikeyev, Yorkina, Popovich, Tereshkova and Solovyova.

Another group photo; this one taken at the Black Sea resort of Sochi in the autumn of 1962. Back row (from left): unknown, Anikeyev and Filatyev. Seated: unknown. Middle row: Yorkina (hiding behind Nikolayev), Leonov and Kartashov. Front row: Khrunov, Gagarin, Rafikov and Solovyova.

Ponomareva, Solovyova and Tereshkova at the Baikonur launch complex.

Tereshkova in final
training for her Vostok 6
mission.

clerk," she later recalled. "As for the degree of our preparation for the flight, we were all running 'neck and neck'" [7].

Valery Bykovsky would be selected over his backup pilot Boris Volynov to join Tereshkova in the joint mission as the prime pilot for Vostok 5. According to researcher Bart Hendrickx, in translating the diaries of Nikolai Kamanin: "The choice of Bykovsky was at least partially based on his weight, which was 14 kg less than Volynov. Installation of extra equipment on Vostok 5 meant that the vehicle was pushing the Vostok rocket's maximum payload capacity. The final decision was made at the cosmodrome and announced on 4 June." Volynov would continue to serve as Bykovsky's backup for the flight [1].

It was hot at the Baikonur cosmodrome as launch day drew near, with the mercury rising to around 42°C in the shade.

Tereshkova and
Bykovsky.

Cosmonaut No. 5

At 2:59 PM Moscow time on 14 July 1963, after an unprecedented number of launch delays due to technical and weather concerns dating back to a week before, Lt. Colonel Valery Bykovsky (codenamed "Hawk") became the fifth Soviet pilot from his cosmonaut group to be launched into space. Even then the launch came close to being scrubbed once again when there was a problem with an arming device related to Bykovsky's ejection seat, and then the R-7 rocket's Block-Ye upper stage failed to respond to a launch signal test. The delay dragged on for so long that Volynov, clad in his spacesuit in the dark blue transfer van, was told to prepare himself to fly after the delay stretched to more than four hours, as Bykovsky might need to be replaced. However, this did not occur, and right at the very end of the launch window that day the lift-off went ahead.

Bykovsky's Vostok 5 spacecraft had soon settled into an elliptical orbit of 122 mi by 146 mi. In the Western media there was informed speculation that his flight might not only last a record eight days, but that it might include a "cosmic rendezvous" with the first space woman. Tass news agency formally identified Bykovsky, then 28 years old, as the pilot of Vostok 5, a married man with a three-year-old son, also named Valery.

While there was wild excitement in the Soviet Union over this latest launch, everyone was eagerly anticipating a bulletin that was expected in the next day or two—one stating that a woman had joined Bykovsky in orbit. In fact, that woman was seated soon after the Vostok 5 launch at a control console watching images of the cosmonaut on a television monitor, and waiting for a transmission slot to become available. When that occurred she sent a radio message up to Bykovsky. "Yastreb, Yastreb [Hawk, Hawk], so you recognise my voice? A warm greeting to you ... I congratulate you on a good beginning."

As she watched, Bykovsky smiled and replied: "I am waiting."

With Bykovsky safely in a good orbit (albeit slightly lower than planned), attention was now focused on Vostok 6. According to Kamanin, the final decision to launch Vostok 6 two days after the other craft was only made after Bykovsky had achieved orbit. Only six days earlier, on 8 June, the favoured option had been to launch Vostok 6 five days after Vostok 5 and have both spacecraft land on the same day.

There was always conjecture that Tereshkova had not been the first chosen to fly the Vostok 6 mission, but this was never the case. However, there was a problem on launch day, two days later, that could have seen her lose that exalted place to her backup. While donning their spacesuits some sealing in the neck area of Solovyova's suit was broken, and the spacesuit had to be quickly changed for that of Ponomareva.

There was no problem in that, but prime pilot Tereshkova was taller than the other two women, and if the problem had occurred in her spacesuit there would have been no backup suit to accommodate her larger frame. Out of necessity the flight would then have gone to her backup Solovyova by default [7]. As it was, all of the women candidates were at the launch site that day, with the exception of Tatyana Kuznetsova.

Stellar duo

Over the next two days, while the world waited for news of a female cosmonaut, Bykovsky patiently submitted standard reports on his condition and the state of his spacecraft as he commenced his lightweight work programme—observing and photographing the Earth as well as the Sun's corona, biological experiments with fruit flies and monitoring the growth of some peas. He would also exchange communications with Premier Khrushchev as part of the flight's immense propaganda value.

As anticipated, two days after the launch of Vostok 5, mounting speculation came to an end with a message broadcast by the Tass News Agency: "On 16 June 1963, at 12:30 PM Moscow time, a spaceship, Vostok 6, was launched into orbit ... piloted, for the first time in history, by a woman, citizen of the Soviet Union, Communist Comrade Valentina Vladimirovna Tereshkova." It added that her call sign would be *Chaika* (Seagull).

At a post-flight press conference in Moscow, Bykovsky said he was well prepared when it came time for the launch of Vostok 6.

> "When the spaceship Vostok 6 was to be put into orbit, I knew in advance at what time. Therefore I had my radio-receiving apparatus switched on at the time, and at the exact time anticipated I heard ground control speaking to Valentina before she was actually in orbit. By that time I was approaching the territory of the Soviet Union, and I was eagerly awaiting the moment when she would finish reporting to ground control that she was in a state of weightlessness and that all was well. I did not interrupt her conversation with the ground, but then, when I saw that she had finished speaking, I cut in. There, actually, our space rendezvous began. It began with congratulations—my congratulating her on her successful launching" [8].

The two spacecraft would actually come within three miles of each other during Vostok 6's first orbit, but this would prove to be the closest distance achieved between the two craft throughout the dual mission. In their post-flight reports Bykovsky stated that he had been unable to see Vostok 6, while Tereshkova said that she felt she might have seen Vostok 5.

Whatever the propaganda reasons for the flight of Tereshkova might have been, her flight created massive headlines around the world, and photographs showed the world's first space woman to be a young, attractive and single female. Those who had become inured to seeing ruggedly handsome men's faces enclosed in space helmets were entranced by the sight of this photogenic young woman staring shyly out from her helmet. Once again, the Russian people were in raptures at this latest spectacular achievement in space.

Despite her continuing denials, it is known that Tereshkova was not well during the early part of her flight, and in fact she would later grudgingly admit to having been physically sick on the third day of her flight—although she would insist it was simply a bad reaction to something she had eaten. Fortunately for Tereshkova the work programme for her flight was light with very few practical responsibilities. One of her major tasks was to practise manual reorientation of the spacecraft, which would be necessary in the event of a manual re-entry, but her initial attempts were not altogether successful. This would earn the ire of Nikolai Kamanin, who would berate her performance in his diaries. Later attempts were far more successful, and it was apparent that not only was Tereshkova feeling ill and suffering from headaches, but she was quite obviously out of her depth in regard to the performance and technical aspects of her mission.

It had originally been planned for Bykovsky to remain in orbit for eight days, but an underperforming upper stage meant that his lower orbit and the effects of gravity would not sustain this flight duration. On the third day of Bykovsky's flight aboard Vostok 5, on 17 June, a preliminary decision was made to bring the cosmonaut back after a five-day or six-day mission. The following day it was further decided to bring both spacecraft down on 19 June.

There also appears to have been communications difficulties for Tereshkova— either because something was wrong with her receiver or because she had mistakenly selected the incorrect channel. Listeners monitoring the dual flight in the West reported calls from ground control to Bykovsky, wanting him to relay messages to Tereshkova, as they could not communicate with her. To which Bykovsky responded: "I have tried myself to contact her without success, but I do not think there is any reason for concern." The later explanation was quite understandable, given Tereshkova's fatigue and illness—she had simply fallen into a deep sleep.

Homeward bound

On 19 June, as scheduled, Vostok 6 began its return journey to the ground on the 48th orbit, but again there would be post-flight criticism of Tereshkova for failing to confirm the successful operation of the solar orientation system. To the bafflement and concern of ground control she would also remain silent throughout the craft's early re-entry phase, and did not report retro-fire or the successful separation of her craft from the service module—a technical difficulty which had plagued earlier Vostok flights.

An automatic ejection from her capsule took place four miles above the ground, and as she parachuted down Tereshkova opened her faceplate in order to get a clear view of her landing site. She had been warned against looking up, in case of falling debris, but she decided to check her parachute and was struck on the nose by a shard of falling metal. Fortunately it did little more than inflict a small cut and bruise.

At 11:20 AM Moscow time Tereshkova would touch down, 385 miles northeast of the city of Karaganda in the Altai region of Kazakhstan. Some surprised collective farm workers had watched her descent and that of her bulbous spacecraft, and ran to the site. Tereshkova would reward them with a smile and some food items from inside her spacecraft, much to the later frustration of physicians, who could not accurately gauge her food and drink intake during the actual flight. After an hour, by which time Tereshkova had changed into a lightweight tracksuit, the first of the rescue teams arrived on the scene. There is a well-known colour photograph of the cosmonaut casually sitting in the Sun in front of the open hatch of her spacecraft, surrounded by a small crowd of curious local onlookers and chatting with a blonde-haired woman.

It was speculated for some time that this might actually have been her backup pilot for the flight; however, the mystery woman later proved to be Lyubov Maznichenko, a renowned parachutist and a member of the Vostok 6 rescue team [9].

Just one orbit later, having surpassed the endurance record set by his colleague Nikolayev with nearly 119 hours in space, Valery Bykovsky would also endure the re-entry phase aboard Vostok 5. "The solar orientation for retrofire worked correctly," he would later report, "and the TDU braking engine fired for 39 seconds." However, there would be a frustrating repetition of the problem encountered by Gagarin and Titov when a restraining strap designed to keep the two spacecraft components together prior to re-entry failed to disconnect as planned, causing the combined descent and service modules to gyrate wildly for a time, until the faulty strap finally burned through with the heat of re-entry.

As he descended, Bykovsky would later state that the force of gravity initially distorted his face and he could not see very well, but as the g-forces abated his eyesight improved and he began to prepare for the ejection process. Through the "evidently burned glass" he could see the ground approaching. "I look and attempt to estimate [the] distance to the earth and the clouds," he recalled. "Vainly."

Valery Bykovsky after his first flight into space.

When it came time for the ejection he heeded Gagarin's words and did not look up after the hatch cover above his head was automatically ejected. Instead, he says he concentrated on the instrument panel until he was explosively ejected through the open hatch. Between his legs he could see the capsule twisting and plummeting towards the ground, trailing ribbons associated with his ejection before its parachute was deployed. Then Bykovsky's own red-and-white parachute was released, billowing above him. The force of the deployment caused him to hit his mouth on the visor of his helmet, but the unexpected impact did not cause any damage. Then his contoured seat also fell away and he was only a few seconds from landing. "I opened my helmet [visor] and inhaled air," he later recalled. "Pleasant steppe air" [10].

Bykovsky touched down without incident some two hours and three orbits after Tereshkova, but nearly 500 miles from her landing site, hitting the ground 335 miles northwest of Karaganda. As Bykovsky waited for the recovery teams more and more local people turned up, arriving on horseback, in tractors, cars and even self-propelled combines. Then the first of the rescue physicians, Vitaly Volovich, turned up on a motorcycle after parachuting down from one of the helicopters that had flown in. As Volovich recalled in his memoirs:

"Bykovsky was already aboard the helicopter which had arrived a short time before us. I climbed in as well.

"Good to have you back, Valery!" Laying out my medical instruments, I began my examination. The cosmonaut was in quite good health, only his pulse was a little rapid. But that was understandable considering the stuffy heat inside the compartment. After the medical check-up, we decided to fly to a regional centre, Maryevka, which was close by" [11].

A wedding and a shooting

Valery Bykovsky's return was cause for celebration by the Soviet people, but it had already been overwhelmed by news of the safe return of Valentina Tereshkova. She would be feted across her nation and around the world. Her immense popularity would lead to Premier Nikita Khrushchev audaciously pushing for the romance between her and Andrian Nikolayev to be fully exploited by gently (but at the same time forcefully) pressuring them to take their affair a step further by getting married. They would be wed in a Moscow registry office on 3 November 1963, amid much fanfare, which was followed by a lavish celebration dinner in a former czarist palace in the Lenin Hills. Khrushchev (and space physicians) were openly keen for the couple to have the first "space baby" born to parents who had both been into space, and their wish was granted just seven months later when a healthy baby daughter Yelena Andrianovna was born by caesarean section.

Tereshkova wanted to fly into space again, but this would never be allowed to happen. Eventually even the so-called "cosmic marriage" would fail, and the couple drifted apart. Before their separation, however, they were involved in a fatal motorcade shooting in Moscow on the bitterly cold morning of 22 January 1969 as the nation celebrated the return of the cosmonauts involved in the dual flight of the

The wedding of Valentina Tereshkova and Andrian Nikolayev.

Soyuz 4 and Soyuz 5 spacecraft. Several cosmonauts were in limousines travelling in convoy which was heading towards the Kremlin where a reception had been organized, when a gunman named Ilyin, angry at being conscripted into the Soviet army, opened fire with two automatic pistols on the limousine he thought was carrying Leonid Brezhnev, killing the driver and seriously wounding a police motorcycle outrider. Brezhnev was not in that vehicle, but four cosmonauts—Tereshkova, Nikolayev, Leonov and Beregovoi—suddenly found themselves under fire and had to fling themselves to the floor of the limousine as bullets sprayed and shattered their windows. They would sustain mostly superficial wounds from the flying glass, but once Ilyin had been arrested they finally continued on to the reception—badly shaken, but grateful to be alive.

THREE MEN AND A VOSKHOD SPACECRAFT

Speculation about a new Soviet space spectacular had begun to mount once again early in October 1964, no doubt fuelled by comments from Vostok space partners Andrian Nikolayev and Pavel Popovich. "A group of cosmonauts are ready for a flight and we will not delay it," Nikolayev was quoted as saying in the Government newspaper, *Izvestia*. There were also strengthening rumours that a second woman

There was intense but unfounded speculation that Marina Popovich would soon fly aboard a Vostok spacecraft.

might be launched into orbit, and the spotlight naturally fell on record-breaking aviator Marina Popovich, wife of the cosmonaut. *Izvestia* reported that she had completed "the higher aviation school and has left now for a distant business trip."

When Popovich was asked whether his wife was preparing for new flights, he tantalizingly replied: "The machine with which she is now familiarising herself is considerably more powerful than the one in which she set a new speed record two months ago."

Despite all of the rumours, neither Marina Popovich nor any other woman would be aboard the next Soviet human spaceflight. Instead, in a highly dangerous gamble to overshadow America's forthcoming Gemini two-man programme, three men without any protective garments would be squeezed aboard what was being touted as a "new generation" Soviet spacecraft named *Voskhod* (Sunrise), for a one-day mission that could so easily have ended in catastrophe and the deaths of all three cosmonauts.

A dangerous gamble

Vladimir Komarov would have to overcome many difficulties before he could make his first spaceflight. Early on in his training he fell ill, and was hospitalized for an operation on a rupture. He was told to take it easy, and this would preclude any parachute jumps for six months—a crucial aspect of the training programme.

After concentrating on the theoretical side of spaceflight for five months, Komarov was able to rejoin his colleagues at the spaceflight training centre and complete exercises on the centrifuge and in the isolation and altitude chambers. Despite earlier fears that he had fallen too far behind in the programme he quickly caught up with his colleagues. His diligence and perseverance were rewarded when he was assigned as second backup to Pavel Popovich for Vostok 4. He then moved up to become first backup when the cosmonaut originally holding that position, Nelyubov, was medically disqualified for a period after failing a centrifuge test.

For a while it looked as if Komarov might get the chance to fly a Vostok himself, when an ambitious flight programme was devised that would consist of three ships flying in space at the same time. Eventually it was decided that only two spacecraft would be launched, and one would carry the first woman into space. It was felt that this alone would create enough of a worldwide sensation. Then came a second devastating blow when an extra-systole, or VPB (ventricular premature beat)—a premature contraction of the heart, independent of the normal rhythm—was detected following a regular post-centrifuge cardiogram. Ordinarily this was a fairly minor condition, but to a space traveller an anomaly such as this could mean a swift end to their career, as U.S. astronaut "Deke" Slayton would experience to his distress.

Once again, Komarov was taken off the training regime, and he did not even attend the launch of Vostok 5 and Vostok 6 along with his fellow cosmonauts. Knowing that a discharge from the cosmonaut corps was potentially imminent, Komarov set about proving that he was physically fit, despite the VPB. A subsequent medical examination concluded that he was otherwise superbly fit, but the condition was still cause for concern by the physicians. Knowing his career was on a very precarious line Komarov began to appeal to the nation's top cardiologists and presented their findings at an official top-level hearing. To his delight the Council agreed that the ailment was probably of very little consequence, and that he should be allowed to resume cosmonaut training.

In February 1964 the Vostok programme came to an end, but the first Soyuz manned launch was still some two years away, so an interim programme was devised to keep ahead of the United States in the Space Race. This new programme was meant to consist of several manned flights, but ultimately only two would be launched to pre-empt NASA's announced plans for Project Gemini. The first would involve flying three men in a single spacecraft on a short mission, while the second would see a cosmonaut egress out into raw space on history's first spacewalk. In order to realize this stop-gap programme, a reluctant Sergei Korolev would first be given a direct order to re-fashion one of the four remaining single-occupant Vostok ships into a craft capable of carrying three cosmonauts. With the first flight of America's two-seater Gemini spacecraft imminent, something spectacular and audacious was needed: something to convince Americans that the Soviet Union still held a substantial technological and psychological lead in the Space Race.

The constraints involved in refitting this "new" craft, christened Voskhod, caused immense problems for the embattled Korolev, but in the end he managed to squeeze in three seats, with one seat positioned slightly forward of the other two. It could only be done at the expense of safety, and it was a particularly hazardous compromise. While the Vostok cosmonauts had used ejection seats to parachute to a safe landing, crewmembers aboard a Voskhod craft would be forced to remain in the descent capsule until touchdown, which could entail a brutally hard landing. But there was another, even more dangerous initiative utilized: in order to save weight and because of a virtual lack of space inside a manned spacecraft, the designers had no alternative but to dispense with the protective spacesuits and helmets for the crew.

By this time, ignoring all the safety factors involved in the flight, arguments had broken out between the various institutions and departments, all of whom wanted to

The first 11 Soviet cosmonauts. *From left to right:* Gagarin, Titov, Nikolayev, Popovich, Bykovsky, Tereshkova, Feoktistov, Komarov, Yegorov, Belyayev and Leonov.

be represented in the crew. In the Vostok programme seats could only be occupied by qualified air force pilots, but with two additional seats available for non-pilot occupancy, Korolev's OKB-1 design bureau, the Academy of Sciences, and the Ministry of Public Health all wanted one of their own onboard the flight. Bitter discussions dragged on for months, but in the end a compromise was reached. It was determined that the crew would consist of Komarov as commander, Konstantin Feoktistov, a deputy of Chief Designer Korolev, and Boris Yegorov, a medical doctor.

The first manned Voskhod spacecraft was launched from the Baikonur cosmodrome on 12 October 1964 atop an improved 11A57 version of the R-7 carrier rocket, returning safely after a flight lasting 1 day and 17 minutes, despite a sincere but denied request from Komarov that consideration be given to extending the flight for another 24 hours. Following the announcement that they had returned and were all well, a relieved Korolev reportedly commented that he couldn't believe the crew had survived and returned from orbit without injury. "Is it really true that the crew has returned from space without a single scratch?" he asked. He had never believed that it was possible to successfully and safely turn a one-man Vostok into a three-man spacecraft [12].

Although it was widely reported in the Soviet press of the time that Komarov had been testing spacecraft systems while his two companions conducted medical and geophysical experiments, it became known in later years that there was precious little room in the spacecraft to do anything other than a simple range of medical and biological tests, and Earth observation exercises. Indeed, the principal task of the crew was to achieve orbit and get back down intact and alive. The propaganda factor was paramount, and anything else that could be done on the flight was quite incidental.

The crew of the first manned Voskhod spacecraft was Konstantin Feoktistov, Vladimir Komarov and Boris Yegorov.

Disappointing from a propaganda point of view, however, was the fact that the crew's safe return was largely overshadowed by an unexpected turn of events in the Soviet Union. After landing the crew was expecting the traditional congratulatory phone call from Premier Nikita Khrushchev, but instead they were whisked aboard helicopters for a return journey to the cosmodrome. On the day they landed Khrushchev had been suddenly overthrown as leader and replaced by a trio of new leaders; Nikolai Podgorny, Alexei Kosygin and Leonid Brezhnev. It would be a grinning Brezhnev, as First Secretary of the Central Committee) who greeted the three cosmonauts following their return to a hero's welcome in Moscow on 19 October.

At a 90-minute press conference held four days earlier back at the launch site, Komarov had stated that the landing of the Voskhod spacecraft, 193 miles northeast of the town of Kustany in Kazakhstan, had been so soft that the spacecraft had left no indentation in the field, and in a monumental stretch of the truth declared that the Voskhod was a "far more comfortable" spaceship than Vostok [12].

TO WALK IN SPACE

With the relatively successful completion of the first Voskhod mission, an equally bold flight would see the Soviet Union achieve yet another space "first"—a tethered man outside of his spacecraft, performing what became popularly known as a "space-

Komarov assisting a sculptor to manufacture a bust of the cosmonaut.

walk" or EVA (extra-vehicular activity). Of course in zero gravity, and with nothing to give traction, a person cannot actually walk in space; it is more a matter of floating and propelling themselves by artificial means.

Training for EVA

First, there were major problems to overcome in achieving this feat. To begin with, the Voskhod hatches were actually bolted in place, and as a consequence were not hinged to allow any occupant of the crew compartment to exit the spacecraft. As well, because the interior of the Voskhod was never expected to be exposed to the vacuum of space, its systems' circuits had not been designed to operate in those conditions, Additionally, it would not only be a difficult task for a spacesuited cosmonaut to depressurize the craft, but to repressurize would consume a far greater amount of oxygen and hydrogen than could be transported into space.

With these obstacles in mind, there was a need to develop some form of lightweight, collapsible and extendable tubular airlock outside the craft that could be inflated when the time came, as well as a hinged main hatch that would swing open to allow a cosmonaut wearing a protective spacesuit and helmet to move into the inflated airlock. After this, the hatch would be secured by the second crewmember and the airlock pressure released. The spacewalking cosmonaut could then open the outer end of the airlock and exit, attached to the spacecraft by a tether. Throughout

The prime crew for Voskhod 2 was Pavel Belyayev and Alexei Leonov.

all of this there was one great danger lurking; once the spacewalking cosmonaut was sealed inside the airlock he would be out of the reach of any possible assistance by the flight commander, and would likely perish if a severe problem such as a torn spacesuit arose.

A training group of six cosmonauts was selected for the Voskhod 2 mission in April 1964, at the same time as those training for the first, three-man Voskhod. Five of them were from the first cosmonaut group—Belyayev, Gorbatko, Khrunov, Leonov and Zaikin, and Pyotr Kolodin from the second group, selected in 1963. The three in line to command the mission were Belyayev, Gorbatko (backup) and Zaikin (support crewmember). Leonov was the prime crewmember for the proposed EVA, with Khrunov acting as his backup and with Kolodin in support. In January 1965 Gorbatko was diagnosed with tonsillitis and was temporarily removed from flight status, which meant that Zaikin moved into the backup role. But there was more misfortune in store for Gorbatko.

"At first everything went well," Gorbatko recalled, "then suddenly the doctors found in me some undesirable changes in the electro-cardiogram. I was confined to hospital for one and a half months. Some people said my 'space career' had come to an end. But the matter was set right: my tonsils were removed and in the cardiogram everything came back to normal."

"I passed through the medical commission once again and was asked to continue training. But I landed awkwardly at the time of a parachute jump and a crack appeared in the bone near the ankle. It was my own fault. By this time I had already jumped about 120 times. I was over-confident. Quite honestly I was afraid, for at that very time the crew for the new spacecraft were just being recruited. I was very much

Belyayev and Leonov trying on the suits they would wear on their flight.

afraid that I would get the sack, but as you see ..." [13]. Eventually Belyayev and Leonov would be named as the prime crew for the mission.

As author Rex Hall and his co-author David Shayler noted in their book, *The Rocket Men*: "Leonov recalled that he first saw the airlock chamber in the summer of 1964 at OKB-1, where it was attached to the side of a Vostok spacecraft. At the same visit to the design bureau, he was to participate in a strenuous two-hour simulation in an altitude chamber, while wearing the suit. The Soviets had adapted a flown Vostok (Popovich's) to a Voskhod configuration for training" [12].

In order to test the integrity of the airlock and hatch system, an unmanned version of Voskhod 2 was launched on 22 February 1965, and up to a point is believed to have been completely successful. The airlock fully deployed as planned, while the hatch system and repressurization functioned without a hitch. However, as the unmanned craft began its third orbit the automatic self-destruct system was triggered by duplicate signals emanating from two ground stations instead of just one, and the Voskhod was blown to pieces. Fortunately, the cosmonauts knew that the manned craft would not be carrying a similar self-destruct system.

Exiting a spacecraft in orbit

On the morning of 18 March 1965, Belyayev and Leonov arrived at the launch pad amid cold winds and flurries of snow, but there was very little concern about the

Alexei Leonov becomes the first person to conduct a "walk in space".

weather and the launch would proceed. Once they arrived at the top of the gantry the two men had the unique task of boarding their spacecraft through the deflated airlock and hinged hatch. The launch took place right on 10:00 AM, and Voskhod 2 settled into a comfortable elliptical orbit of 107.8 mi by 309.2 mi above the Earth. They had beaten America's first Gemini launch into space by just five days.

Once they had completed their early tasks the two spacesuited cosmonauts immediately began to prepare for the EVA, which included the inflation and extension of the airlock and Belyayev assisting Leonov to don his life support system backpack. Once all the preparations had been made the inner hatch was opened by Leonov, who bade farewell to his commander and slipped into the airlock, where he attached his 17.5-foot tether. Belyayev then closed the inner hatch. A little over an hour and a half after being launched aboard Voskhod 2, at 11:35 AM Moscow time, Leonov opened the outer hatch at the end of the airlock and made his way out into the void. He would later describe this by saying he "popped out of the hatch like a cork."

As he floated out to the end of his tether, Leonov took in the spectacular view below, looking across the Mediterranean from the Straits of Gibraltar to the Caspian Sea. Ghostly images of his feat would be presented on Russian television screens just 30 minutes later. They showed Leonov emerging from the spacecraft, waiting for the official go-ahead order from Belyayev and then floating out of the airlock, moving from a horizontal position to a vertical one while holding onto a handrail attached to the exterior of the craft. He would make swimming motions, and by pulling on the tether connecting him with Voskhod 2 would be able to turn somersaults. He would also inspect the exterior of the Voskhod and take photographs with a camera he had mounted on a hatch ring on the end of the airlock.

Ten minutes after exiting the airlock, Leonov's time was up, and Belyayev told him it was time to get back inside the spacecraft. His spacewalk time at an end, Leonov retrieved the camera and pushed it down to the airlock for later retrieval, which was a far more exerting exercise than he'd imagined. He pushed his legs down into the airlock, but discovered to his mounting horror that his spacesuit had ballooned so much that he was unable to squeeze back inside. The more he tried, the more his visor fogged up, his body temperature rose, and his heart beat increased. He began panting hard with the exertion.

Leonov realized there was only way he was going to be able to force his way back into the airlock, and that was to reduce the pressure inside his spacesuit—a perilous operation that could kill him in moments. Without seeking permission from Belyayev or the ground he began to bleed pressure from his suit, but even as he reached dangerous levels he could still not squeeze into the airlock. Summoning up his courage he reduced the pressure right down, knowing that if he could not get back into the airlock he was a dead man anyway. Finally, his suit had lost enough volume for him to slide into the airlock and secure the hatch behind him. By this time his body temperature had climbed to critical levels, he was fatigued to almost the point of collapse, and perspiration had filled the legs of his suits, causing them to slosh whenever he moved. In *The Rocket Men*, authors Hall and Shayler offer the startling fact that: "More than thirty years later, Leonov revealed that he had, in his helmet, a suicide pill which he could take if he had been stranded outside, leaving Belyayev to return without him!" [12].

Leonov would finally return to his couch in the crew compartment, to the obvious delight and relief of Belyayev. The airlock was jettisoned, and the crew could now begin what was a far less exciting mission programme on a flight of just one day's duration.

Hazards to overcome

There would be even more dramas ahead when the autopilot landing system failed and Belyayev noticed that the Voskhod's attitude was incorrect for the retro-burn. Taking manual control he shut the system down, and the two cosmonauts would complete an extra orbit while he consulted with Korolev and the ground controllers. Following instructions, he then attempted to initiate a manual orientation via the Vzor optical apparatus for re-entry, but found it impossible while clad in his spacesuit and strapped firmly into his couch.

One of the problems for the crew was the curious cabin layout. The Vzor optical apparatus and the hand controller were located to the left of Belyayev's seat, rather than in front of him, which meant that the two men could neither peer through the sight nor operate the hand controller while fastened in their couches. This meant they had to unbuckle themselves and leave their seats in order to perform these tasks. As well, Belyayev had to remove his helmet so he could move his head around. He then lay down over both couches, the only way in which could use both hands to operate the manual controls. Meanwhile, Leonov manoeuvred himself under his seat so he could use his arm to prevent Belyayev from floating upwards and blocking the optical device. Following this successful procedure they could fire the braking rocket, but first they had to make their way back into their couches and strap themselves in once again—not an easy task in microgravity. However, it then took another 46 seconds to

Belyayev and Leonov after
their history-making flight
aboard Voskhod 2.

restore the centre of gravity before the braking engine could be manually fired. These delays would eventually lead to them overshooting their intended landing zone by a very wide margin.

Belyayev and Leonov had earlier been hoping to touch down in the flat steppes of Kazakhstan, but the orientation problems meant that they ended up missing the intended landing zone by some 1,240 miles. Voskhod 2 came down in an isolated snow-covered forest in the northern Urals, crashing through treetops and ending up firmly wedged between two large trees, suspended several feet off the ground. The two men spent an uncomfortable, freezing-cold day and night in this precarious situation, with temperatures dropping to −5°C, and listening to the nearby howling of hungry packs of wolves.

Meanwhile, across the Soviet Union, concerns mounted as citizens waited for news of the safe recovery of the two cosmonauts, but all they heard as the hours went by was patriotic music which they knew could be a prelude to bad news. Finally, at dawn, a ski patrol homed in on their signal beacon and eventually located the capsule. Two very relieved cosmonauts were then helped down from the spacecraft by around 20 rescuers, but as it was considered far too hazardous an operation to airlift them from the dense pine forest by helicopter they had to spend a second night in the forest along with their rescuers before they were fitted with skis and guided through the forest to a helicopter waiting in a cleared area.

Quite naturally, Leonov had become the focus of an outpouring of public adulation during and after his flight as a result of his historic walk in the cosmos, and mission commander Belyayev seemed quite content to take a back seat.

A death too soon

Sadly, Pavel Belyayev would never make another space-flight, although he participated in the Soyuz and Almaz programmes, and was assigned as one of a team of cosmonauts training for a circumlunar flight scheduled

Pavel Belyayev.

Cosmonauts Beregovoi, Leonov and Filipchenko paying tribute at Belyayev's grave.

for 1968. However, that programme was cancelled following the repeated failure of the N-1 lunar booster, and the continuing success of America's Apollo programme.

In December 1969 Belyayev was suffering from an increasingly acute illness that he decided to keep to himself, fearing he might be removed from flight status. Finally, he had to seek attention due to the pain and was diagnosed with a bleeding stomach ulcer. He entered hospital and even though it was a relatively simple operation involving the partial removal of his stomach, complications set in following the surgery and he developed peritonitis and pneumonia. Colonel Pavel Ivanovich Belyayev, the tenth Soviet cosmonaut to fly into space, died in hospital on 10 January 1970.

Sadly enough, Belyayev's death would cause an unexpected problem of protocol for the Soviet government, which had to decide if the decorated cosmonaut should be given the very distinct honour of burial in the Kremlin Wall. He had died of natural causes, it was argued, and as the tenth cosmonaut to fly into space was his achievement of sufficient significance to warrant such recognition? A decision was finally made that he deserved to be inurned in the Wall, but in an extraordinary turn of events his widow Tatyana flatly refused to allow this, saying it was her right to have him buried in a place where she could visit him, unhindered and whenever she wanted, which was not permitted at the Kremlin. Her request was finally granted, and her husband's body was laid to rest in the Novodevichy cemetery in Moscow, buried alongside prominent politicians, artists, actors, scientists and others who had carved names for themselves in Soviet society.

These days Belyayev's grave is easily identified by a life-size bronze image of the cosmonaut wearing a spacesuit, with a helmet tucked under his arm.

REFERENCES

[1] Bart Hendrickx, "The Kamanin Diaries 1960–1963," *Journal of the British Interplanetary Society*, **50**, 33–40, 1997.

[2] David J. Shayler and Ian Moule, *Women in Space: Following Valentina*, Springer/Praxis, Chichester, U.K., 2005.

[3] Evgeny Riabchikov, *Russians in Space*, Doubleday & Co., New York, 1971.

[4] Valentina Tereshkova, *Stars Are Falling*, Progress Press, Moscow, 1964.

[5] Valentina Ponomareva, interview with Bert Vis and Rex Hall, Star City, Moscow, 14 April 2000.

[6] Michael Cassutt, *Who's Who in Space* (International Space Year Edition), Macmillan, New York, 1993.

[7] Yuri Zaitsev, "The story of women in space," *Space Race News*, RIA Novosti, Moscow, 2007.

[8] Valery Bykovsky, Moscow University press conference, Moscow, 25 June 1963.

[9] Francis French and Colin Burgess, *Into that Silent Sea: Trailblazers of the Space Era 1961–1969*, University of Nebraska Press, Lincoln, NE, 2007.

[10] Grigori Reznichenko, *Cosmonaut No. 5*, Politizdat, Moscow, 1989.

[11] Vitaly Volovich, *Experiment: Risk*, Progress, Moscow, 1986.

[12] Rex Hall and David J. Shayler, *The Rocket Men: Vostok & Voskhod, the First Soviet Manned Spaceflights*, Springer/Praxis, Chichester, U.K., 2007.

10

A tragedy, and Gagarin's final flight

The Vostok and Voskhod spacecraft had served the Soviet space programme well, but as plans advanced beyond the limitations of these craft for future projects, including manned lunar flights, it meant that a whole new generation of spacecraft would have to be designed and flown. Vostok and its improvised variant Voskhod lacked the capacity for achieving rendezvous and docking, but the new craft—christened Soyuz—would not only serve the Soviet Union well, but almost four decades on it continues to fly into space, albeit greatly modified.

THE SOYUZ SPACECRAFT

Evaluating options for a multipurpose craft that would eventually replace Vostok (and Voskhod) had already been taking place since 1958 at Korolev's OKB-1 design bureau. A vehicle was needed that would support Earth-orbital operations, have the capacity to rendezvous and dock with large orbiting space laboratories in orbit and even to carry cosmonauts on circumlunar missions.

By 1962, development work on the new-generation spacecraft was becoming more defined, and the designers had come up with a vehicle known as Product 7K, or, more simply, Soyuz. Eventually the spacecraft would have three major components. At the rear end would be the Instrument (or Propulsion) Module, which housed propellant tanks, on-orbit storage batteries, and approach and orientation engines. On the exterior of the Instrument Module would be two sets of gull wing–like solar battery panels and the larger of two thermal control units. The recoverable, bell-shaped Descent Module (DM) was situated between the Instrument Module and the Orbital Module (OM). The DM would be occupied by the crew for launch, during orbital manoeuvring, docking and undocking procedures, and for re-entry and landing. Finally, the bulbous OM would be used for orbital operations, and where required would be fitted with a docking apparatus. A hatch in its side would be used for crew boarding on the launch pad, as well as for orbital EVA egress. Fitted with a protective ascent shroud and rocket-powered launch escape system, the Soyuz assembly would be launched aboard the R-7 booster [1].

Komarov's second assignment

On 14 January 1966 the Soviet space programme was delivered a monumental blow when Chief Designer Sergei Korolev unexpectedly died at the age of 59. He had entered the Kremlin hospital nine days earlier for a relatively minor operation to remove a bleeding polyp in the straight intestine, but the operation had been delayed by the need for more tests. When the operation finally began a fist-sized tumour (later found in the biopsy to be malignant) was discovered in the abdominal cavity, and was removed by means of an electronic scalpel. However, severe internal bleeding resulted, and just over four hours later, despite the best efforts of the surgeons, Korolev's pulse weakened and stopped. He could not be revived [2].

Korolev's death removed a uniquely dominant and almost irreplaceable element from the management of the Soviet space programme. He would be given a state funeral four days later. The urn containing his ashes would be placed with honour in a niche in the Kremlin Wall, and covered by a marble plaque inscribed with his name, date of birth and death. He would eventually be succeeded as Chief Designer by his First Deputy, Vasily Mishin.

The prime and backup pilot for the first Soyuz flight, Vladimir Komarov and his understudy Yuri Gagarin.

Losing Korolev could not have come at a worse time for Soviet space plans. Mishin was under a lot of political pressure to get things moving, but he first had to assess Korolev's plans for up to four more Voskhod missions before Soyuz—still plagued by development problems—was ready to fly. The most immediate flight, Voskhod 3, was a long-duration mission scheduled to fly in the early part of March 1966, which would be crewed by Boris Volynov and Georgi Shonin. However, the flight kept being put back to June and then July, and much to their frustration the cosmonauts involved were told to take a short holiday while further assessments of their mission were made. A November launch was then proposed, but by this time a decision had been made and the Voskhod programme was terminated after just two flights.

Although the Soyuz spacecraft was still far from ready, the push was on to get it into space to counteract the ongoing publicity being generated by NASA's highly successful Gemini programme, which was sending manned craft into orbit every two months on average.

By this time, unofficially, Vladimir Komarov had been assigned as the prime crewmember for Soyuz 1, with Yuri Gagarin set to train as his backup pilot. The assignment meant that Komarov would become the first cosmonaut to fly into space twice. Valery Bykovsky, Alexei Yeliseyev and Yevgeny Khrunov would form the prime three-man crew for the second Soyuz mission.

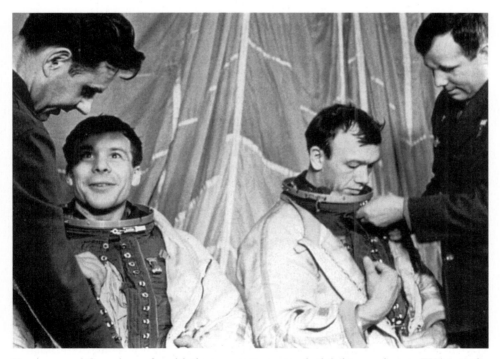

Komarov and Gagarin confer with the two cosmonauts scheduled to conduct an EVA transfer from Soyuz 2 to Soyuz 1, Khrunov and Yeliseyev.

The new Chief Designer

With the sudden loss of Korolev there came a shift in the philosophy of appointing new candidates to the cosmonaut team. As the Soyuz spacecraft was a far more complex piece of machinery and—unlike NASA's spacecraft—with most functions still controlled by automatic means rather than by the pilot-cosmonaut, it was felt that the Soviet space programme might be better served by introducing engineers who could monitor the systems rather than the previous reliance on pilots, who generally had very little engineering experience.

Because the Soyuz could carry up to three men, it was decided by Mishin to introduce more highly qualified specialists into the cosmonaut cadre; specialists who could look at design solutions in-flight, but could also be directly involved in the design and ground testing of the spacecraft. Mishin therefore began taking practical steps toward changing the composition of the cosmonaut team. In May 1966 the design bureau OKB-1 organized the implementation of a flight methods department for the training of civilian cosmonaut candidates, which immediately brought Mishin into direct confrontation with Air Force officials, who wanted to retain their monopoly on the selection and training of cosmonauts.

By using the colossal influence of his position as the nation's new Chief Designer, Mishin was able to call in some favours from higher up, resulting in him challenging the authority of the Air Force by saying that if a compromise could not be reached, all future cosmonaut selections would be restricted to engineers and scientists, and the

The prime crews for Soyuz 1 and 2 photographed at the Baikonur launch complex. From left, Vladimir Komarov, who would pilot Soyuz 1, and the three planned Soyuz 2 crewmembers Valery Bykovsky, Yevgeny Khrunov and Alexei Yeliseyev. Mishin is on the right in dark glasses.

Air Force's cosmonaut training centre would be closed down. His bluff worked, and the planned compromise was reached. The training centre would continue its work, but each Soyuz crew would typically comprise a pilot, a civilian engineer, and a flight researcher [3].

Troubled mission

On 23 April 1967 the two-year hiatus between Soviet space missions came to an end when Tass news agency reported the successful launch of Soyuz 1, carrying veteran cosmonaut Vladimir Komarov on his second mission. What would really capture the attention of Western observers was not so much Komarov's second flight, but the name given to his spacecraft—*Soyuz*. The word means "Union", and though this could simply have been a reference to the U.S.S.R., many were convinced that it was also a reference to a possible docking in orbit with a second vehicle. What was also remarkable was that the spaceship was named Soyuz 1—the first time a new-generation manned Soviet spacecraft had carried a number designation. Gagarin's spacecraft had simply been called Vostok, and Gherman Titov's Vostok 2. The same applied for Voskhod and Voskhod 2. It was thought that giving this new craft a number designation might have some significance, and they were right: Soyuz 2 was already on the launch pad.

The fears held by many of the cosmonauts, including Komarov and Gagarin, about launching Soyuz before all the new craft's problems had been overcome would soon be realized when things began to go wrong aboard Soyuz 1. Shortly after lift-off Komarov had begun to experience serious difficulties, and resolving these quickly became his first priority. One of the two solar panels intended to supply power to the spacecraft had failed to deploy shortly after the spacecraft reached orbit. Then, despite several attempts, Komarov was unable to manually align Soyuz 1 after the automatic orientation system failed.

As a result of these problems it was felt that the launch of Soyuz 2 the following day with its three-man crew should be either postponed or abandoned. That spacecraft was planned to dock with Soyuz 1, after which two of the crew (Yeliseyev and Khrunov) would perform a spacewalk over to Komarov's ship. According to schedule, all three cosmonauts would eventually land aboard in Soyuz 1 while Bykovsky made a solo return to Earth. Amazingly, in light of all the problems Komarov was struggling to correct, the recommendation to abort the mission was ignored and preparations for the second Soyuz launch continued.

Only hours later, when it became obvious that Komarov would not be able to rectify the orientation situation, was a decision finally reached to cancel the launch of Soyuz 2 [4].

Loss of a cosmonaut

Over the next 27 hours a frustrated Komarov vainly worked to gain control over his malfunctioning spacecraft, but despite his best efforts he finally received confirmation that Soyuz 1 would have to return to the ground. A first attempt failed when the

The shattered, fire-gutted remains of the Soyuz 1 descent module.

Soyuz craft could not be correctly aligned for re-entry, but a second attempt on the 19th orbit was successful, and Soyuz 1 began to plunge back into the atmosphere.

It was after the craft had safely negotiated the ferocious heat of re-entry that other things began to go horribly wrong for Komarov when his main parachute failed to deploy as a result of what was later found to be a design fault in the parachute container. This undiscovered flaw meant that a smaller drag chute, which had deployed and opened as planned, had insufficient force to drag out the main parachute. When the main chute failed to deploy a reserve chute was brought into operation for just such a contingency. Unfortunately the small drag chute trailing behind the spacecraft became inextricably tangled in the reserve parachute. These problems were totally unrelated to any of the difficulties encountered while in orbit.

Komarov, strapped helplessly in the plummeting Descent Module, would have quickly realized that there had been no reassuring thump of sudden braking of a parachute blossoming out and biting into the thickening air. He would have known that in moments his spacecraft was about to slam into the ground, and he was doomed.

Soyuz 1 hurtled down and hit the ground at around 120 miles an hour. Small braking rocket engines, normally designed to fire moments before touchdown, exploded on impact. The shattered, flattened wreckage erupted into flames, in a conflagration so fierce that the aluminium outer sheet melted and pooled on the ground.

Several years later, graphic film footage taken shortly after the crash by photographers accompanying the rescue teams was released. It showed the still-burning remnants of the spacecraft and followed the desperate attempts of devastated rescuers to extinguish the fire. Later, when they were able to finally approach the mangled spacecraft, the recovery crews could do little but extract what they could find of Komarov's charred remains for solemn transportation back to Moscow [4].

A nation mourns

Vladimir Komarov's tragic death would prove a tremendous blow to the Soviet space programme, while the entire nation fell into a shocked, collective sadness.

The cosmonaut's remains were inurned in the Kremlin Wall with full military honours on 26 April. At the time, due to secrecy surrounding the mission, it was not known that Yuri Gagarin had actually been Komarov's backup for the mission. Nor could Gagarin or anyone else have known that less than a year later, another solemn funeral procession would take place a few metres away by the Kremlin Wall, this time for the nation's first, much beloved cosmonaut. As author Colin Burgess wrote in his book, *Fallen Astronauts*:

> "The usual shroud of secrecy surrounded the flight of Soyuz 1, which brought forth an equally usual string of rumours about Komarov's death. It was said that his wife Valentina had been brought to mission control to bid an emotional, tearful farewell to her doomed husband. Soviet Prime Minister Alexei Kosygin was said to have talked with Komarov, telling the cosmonaut that his country was proud of him and his forthcoming sacrifice for the space programme. During the descent module's re-entry American listening stations were reported to have listened in to Komarov's pitiful cries while he plummeted to Earth, cursing and renouncing a government that had ordered him to carry out a flight in a trouble-plagued craft that was launched well before it was ready.

The funeral procession for Komarov winds a solemn path to the Kremlin Wall.

Serious observers of the Soviet space programme are absolutely certain that these rumours are all false. For starters, the programme's mission control was situated in Yevpatoriya in the Crimea, while Valentina Komarov would have been at home in Moscow. In addition, mission control would have lost contact with the cosmonaut shortly before the descent module separated from the other two spacecraft modules. This was a perfectly normal occurrence for any returning spacecraft: the ionized air surrounding it causes a radio blackout lasting for several minutes. In 1992 the author spoke with an unflown cosmonaut named Alexander Petrushenko, who had literally been the last person in contact with the doomed cosmonaut. During this exchange, Komarov simply reported to Petrushenko that he had carried out the final correction manoeuvre and was preparing for his return to Earth'' [5].

Komarov's widow Valentina lays a wreath of flowers against his statue on the Avenue of Cosmonauts in Moscow.

Ironically, Komarov's early return from space, and his death, probably saved the lives of the three cosmonauts preparing to fly into space aboard Soyuz 2: Valery Bykovsky, Alexei Yeliseyev and Yevgeny Khrunov. The latter two men, having transferred by EVA, would undoubtedly have perished along with Komarov. As well, subsequent investigations revealed the frightening fact that their Soyuz 2 ship's parachute container carried the same design fault as that of Soyuz 1. Given similar problems to those encountered by Komarov, Bykovsky might also have been killed on his return to Earth.

MARINA POPOVICH TRIES AGAIN

Marina Popovich possessed a restless personality. A pilot in her own right and the wife of a cosmonaut, she had applied for the women's group in 1962, but was excluded because Colonel Nikolai Nikitin was only interested in having proficient parachutists in the squad. No female pilots were being considered, although one woman who made it through was Valentina Ponomareva, a trained parachutist who had also learned to fly at a local aero-club in Moscow.

In 1964 Marina enrolled in the air force test pilot school in Akhtubinsk, 800 miles east of Moscow and close to the Kazakhstan border, training under Stepan Anastasovich Mikoyan, the renowned pilot nephew of aircraft designer Artem Mikoyan (the "Mi" in MiG). Stepan Mikoyan was the commander of Soviet aviation at the Research and Flight Test Institute of Chkalovskaya and Akhtubinsk. Having graduated successfully, Marina Popovich then went on to fly MiG-15, MiG-17 and MiG-21 fighters, and logged more than 6,000 hours flying time, eventually rising in rank to colonel in the Soviet Air Force.

A further application

Also in 1964, Marina applied once again to become a cosmonaut, in a group that would be recruited from experienced test pilots. At that time the criteria stated that an applicant had to be under 35 years of age, have more than 1,500 hours flying time, and hold a degree in higher education—all of which she met. The candidates had to complete 20 crucial exams, 19 of which required a pass mark of 5. The other subject, meteorology, only required a pass mark of 4. She then underwent 40 gruelling days of tests, only to be told at the end, without any explanation, that she had been unsuccessful. When she demanded a reason for her exclusion, it was revealed that she had not been considered because she had a child, Natalya, then just eight years old. Pavel was absolutely furious, and demanded to know why she had had to endure 40 days of testing if it had already been decided beforehand that she was unsuitable for this reason.

Popovich was actually becoming increasingly perturbed and perhaps even a little jealous of his wife's flying aspirations (she even had more flying hours than him when he became a cosmonaut), and this not only caused disputes, but he was making his strong views on the subject known in the cosmonaut community. Marital trouble

Marina and Natalya Popovich on a boat cruise with her husband and his fellow cosmonauts Nikolayev and Gagarin.

finally erupted on the evening of 11 April 1966 when Gagarin, Gorbatko, Nikolayev and Popovich and their respective wives were out for the night entertaining some delegates to the 23rd Party Congress from Kiev. Late in the evening, Marina saw her husband Pavel embracing Gorbatko's wife and angrily confronted him. Words were exchanged, and then Popovich unwisely lashed out at his wife. Her brother was also in attendance, and he punched Popovich in the face, giving him a black eye. Once the wives of some other cosmonauts heard about the incident they banded together and wrote a strong letter to Popovich's superiors denouncing his actions, not only in hitting his wife, but for trying to force her to abandon her own career in aviation.

Eventually, despite the birth of their second daughter Oksana in 1968, the marriage broke down and the couple were divorced. While Pavel continued with his cosmonaut career and would later remarry, Marina left Moscow after being invited by legendary aircraft designer Oleg Antonov to join his company's design bureau as a test pilot. In this capacity she went on to hold over 100 world flying records (the vast majority of which she still holds today), and would fly 40 different types of aircraft; six of which she flew as a test pilot. She would also walk away from six bad crashes. Marina was the first Soviet woman to fly faster than Mach 1, reaching a speed of 1,125 mph, and the third woman in the world to achieve this supersonic milestone after American Jacqueline Cochran and French aviatrix Jacqueline Auriol. Like Pavel, she would also marry again, this time to Boris Aleksandrovich Zhikharev, a major general involved in aviation. Today, with a PhD in Flight Technology from Leningrad University, she lives in Star City, but travels the world giving talks as a self-proclaimed expert in Soviet UFO and related paranormal subjects [6].

GAGARIN TO FLY AGAIN

On 25 May 1961, six weeks after his historic orbital flight, Yuri Gagarin had been appointed commander of the first cosmonaut detachment. It was a position he would hold until April 1965 when he handed the responsibility on to Andrian Nikolayev. Sergei Korolev had been known to be highly critical of the protracted public relations exercises forced on Gagarin, which would continue to a marginally lesser degree after this appointment. While grudgingly admitting it was very useful in propaganda terms, Korolev also saw it as a waste of his favourite "little eagle" on nugatory social events and official functions that took him ever farther away from his former proficiency as a skilled pilot and cosmonaut.

With many months of tours, publicity, meeting dignitaries and a tiring schedule behind him after his flight, Gagarin was ready for some real work once again. On 1 September 1961 he undertook studies at the engineering faculty of the N.E. Zhukovsky Air Force Engineering Academy. He was also delighted to be given the responsibility of heading a training programme for the second, and female, group of cosmonauts, along with Nikolayev. As good friends, the men would work well together, and Gagarin was simply happy to be back in his element once again, doing some useful work.

On 12 June 1962, Gagarin was promoted to the rank of lieutenant colonel while preparing a demanding but structured training programme for the five women, knowing that one of them would be in line to fly into space after the next planned venture—a tandem mission between Vostok 3 and Vostok 4.

Two months later, Gagarin served as prime CapCom during the historic joint mission, flown by Nikolayev and Pavel Popovich, launched on 11 and 12 August, respectively. Although he was still called upon occasionally to appear at a number of functions, Gagarin's training programme for the women worked satisfactorily. Valentina Tereshkova would eventually be selected to fly aboard Vostok 6, launched

Yuri Gagarin had never lost hope of flying into space on a second mission.

Gagarin with his two
daughters Lena and Galya.

on 16 June 1963 on a tandem flight with Valery Bykovsky on Vostok 5, and she would
return a national hero having completed 48 orbits of the Earth.

Gagarin would be further promoted to the rank of full colonel on 6 November
1963. The following month, on 20 December, Gagarin received an appointment as
Deputy Chief of the Cosmonaut Training Centre for spaceflight training under Major
General Nikolai F. Kuznetsov (who had been appointed commander of the TsPK
training centre on 2 November 1963), and was also elected a member of the Soviet
Supreme.

However, his fame and immense propaganda value, while it had made him a
beloved international icon of the Space Age, would become an onerous liability to
Gagarin. On 11 June 1964 he was officially grounded. The world's first spaceman was
no longer permitted to fly aircraft.

A mischievous Gagarin gives
Tereshkova's ear a playful
tweak as they pose for
photographs.

Restless hero

While balancing an onerously heavy workload of public appearances and duties
related to the Soviet Supreme and the cosmonaut training centre, Gagarin grudgingly
accepted his lot for a while. As more and more Soviet cosmonauts followed his path
into the cosmos, however, he grew increasingly restless, complaining to those who
would listen that he should be permitted to fly into space again. A single orbit was not
how he wished his spaceflight record to stand, and he dreamed of going back into
space. In a bold move, he submitted an official request for another flight to Kamanin,
but his application was set aside. The space programme chiefs were initially loath to
risk the life of a true national hero on another risky flight, but eventually, and with
considerable reluctance, they gave in.

In October 1966, with his grounding orders rescinded, a delighted Gagarin was
officially assigned as backup pilot to his good friend Vladimir Komarov for the Soyuz

Training for a spaceflight that would never come.

1 mission, then scheduled to rendezvous in orbit with Soyuz 2, launched a day later. The latter craft would be carrying three cosmonauts: Valery Bykovsky, Yevgeny Khrunov and Alexei Yeliseyev. While Bykovsky remained on Soyuz 2, Khrunov and Yeliseyev would carry out a planned spacewalk and transfer to Komarov's Soyuz craft for the journey home. If successful, it would be regarded as a significant feat and an indisputable milestone in spaceflight history. Gagarin's backup role for Komarov would place him squarely in line to command the follow-up Soyuz 3 mission.

Unfortunately, serious problems developed soon after the launch of Soyuz 1, and the launch of the three-man Soyuz 2 was scrubbed while every effort was made to get Komarov back safely. Sadly, it was not to be: Komarov died on impact when his plummeting spacecraft slammed at high speed into the Kazakh steppe.

Following the tragic loss of Komarov, all Soviet manned space activity came to an abrupt end. Gagarin would also learn that his friend's death had other ramifications, when he was grounded once again from all flying activities. This time, his

A MiG-15 UTI jet trainer similar to the one in which Gagarin and Seryogin would lose their lives.

superiors were adamant; due to the dangers involved he must never be permitted to make another spaceflight. His life was regarded as far too valuable by the Soviet government to run the risk of losing it in a similar catastrophe.

For Gagarin it meant the crushing end to a dream. Shunted back into an administrative desk job he was desperately unhappy, but persevered with his attempts to salvage some pride by getting back to active flying in jet aircraft. Amazingly enough, his superiors finally concurred with him that the ban could be relaxed a little.

On 17 February 1968, Gagarin presented his diploma thesis at the Zhukovsky Air Force Engineering Academy, clearing the way for him to undertake flight training. On 13 March 1968, 11 days after his graduation from the Academy, the stand-down order was rescinded for a second time. To his unbounded relief Gagarin was given permission to resume flying, although he was only allowed to take to the skies if accompanied by an experienced instructor who could gauge his proficiency. Almost immediately he began refresher training in MiG aircraft, although his other duties meant he could not fly as often as he wished. He would perform 18 flights from 13 to 22 March in the two-seat MiG-15 UTI trainer, logging a total of seven hours flight time.

A subsonic trainer fighter with a single turbojet engine, the MiG-15 UTI was equipped with two pressurized cabins: the first was for the trainee pilot, and the second for his instructor. Located in the forward section of the fuselage, both cabins were fitted with ejection seats and canopies. The canopies could be ejected independently by either crewmember in the event of an emergency by means of an explosive charge, or manually if the charges misfired, by opening the restraint locks. The instructor, sitting behind his pupil, could control and correct any actions from the rear cabin.

The fatal day

There was near waist-deep snow on the ground on the freezing cold morning of Wednesday, 27 March 1968, as Gagarin readied himself for yet another training flight from the Chkalov Air Force Base, near the cosmonaut training centre. Due to his work and travel commitments he hadn't flown for several days, and was looking forward to taking the controls again.

It would be a routine but structured training session aboard the MiG-15 UTI with his 46-year-old pilot instructor, Colonel Vladimir Seryogin, a senior test pilot and highly decorated military hero who had flown more than 200 combat missions during the Second World War. The call sign of their aircraft that cool, overcast day was 625, and though it would have been noticed by the pilots, no mention seems to have been made of the fact that it was fitted with two expendable 260-litre wing-mounted fuel tanks, an add-on feature giving an aircraft far more flight endurance, but one also known to make the MiG-15 unstable in poor conditions.

The ill-fated day had already begun badly for Gagarin. First, his car broke down, which meant he had to catch a bus to the air base. Then he realized he had forgotten his gate pass, so he had to return home to pick it up before returning to the Chkalov airfield. It was reported that his superstitious colleagues chided him about having a bad-luck day, and that he should reconsider flying. But Gagarin chose to ignore their well-intentioned advice.

Taking off into strong gusting winds at 10:19 AM, Gagarin and Seryogin flew out between two thick cloud layers, steadily gaining altitude as they headed at high speed

All too soon the world would lose history's first spaceman in a fatal airplane crash.

to the designated training area where Gagarin would take over the controls and complete a few practice manoeuvres. At 10:31 the session was abruptly terminated. Using their call sign of 625 they requested an unexpected change of course to take them back to base, which was granted by air traffic controllers. It was the last ever transmission from the MiG.

It did not take long for the air traffic controllers to become concerned. The MiG's transponder signal had suddenly vanished from their screens and there was no response to repeated calls.

Gagarin's colleague Alexei Leonov had been training a group of cosmonauts on lunar landing techniques that badly overcast day, which involved making parachute jumps from a helicopter flying near Kirzhach airfield, some 80 miles northeast of Moscow. As local weather conditions deteriorated even further and persistent rain turned to sleet, Leonov wisely called off the exercise and requested permission to return to base. Just as he was waiting for a response he and the trainees heard two dull explosions in the distance, a couple of seconds apart. He reported this and, fearing the worst, suggested that a search helicopter be immediately sent to the area. The alarm was raised and a helicopter was hurriedly despatched to the thickly snow-covered site to look for any sign of a missing aircraft or parachutes in the trees. Eventually, two IL-14 and four Mi-4 helicopters were in the air, searching Kirzhach, Pokrov and the outlying areas of eastern Moscow [7].

The controllers' worst fears were realized two hours later when it was reported that a pall of smoke in a birch forest near the village of Kirzhach had led one of the helicopter pilots to a 20-foot crater and the wreckage of a MiG fighter strewn amid the trees, two miles from Novoselovo village. The news got even worse: the two pilots were assumed to be dead, and one of them was identified through flight records as Yuri Gagarin.

As a devastated Nikolai Kamanin would record in his diary, he was at the crash site when a breakfast meal ticket made out to Gagarin was discovered in a piece of torn clothing. "Then we found Gagarin's wallet with his personal identification, driver's licence, 74 roubles and a photo of Sergei Pavlovich Korolev on the front" [8].

Doctors at the impact area would also have the gruesome task of trying to identify the two bodies from whatever body parts could be found. Seryogin was eventually identified from fragments of his scalp, ears and a foot, and Gagarin by a dark mole behind the remains of an ear. As well, a part of the cosmonaut's left hand and some fingers (which later made fingerprint identification possible) were located on an engine control handle.

Requiem for a cosmonaut

Almost immediately people began asking what had caused the crash that took the young cosmonaut's life, but meagre information subsequently released by Soviet officials only served to fuel speculation, giving rise to a string of theories and unfounded rumours of a cover-up at the highest levels.

Reports began to circulate hinting that a technical malfunction, even sabotage, had brought down the MiG, while others suggested the cause was nothing more than

A distraught
Tereshkova tries to
console Gagarin's
widow, Valentina.

an unfortunate bird strike. Further speculation centred on the discovery of fragments
of a weather balloon near the crash site, indicating that the aircraft may have flown
into it, bringing both down. Articles were written stating that the crew had received
incorrect weather data, in particular the amount of cloud cover, which may have
caused them to believe they were flying at a much higher altitude.

Unfortunately, a 1968 government commission's findings into causes of the crash
were promulgated in haste in its eagerness to return a quick and valid verdict. Headed
by Igor Kuznetsov, then a member of the State Institute of Exploitation and Repairs
of Air Force Aviation Equipment, their investigation at the crash site was detailed
and thorough, involving some hundreds of investigators. Cosmonauts Alexei Leonov
and Gherman Titov were also assigned to this team. Some 90% of the aircraft was
eventually located by soldiers meticulously combing the woods within a three-mile
radius of the crash site, extended to nine miles north of the crater.

Despite the thoroughness of this investigation, *Pravda* later stated in an article
that "much of the relevant data for the enquiry was lacking, while some facts were not
looked at hard enough" [9].

To everyone's chagrin most of the findings were kept secret, with Brezhnev
forbidding the investigation team to publish their findings and conclusions, based
on the mistaken premise that it might "unsettle the nation". The 30-volume report
was subsequently consigned to the state archives, and only a brief summary of its
findings would be released.

At first, crash investigators were deeply puzzled by the absence of the parachute
packs belonging to the two pilots, which could not be located. Three days later,
however, KGB agents located the missing packs which had been hidden beneath a
pile of manure in a neighbouring village. It seemed that looters had found the packs

An obelisk marks the place
where the MiG training jet
slammed into the ground,
killing Seryogin and Gagarin.

and hidden them, probably with the intention of using the silk material for domestic purposes, but they had been totally unaware of the identity of one of the pilots.

As a result of the ambiguity of the special committee's findings, dark, unsubstantiated rumours would begin to circulate on the principal cause of the crash. One of those sadly lingering to this day infers that one or both pilots were intoxicated when they took to the skies, with Gagarin consuming a bottle of vodka before the flight. Similarly, he is said to have attended a birthday party two nights earlier at which he drank so much vodka that he was still under its effects on the day of the crash.

Another bizarre story claimed that the accident was caused by the pilots chasing and trying to shoot a moose after spotting it from the air, while yet another was based on the fanciful notion that the KGB had murdered both men at the behest of Soviet leader Leonid Brezhnev. At the root of this particular story was the indisputable fact that Gagarin had always been seen as a protégé of Nikita Khrushchev. Once Khrushchev had been toppled from power, Gagarin is also said to have fallen from grace with the new Kremlin hierarchy. It was even suggested he had been killed under instruction by KGB agents because his voracious consumption of women and alcohol had openly flourished since the Vostok flight, and he was proving something of an embarrassment to the new regime.

Ultimately, even though it was compiled in relative haste, the official report offers the most credible scenario. It concluded that the most likely cause of the crash was the MiG going into a violent spin after flying into the trailing vortex of another jet fighter. The question of alcohol as a contributing factor was dismissed by the investigators, who stated that Gagarin had passed two medical examinations before take-off, and a post-mortem examination of body parts found no alcohol in his system.

Another investigation

More than 16 years after the event, another team of Soviet experienced investigators, led by the eminent aviation engineer Lt. General Sergei Mikhailovich Belotserkovsky, was officially commissioned to begin an in-depth study into possible causes of the crash.

Belotserkovsky, who died aged 80 in the summer of 2000, was a professor who had lectured Gagarin and the other cosmonauts on engineering and scientific training at Moscow's Zhukovsky Air Force Academy. One unnamed member of the first cosmonaut group once wrote of him: "Professor Belotserkovsky worked with us for

Gagarin with Belotserkovsky at the Zhukovsky Air Force Academy.

many years. We, and especially Nikolayev, Leonov and Volynov, who defended their theses under his leadership, really consider him our teacher. But Gagarin was the most intimate and dear disciple" [10].

The Belotserkovsky report, prepared over several years, would state that a minute after Seryogin and Gagarin had left the runway a pair of much faster MiG-21 jets had also taken off, rapidly overtaking the smaller MiG-15 and leaving it behind. The next aircraft in the flight line that morning was another MiG-15, call sign 614, who could not see Gagarin's aircraft (625) in the murky conditions. The investigative team reported several air safety violations had occurred that day, and the two MiG-15s were brought "in dangerous proximity to each other" less than 1,640 feet apart.

According to Moscow's Novosti News Agency in a 1989 article, "[the pilot's] manoeuvre to prevent dive and spin while levelling off—the downward deflection of the aileron on a dropping wing to prevent the aircraft from going down in a spin—led to wing stall and spin. The entire accident happened in a moment" [11]. Still diving and unable to stabilize the aircraft, with evidence later showing that Seryogin was in fact the last person to control the MiG, the pilots crashed through some trees and into the ground before they could level out and recover some altitude.

The Novosti article was backgrounding a computer re-enactment of the crash that had been put together by Professor Belotserkovsky's team based on information they had gathered over a painstaking investigation. Having simulated the prevailing weather conditions on that day, and the position of all aircraft involved, their report concluded that: "Either Gagarin's Mig-15 fell into the vortex wake of another aircraft or else it banked sharply to avoid hitting another aircraft or an instrument probe [weather balloon]. The aircraft went into a spin characterised by maximal energy loss. Subsequently it recovered from the spin after which the aircraft ploughed into the ground."

In conclusion, Novosti reported that, "The commission's findings unequivocally showed that careless flying or lack of discipline on the part of the crew, as well as ground-control negligent attitude or someone's malicious intent, were ruled out as possible causes for the disaster" [11].

Despite an inordinate amount of research, the task group's findings would be viewed as unconvincing, giving rise to even more scepticism and rumours that were both fantastic and plausible in the public perception. One bizarre but popular story that did the rounds involved the sighting of a UFO in the area which had emitted paralysing mind rays, disabling the pilots, while another iterated that a Bulgarian psychic named Vanga had declared the reported death of Gagarin to be a monstrous conspiracy, and that the cosmonaut was actually alive and residing under cover in the United States.

Alexei Leonov is adamant that culpability for the accident actually rests with the pilot of another aircraft in the area that day. At a press conference held at the Central House of Journalists in Moscow on 24 March 2003, Leonov said the unnamed pilot "failed in his brief for the flight," and did not fly the appropriate profile that day. It was, he believed, the wash from this aircraft's afterburners that sent the MiG-15 into its fatal spin.

In his 2004 book, *Two Sides of the Moon* (co-authored with astronaut Dave Scott), Leonov wrote more about his personal belief as to the cause of the crash:

"At the time of the accident, it was known that a new supersonic Sukhoi SU-15 jet was in the same area as Yuri's MiG. Three people who lived near to the crash site confirmed seeing such a plane shortly before the accident. According to the flight schedule of that day, the Sukhoi was prohibited from flying lower than 10,000 metres. I believe now, and believed at the time, that the accident happened when the pilot jet violated the rules and dipped below the cloud cover for orientation. I believe that, without realising it because of the terrible weather conditions, he passed within 10 or 20 metres of Yuri and Seryogin's plane while breaking the sound barrier. The air turbulence created overturned their jet and sent it into the fatal flat spin" [7].

Fellow cosmonaut Viktor Gorbatko, meanwhile, disagreed with Leonov's conclusions, saying that like Gherman Titov he felt it was possible that Seryogin and Gagarin had actually run into a weather balloon. "There can be many theories of Gagarin's death," Gorbatko told journalists, "but the one conclusion is that 35 years on we can't come to grips with the irreparable loss" [9].

In April 1974, one-time cosmonaut Valentin Varlamov was interviewed by respected journalist and space historian Yaroslav Golovanov, and their conversation turned to the deaths of Gagarin, Belyayev and Komarov. "I saw death much," Varlamov stated sadly. "I lost three close friends. It was long ago, and at times I am still pained with the memory of their faces. And [Gagarin] I can't forget. Now he stands before me, I see him; for me he did not perish ... I don't belittle the merit of the other guys. There were many outstanding guys among us. But nobody, no way, could replace Yuri; everybody says this. Probably I could talk about him a lot, but I knew him too well to do so" [12].

Conclusions drawn

Unknown to many at the time, the KGB's counter-intelligence department had also conducted its own secret investigation into the crash. They found the ground staff that day were inattentive and guilty of actions amounting to "a dangerous violation" of standing orders. They had failed to notice that the MiG had wing-mounted fuel tanks, which should not have been permitted on an aircraft involved in the training exercises listed for that flight, and definitely not in inclement weather. The report, finally released in 2003, also stated that the officer in charge of air traffic control that day, identified only as Colonel *Y*, had supplied Gagarin with an out-of-date weather briefing.

The KGB report concluded that, based on the weather briefing, Seryogin would have assumed the spinning MiG to have been some 820 metres (2,700 feet) above the ground before coming out of the clouds, allowing plenty of time for a recovery to level flight. However, the cloud base at the time was actually somewhere between 275 m and 365 m (900 ft to 1,200 ft). Altogether, the MiG had fallen approximately

3,650 m (12,000 ft) in one minute before it hit trees and crashed near the town of Kirzhach. The angle at which the MiG finally hit the ground suggested that they had very nearly managed to recover from their fatal dive.

"One can conclude," the Soviet newspaper *Pravda* later stated when reporting on the KGB findings, "that 625 got on the tail of 614 and was following it. Finding itself in the trailing vortex of the aircraft in front, the plane piloted by Gagarin and Seryogin got into a spin." It is thought that the pilots may have succeeded in controlling the spin, but they were too low, could not see the horizon, and were most likely disoriented in the prevailing thick cloud.

"The crew's actions aimed at regaining level flight were correct in the highest degree," the report stated. "The pilots retained their capacity for work to the end of the flight, skilfully and efficiently piloting the aircraft" [9].

Years after his death, doubts still linger as to the cause of the fatal accident that took the life of Yuri Gagarin.

A wall of secrecy

In March 2005, almost four decades after the fatal crash, the principal investigator of the task group given the job of determining the cause of the crash in 1968, Igor Kuznetsov, then an aviation equipment specialist at the Defence Ministry's Scientific Research Institute, went public with demands for a fresh resumption of the investigation. He was quoted in the *Moskovsky Komsomolet* newspaper as saying that new evidence using modern methodology had come to hand.

Together with 30 eminent colleagues, experts in their fields, Kuznetsov personally petitioned President Vladimir Putin in 2007, asking him to sanction a fresh investigation into the incident based on their findings. To his frustration, the Kremlin vetoed the suggestion, and said it had no interest in questioning what they referred to as the "original findings" of the KGB investigation. "What original findings are they talking about?" Kuznetsov demanded of this latest rebuff. "There were no original findings; just speculation. At the time we faced a wall of secrecy designed to stop us finding anything that might damage the Soviet reputation," he stated. "But now so many files have been de-restricted that I believe it would be possible once and for all to find the real reason for the death of the first man in space" [13].

Kuznetsov claims that both pilots lost consciousness after a cockpit vent panel in Gagarin's cabin was accidentally left partially open on the ground, which meant that the cabin of the MiG-15 would not have been hermetically sealed after take-off, resulting in a dangerously low level of pressurization. This anomaly was apparently recorded on a cockpit pressure gauge. Part of the evidence presented showed that Russian-built MiG-15 aircraft do not have these vent panels, but the MiG Seryogin and Gagarin were flying that day had been manufactured in Czechoslovakia, where the panels were a design inclusion.

The pilots, Kuznetsov continued, would only have realized a lack of pressurization after they had reached their recorded altitude of 13,000 feet, by which time they would both have been unknowingly suffering from prolonged hypoxia, or oxygen deprivation, severely affecting their judgement. He believes they tried to abort the training run and over the next six minutes tried to descend to a safe altitude. However, as reported by the online Mosnews.com, "For five out of the six minutes, the men experienced increased *g*-loads. The cabin pressure mounted like an avalanche. It is possible that in five seconds, the men suffered aerodynamic shock. The pressure was rising at the rate of 14 mm of mercury per second because the rate of vertical descent was 140–150 metres per second. The pilots probably lost consciousness and failed to regain control of the aircraft" [14]. The MiG-15 was also notoriously flawed: at transonic or supersonic speeds the jet's elevators became ineffective.

Pressure suits and sealed helmets would undoubtedly have assisted the two men to remain conscious, but in those days Soviet pilots only wore leather jackets and crash helmets. In addition, as Kuznetsov would emphatically state: "In 1975, our medics prohibited pilots from descending faster than 50 meters per second. But it was 1968 and Gagarin's aircraft was losing altitude at 145 meters per second—three times faster than the accepted norm."

"I can't say who lost consciousness first. What's important is that they didn't try to pull out of the nosedive at 2,000 meters ... Somewhere between the altitude of 4,100 and 3,000 meters they either lost consciousness or found themselves in a pre-fainting state. That's what would happen in a non-hermetic cabin."

No fault could be attributed to Seryogin and Gagarin for the open vent according to Kuznetsov, who opined that it could have been left open either by a technician or a pilot on the aircraft's previous flight. He was equally critical of a lack of attention by the ground staff that day (none of whom would ever face charges), and added that the two pilots had both "acted strictly according to regulations" [13].

REFERENCES

[1] Rex D. Hall and David J. Shayler, *Soyuz: A Universal Spacecraft*, Springer/Praxis, Chichester, U.K., 2003.

[2] Asif Siddiqi, *Challenge to Apollo: The Soviet Union and the Space Race, 1945–1974*, NASA SP-2000-4408, NASA, Washington, D.C., 2000.

[3] Slava Gerovitch, "Human–machine issues in the Soviet space programme," in Steven J. Dick and Roger D. Launius (eds.), *The History of Spaceflight*, NASA SP-2006-4702, NASA Office of External Relations History Division, Washington, D.C., 2006.

[4] David J. Shayler, *Disasters and Accidents in Manned Spaceflight*, Springer/Praxis, Chichester, U.K., 2000.

[5] Colin Burgess, Kate Doolan and Bert Vis, *Fallen Astronauts: Heroes Who Died Reaching for the Moon*, University of Nebraska Press, Lincoln, NE, 2003.

[6] Interview with Marina Popovich conducted by Rex Hall and Bert Vis, Star City, Moscow, 14 April 2007.

[7] David Scott and Alexei Leonov, *Two Sides of the Moon*, Simon & Schuster, London, 2004.

[8] Bart Hendrickx, "The Kamanin Diaries 1960–1963," *Journal of the British Interplanetary Society*, **50**, 33–40, 1997.

[9] Pravda Online, *The Day First Man in Space Died Ironic Death*, 28 March 2003. Website: *http://english.Pravda.ru/main/2003/03/28/45173.html*

[10] Wayne Jackson (ed.), "Tribute to Professor Sergei Mikhailovich Belotserkovsky (April 1920–August 2000)," in *Wake Vortex Predictions: An Overview*, Transportation Development Center, Transport Canada, March 2001.

[11] Anthony Curtis (ed.), Space Today Online Questions, *Hero's End*. Website: *http://www.spacetoday.org/Questions/Questions.html*

[12] Yaroslav Golovanov, "They were the first," serialized in *Izvestia*, 2–6 April 1986. From translation by Jonathan McDowell.

[13] Ed Holt, "Inquiry promises to solve Gagarin death riddle," *Scotland on Sunday*, 3 April 2005.

[14] Mosnews.com online, "Russian expert demands resumption of Gagarin death probe," 28 March 2005. Website: *http://www.mosnews.com/news/2005/03/28/gagarin.shtml*

11

Pushing the limits

Following a lacklustre solo test flight aboard Soyuz 3 by Georgi Beregovoi, who clearly was not up to the complexities of the mission, the Soviet Union next launched Soyuz 4 and Soyuz 5 into orbit on a mission to link up in space and effect a partial crew transfer by EVA. Vladimir Shatalov was launched first aboard Soyuz 4 on 14 January 1969, followed into orbit the next day by the three-man crew of Boris Volynov, Yevgeny Khrunov and Alexei Yeliseyev on Soyuz 5. All four cosmonauts were making their first flights into space. Just three orbits after Soyuz 4 and Soyuz 5 had linked up and an EVA crew transfer involving Khrunov and Yeliseyev had been successfully completed, it was time to undock the two vehicles. Once they had disengaged, Shatalov fired small thrust rockets and Soyuz 4—now with three cosmonauts onboard—slowly slipped away from its sister craft, leaving Volynov as the sole occupant of Soyuz 5.

A TERRIFYING ORDEAL

Mission planning called for Shatalov's crew to return first, so early the following morning the de-orbiting rockets were fired on Soyuz 4 and the spacecraft began its inexorable descent into the atmosphere. The Orbital Module and Instrument (or Service) Module were explosively discarded from either end of the Soyuz 4 Descent Module, which then commenced a controlled ballistic re-entry back to Earth. Few problems were encountered, and the three cosmonauts successfully touched down at 9:53 AM Moscow time, landing under their main parachute in freezing cold temperatures some 25 miles northwest of Karaganda in the steppes of Kazakhstan. Soon after, the Tass newsagency reported the safe return of Soyuz 4 and its three-man crew, and stated that Soyuz 5 was "continuing its flight under Colonel Boris Volynov's command," adding that "all was well" onboard.

From left, the crew of Soyuz 5: Volynov, Yeliseyev and Khrunov, together with the commander of Soyuz 4, Vladimir Shatalov.

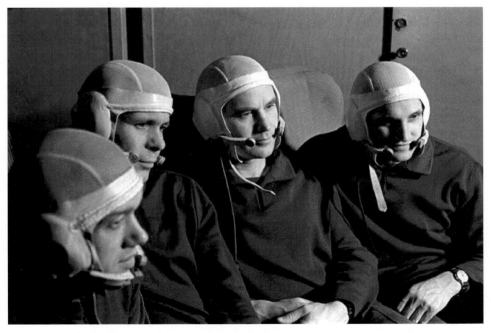

Yeliseyev, Khrunov, Shatalov and Volynov prior to the tandem flight.

By 18 January, the three-day flight of Soyuz 5 was nearing an end and Boris Volynov began stowing gear in readiness for his own re-entry and landing. While it had been a crowded ascent with Khrunov and Yeliseyev along for the ride, there was now plenty of room for the sole cosmonaut as he floated around inside the Descent Module, dressed only in a comfortable lightweight woollen track suit in which he would make the re-entry. There was no protective spacesuit or helmet for him to wear. Volynov could not know it at the time, but he would shortly become the very last person from any nation in that century to return alone from space.

Volynov had already received a report on conditions in the landing area, alerting him to the fact that the ground temperature was expected to be a chilly five degrees lower than that experienced by Shatalov's crew, but he could also expect clear skies and favourably light winds. Over a nine-minute period following orbital sunrise on the next-to-last revolution of the Earth he practised a manual reorientation of Soyuz 5, afterwards advising ground controllers that he had not been able to fully complete the task in the allotted time. Nevertheless, he was given instructions to commence manual de-orbiting procedures on the following orbit, but was routinely given the commands necessary to initiate an automatic orientation should he not be able to complete this procedure manually.

Return journey

As the time approached for the planned de-orbit firing, Volynov secured himself tightly in his couch. Then, as Soyuz 5 passed over the Gulf of Guinea at 10:20 AM Moscow time, he began the landing sequence. Retrofire took place with the main engines firing perfectly for just under two and a half minutes. Then, in accordance with the flight plan, Volynov explosively jettisoned the Orbital Module at the front of his spacecraft. Following this, he would similarly jettison the Instrument Module containing the rocket engines and power supplies. This was situated behind him, attached to the heat shield end of the Descent Module. When the separation procedure was initiated Volynov heard the explosive bolts fire, but a few moments later when he glanced out of his window he noted with alarm that the whip antennas were still extending from the Instrument Module's solar arrays. This told him that the separation had not been effected, and the Instrument Module was still in place. He also realized he was in trouble; in just half an hour he would either be safely on the ground or dead. Later, it would be found that the explosive bolts responsible for separating the two components were defectively underpowered, but that meant little to Volynov, as he was now unable to abort or control the re-entry phase.

As the combined modules began to touch on the fringes of the atmosphere, the awkward mass of the Instrument Module with its flared base caused the spacecraft to automatically seek the most aerodynamically stable attitude, and it began to tumble, something the pilot could not control. Eventually the craft stabilized itself, but with the substantially unprotected nose of the Descent Module pointing forwards. This meant that the craft's access hatch was facing into the descent instead of the heat shield with its protective six-inch layer of ablative material. As well as soaking up some of the heat, the curved shield was designed to form a thick shockwave in front of the module that would both deflect much of the ferocious heat and help to slow the craft. Volynov knew the forward end of the Descent Module could never withstand the searing heat of re-entry and reported his dire situation to controllers aboard a Soviet tracking ship, although there was nothing anyone could do to assist him. He desperately needed the Instrument Module to separate or he and his craft were doomed.

Disaster looms

As Soyuz 5 continued its terrifying back-to-front ballistic descent into the atmosphere an intense pulse of searing heat engulfed the nose of the spacecraft, and all voice contact with the ground was lost. Then Volynov noticed a pungent smell of burning rubber as the gasket sealing the hatch began to melt. The hatch had only been designed to provide crew access between the Orbital and Descent Modules, and its single inch of insulation offered little protection against the massive wave of super-heated gases now building up against it. Even had he been wearing an insulated pressure suit, this would not save him from a horrifying death. Eventually the heat would exceed the surface temperature of the Sun.

While he could still move his arms, Volynov grasped any handwritten notes detailing the rendezvous and thrust them into his flight log. He bound the log tightly with string and tried to jam it somewhere where he thought it might survive an explosive impact with the ground. He also managed to record some brief recollections of the flight and his problems into a small tape recorder as Soyuz 5 continued to plunge ever deeper into the atmosphere.

Soon after, there were even more terrifying moments when the Instrument Module's propellant tanks began to explode, and the resultant pressure wave forced the front hatch inwards and upwards as it strained against the pressure that threatened to tear it off. Volynov repeatedly muttered "No panic, no panic" as the g-forces began to build. During a normal re-entry hydrogen peroxide thrusters would fire to produce sufficient lift to the capsule in order to reduce these g-forces, but as Soyuz 5 plummeted earthward Volynov's instruments confirmed that even though these valves were open, all of the craft's propellant had been consumed earlier when the onboard computer had unsuccessfully tried to orientate the Descent Module. Volynov sensed that the end was rapidly approaching as the stench of burning rubber grew ever stronger, and he could feel the craft's interior heating up as Soyuz 5 hurtled towards what seemed inevitable destruction.

Under normal circumstances at this stage of the re-entry, Volynov would have been pushed back against his couch with the build-up of g-forces behind him. However the reverse ballistic re-entry meant that he was instead being pressed hard and painfully forward against his harness straps. Meanwhile the cosmonaut had little option but to watch and wait as the Descent Module's automatic orientation system sought to cope with the mounting problems and the brutal heat outside. "I looked out of the window of the capsule and saw the flames," he would later recall, "and I said to myself, 'This is it; in a few minutes I'm going to die.' It's hard to describe my feelings;

there was no fear but a deep-cutting and very clear desire to live on when there was no chance left" [1].

As the combined modules plunged ever deeper into the unforgiving atmosphere, the very heat that might soon take his life now began to work in Volynov's favour. Stress, and rapidly increasingly atmospheric friction on the struts holding the two components, caused them to rapidly overheat. Finally, some 50–55 miles above the ground (as Volynov recalls), the struts reached a failure point where they began to break down. Within moments they disintegrated entirely, allowing the Instrument Module to tear free and fall away. It would quickly burn up and vaporize in the intense heat.

Volynov knew something had happened, but could only imagine Soyuz 5 was breaking up. Then, suddenly, his spacecraft lurched and automatically rotated into the most aerodynamically stable alignment, this time with the heat shield facing forward into the conflagration. It had been a close thing—within seconds the situation would have become unrecoverable.

Out of the frying pan . . .

With this catastrophe averted just in time, a relieved Volynov now began to harbour mounting concerns about the integrity of his charred craft's parachute system, which had also been subjected to severe heat. As well, the Descent Module was spinning around its longitudinal axis. The appalling death of his friend and colleague Vladimir Komarov under similar circumstances only two years earlier invoked sharp memories, inevitably increasingly his anxiety.

Approximately six miles above the ground, with the craft still spinning, the hatch of the parachute container was automatically jettisoned. Soon after, reassuringly, the small drogue parachute popped out as planned, but then his main landing parachute did not fully deploy as the shroud lines began to twist around each other. Fortunately they managed to become untangled in time and Volynov felt a reassuring thump as his parachute suddenly blossomed and bit into the air. However, the delayed opening of the main chute meant that Soyuz 5's descent was now much faster than it should be.

Soon after, as planned, the heat shield was explosively jettisoned, exposing soft-landing retro-rockets in the base of the Descent Module, which were programmed to automatically fire 1.5 metres from the ground. Moments from touchdown they fired, but the spacecraft was still oscillating from the earlier parachute failure, causing the vehicle to slam heavily into the snow-covered, frozen ground.

The unexpectedly heavy impact propelled Volynov forward against his harness with such force that it gave way, resulting in him smashing his jaw and shoulders against the instrument panel, snapping off several upper teeth at the roots.

Soyuz 5's re-entry problems had caused the cosmonaut to land some 400 miles off course, and as he sat in excruciating pain trying to assess the damage to his face he could hear the hot exterior of his charred spacecraft popping and hissing in the freezing cold outside. Bruised and bloody, Volynov knew he needed some urgent

290 Pushing the limits

medical attention, but he was also aware of the fact that he was only wearing a thin woollen tracksuit and it was somewhere around 38°C outside.

There were later misleading (and often repeated) reports that as he sat there he saw some smoke from a nearby farmhouse chimney and made his way across to seek shelter and assistance. However, Volynov emphatically denied this in an interview with Bert Vis in London in May 2001, saying it was completely deserted and freezing cold where he landed, emphasizing that to have left the spacecraft would have been suicidal. "There was nobody to help. It was two hundred kilometres from Kustanay [Kazakhstan] ... no settlements at all ... no farm ... maybe sixty or eighty kilometres there was nothing. I couldn't go anywhere as it was minus 38 degrees and I was wearing that suit. I waited inside the capsule" [2].

It took almost an hour for a rescue aircraft to locate and reach the Soyuz 5 spacecraft, although Volynov could see the aircraft in the distance through his window, doing searching sweeps of the area. They finally spotted his parachute and flew low over the downed spacecraft. Four recovery parachutists then leapt out in order to check on his condition and offer initial assistance. Radio links were quickly established and, soon after, the helicopters arrived. There was intense relief and exultation back at the Mission Control Centre when news came through, confirming that the cosmonaut was alive and safe. Those monitoring his flight had all but resigned themselves to another tragedy similar to that of Komarov aboard Soyuz 1, and had even begun glumly taking up a collection for his family [3].

Boris Volynov. He would not fly into space for seven years after the dramas of Soyuz 5.

A few days later, still sporting his smashed teeth and badly injured jaw, Volynov was awarded his Hero of the Soviet Union medal at the Kremlin, after which he underwent several months of corrective surgery and recuperation in hospital. His injuries had been so severe that he was effectively grounded from consideration for any future space missions for two years. In fact, he was later informed by doctors at the cosmonaut training centre that he might never fly again, as there could be what they uncomfortably termed "a psychological barrier" due to the trauma of his experiences.

Nevertheless, Volynov would fly into space again some seven years after his harrowing return flight aboard Soyuz 5, this time as the commander of Soyuz 21 in July 1976. However, it would prove to be his final space mission.

Yevgeny Khrunov and
fellow Group 1 cosmonaut,
Dmitri Zaikin.

Khrunov

The Soyuz 4/5 mission would prove to be Yevgeny Khrunov's first and only space flight. In July 1969 he was named as backup commander for Soyuz 7, replacing Anatoli Kuklin who had failed a centrifuge test, but shortly after he was involved in an automobile accident and was also replaced on that crew. He would later train for an Almaz military space station mission, and still later linked up with the Interkosmos programme. In 1980 he and Cuban cosmonaut José López-Falcón trained as the backup crew for the Soviet–Cuban mission, but this was flown by the prime crew.

Khrunov was then given his next flight assignment as backup commander of Soyuz 38 along with Dumitru Prunariu on the Soviet–Romanian Interkosmos flight, but he reportedly declined the assignment and unexpectedly left the cosmonaut team on 25 December 1980. It was also heavily rumoured that he had been asked to resign for an unexplained "violation of regulations" [4].

Khrunov next undertook work for the 30th Scientific and Research Institute of the U.S.S.R. Ministry of Defence, appointed as a senior scientific associate of the 120th Laboratory of the 46th Department of the 1st Directorate of the Institute. For six years from 1983 he then worked in the Main Technical Directorate of the U.S.S.R. State Committee for Foreign Economic Cooperation, first as deputy chief and later chief of the directorate.

Retiring from the Air Force with the rank of colonel in 1989, Khrunov then participated in the effort to clean up the site of the Chernobyl nuclear power station accident.

THE "TROIKA" FLIGHT

Georgi Shonin was well liked in the cosmonaut corps. Yuri Gagarin, who had served with him in the same fighter unit prior to their selection as cosmonauts once said of

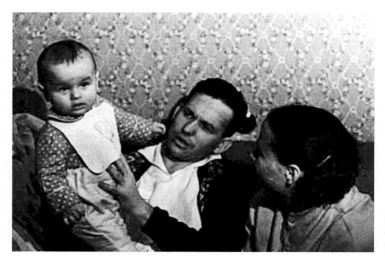

Shonin with his
wife Lydia and
baby son Andrei.

his friend that he had excellent qualities as a man and a pilot. "He has strong will-power, is intelligent and persistent in study and training. Normally he worked in heavy doses. We observed several times that if the situation and programme of study demanded it he could stretch himself to the fullest extent" [5].

Andrian Nikolayev had also become good friends with Shonin during their early cosmonaut training. "During my first day on joining the detachment I got acquainted with Shonin. A calm, modest, hard-working and considerate person. We passed our bachelors' life together. I never thought he would get married before me. He kept away from girls. But when he went on leave he sent me a 'Getting married telegram'. I, of course, congratulated him, but I thought in my heart, 'Did the boy not hurry things up?' Later I was reassured. I looked at the life of the young couple and was certain that they had a good family life. I visited them very often, as I visited the Bykovsky couple."

"A little before my flight [on Vostok 3], Georgi's son was born. At that time he told me: 'You know, we want to call our son Andrei in your honour, to honour our friendship. I was deeply touched by such affection. I thanked him heartily. Thus my little namesake appeared in the 'Star City'." [5].

A long way to space

Georgi Shonin had been the original backup pilot for Valery Bykovsky on the Vostok 4 mission, but he was quickly replaced by Vladimir Komarov when he developed an unexpected intolerance for high-gravity loads on the centrifuge. It seems he later overcame this problem and managed to adapt to its rigours, but he had been perilously close to washing out of the cosmonaut detachment. "During this training I worked very hard and so I got some trouble with my heart," Shonin told interviewer Bert Vis in August 1993. "Some of the physiologists said, 'He must wait.' And so

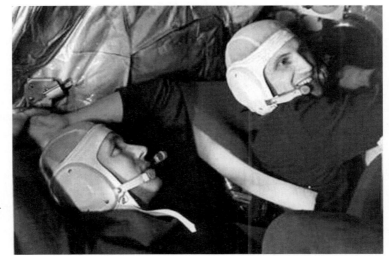

Shonin (left) in
training with
Boris Volynov for
the cancelled
Voskhod 3
mission.

I went to the Crimea for some rest instead of going to Baikonur. That was all okay
and I returned to training again."

Then, in 1964, Shonin began two years of training as prime crewmember along
with Boris Volynov for the 15-day Voskhod 3 mission, only to have the mission
cancelled before it was flown. "At first the flight was planned for the autumn of
1965," Shonin told Vis. "We started flight preparations in 1964; in the summer. At
first, the flight was to be flown in the autumn of 1965 . . . then . . . in the winter of 1966.

The Soyuz 6 crew:
Valery Kubasov
and Georgi Shonin.

Then Korolev died and the launch was cancelled in the spring of 1966. And in May the flight was cancelled altogether.

"At that time there were a lot of people who were against such a flight," Shonin continued, "and I think that it was a great risk to fly in space for twenty days on board Voskhod. The first long American flight, Borman and Lovell in Gemini, they had great problems after fourteen days in space, when they returned. Our flight had to be longer, but we had ... not very good conditions." When pressed by Vis on this issue, Shonin stated that the Voskhod spacecraft was just too small for comfort. "Cramped! For twenty days! I don't know how we could return to earth after that flight. Little space" [6].

The four men—Shonin, Volynov and their backups Shatalov and Beregovoi— were subsequently reassigned to the Soyuz programme. Shonin, who by then had come to be regarded as the most outstanding cosmonaut in flight training, would first serve as backup commander to Shatalov on Soyuz 4 before being assigned to Soyuz 6, along with engineer Valery Kubasov.

Three craft in orbit

Soyuz 6, the first manned Soviet flight in nine months, would actually be the first of a trio of Soyuz craft that would be launched into space, carrying a record total of seven cosmonauts. The *troika* [threesome] mission, as it was known, would practise manoeuvring in formation, test a common flight control system and carry out several important tasks, including an attempt to weld different materials in weightlessness using different techniques. Although they did not allude to the possibility, it was obvious that the Soviets were keen to practise docking procedures and techniques with the three Soyuz spacecraft, constructing a rudimentary space station in orbit. Shonin confirmed with Bert Vis that it was also planned to serve as a test flight for procedures on the L-3 lunar programme.

"This new flight is important in the sense that it will involve the launching of three spacecraft," stated Alexei Yeliseyev pre-flight. He would launch aboard Soyuz 8. "All three of them will carry out manoeuvres and rendezvous. This simultaneous work may give rise to unforeseen difficulties which will have to be overcome. And in overcoming them, we bring closer the time when orbiting stations are created" [7].

In an article published on the eve of the Soyuz 6 launch, the weekly Soviet magazine *Nedelya* devoted its entire centre spread to a generalized article about manned orbiting laboratories, with artists' sketches depicting cosmonauts constructing an interplanetary craft based on an orbiting platform. The article also commented: "Man must build himself a house wherever he goes: on the tundra, in the forests, in the mountains, on the bottom of the ocean, and now in space." Unfortunately the *troika* flight would fall short of that grandiose plan [8].

At 2:10 PM on 11 October 1969, a cold day that begun with swirling rain and strong winds across the steppes, Soyuz 6 lifted off with Georgi Shonin and Valery Kubasov onboard. Soon after, Moscow television interrupted its regular programmes to show videotaped footage of two smiling cosmonauts arriving at the Baikonur cosmodrome, dressed ready for their flight in leather jackets and fatigues. Having

Soyuz 7 would be manned by Viktor Gorbatko, Anatoli Filipchenko and Vladislav Volkov.

exchanged greetings with space officials they climbed the ladder to a lift that would carry them up to the Soyuz 6 spacecraft, stopping to turn and wave to workmen. Then, as the booster rocket thundered into the cloud-covered Kazakh skies, the busiest week in Soviet space history since the launch of the first Sputnik had begun.

Less than a day later the second spacecraft, Soyuz 7, lifted off from a different launch pad at the cosmodrome, carrying an all-rookie crew of Anatoli Filipchenko, Vladislav Volkov and research engineer Viktor Gorbatko into orbit. Gorbatko was actually fortunate to be onboard, as he had broken his leg during a parachute jump in March and it was feared he would fall behind in the training. "But I started training with plaster around my leg," he recalled. "Andrian Nikolayev played a great role in this; he said that I would be ready for the flight and be as prepared as everybody else. Even though the doctors wanted to remove me" [9].

A day after the launch of Soyuz 7, on 13 October, the flagship of the troika fleet, Soyuz 8, thundered into orbit from the same launch pad as Soyuz 6, with Vladimir Shatalov and Alexei Yeliseyev both making their second flights into space.

The principal plan was to have Soyuz 8 rendezvous and dock with Soyuz 7, duplicating the earlier link-up between Soyuz 4 and Soyuz 5. It would be Shonin's task aboard Soyuz 6, which was not fitted with any docking apparatus, to manoeuvre close enough to its two sister ships to record the docking procedure on film. Unfortunately the Igla rendezvous system onboard Soyuz 8 failed to lock onto the system on Soyuz 7, and the crews requested permission for a manual docking. By the time this

The Soyuz 8 crew—Shatalov and Yeliseyev—prior to their mission.

had been granted, however, the two ships had drifted too far apart. The crews were advised that another attempt would take place the following day, 15 October.

With the Igla system still malfunctioning the next day, Shatalov began an intricate manual approach to Soyuz 7, as described by authors Rex Hall and David J. Shayler in *Soyuz: A Universal Spacecraft*.

"The cosmonauts therefore had to rely on visual clues, and there was little chance of success. In addition, Yeliseyev, onboard Soyuz 8, found it difficult to locate Soyuz 7 through the OM [orbital module] portholes against the Earth ... After four small burns of the DPO [approach and attitude control thrusters], Shatalov was still unaware of his range from Soyuz 7 or the orientation of his spacecraft, and was once again forced to abandon the attempt to dock ... consequently, Soyuz 8 and Soyuz 7 passed by each other." [Tass news releases afterwards indicated a closest approach of 500 metres, but did not mention a failed docking attempt] [10].

Two further manual attempts at manual docking by the flight-experienced and highly trained Shatalov and Yeliseyev the following day also failed, much to the frustration of the crews, ground controllers and Soviet space chiefs, and any further docking activities were called off. Even Soviet commentators, generally masters of the cryptic statement, were forced to admit that the *troika* flights had been more of an experimental trial run than an actual attempt to create a rudimentary space station in orbit. Instead, they placed a heavy emphasis on the highly successful "cold" welding experiments carried out onboard Soyuz 6.

The experiment, prepared by scientists at the Paton Institute, had taken place on 16 October. The Vulkan welding outfit was located in the Orbital Module, while the control panel was in the cosmonauts' cabin. On Soyuz 6's 77th orbit Shonin had depressurized the OM, creating a vacuum. Then, using remote control switches, Kubasov turned on the welding unit, initiating three welding processes: low-pressure compressed arc, electron beam and fusible electrode. First of all. he performed some automatic welding using a short arc under low pressure. He then switched on the automatic equipment for welding with an electron beam and a consumable electrode. A special indicator panel on the control board enabled Kubasov to monitor the work of the Vulkan unit, while data on the experiment was recorded in addition to being transmitted to ground control. The resultant samples would later be returned with them to the ground.

A bold journey ends

On 16 October, at 12:52 PM Moscow time, Soyuz 6 touched down onto the frozen, snow-covered barren steppe around 113 miles northwest of Karaganda in the Central Asian Republic of Kazakhstan, having completed 80 orbits of the Earth. When Shonin and Kubasov finally emerged from their Descent Module it was cold and windy, but they didn't really mind as they were back on the good Earth. As well as some local herdsmen, a search party and cameramen were there to greet them, talking excitedly and hugging the cosmonauts as a rescue helicopter landed nearby.

Just 26 minutes less than a full day later, Soyuz 7 also touched down in the midst of wind, sleet and snow flurries, this time 98 miles northwest of Karaganda. Once again, some local people were the first to arrive at the scene, broadly smiling as the three cosmonauts clambered out of their blackened spacecraft, dressed in heavy jackets, long fur-lined boots and leather caps. The ebullient Volkov hugged his comrades and then the ecstatic villagers, while Gorbatko signed a couple of autographs on cigarette packets held out to him. Meanwhile Filipchenko gave the exterior of their craft a thorough inspection as rescue teams began to arrive on the scene [7].

Meanwhile the crew of Soyuz 6, Shonin and Kubasov, were preparing to show Soviet scientists sample metals joined in the Vulkan space welding experiment, described by Tass as "unique". The excited scientists hailed the experiment, in which the molecular structures of metals are broken down and rebuilt, as a major step towards building orbiting space stations. "It is still too early to sum up the results (of the welding tests), but it has already been proven that welding of metals is possible in near-Earth space," wrote welding expert Academician Boris Paton of the Paton Institute in *Pravda* [11].

The following day, at 12:10 PM on 18 October, the final ship touched down in a raging blizzard, right on target and just 90 miles north of Karaganda. All three spacecraft had completed 80 Earth orbits. Exhausted and unshaven, Shatalov and Yeliseyev were nevertheless excited to be back home, and pleased to realize that all three craft had landed so close to each other. Meanwhile, Western observers were left scratching their heads over a triple mission that had begun with so much potential and yet had ended meekly with so few obvious results.

At a later ceremony in the Kremlin's Palace of Congresses, the seven cosmonauts each received high honours, with Shatalov and Yeliseyev receiving a second Hero of the Soviet Union award. The mission may have proved a major disappointment (despite being loudly hailed by the Soviet press as a great achievement), but no fault for this could be laid at the feet of the cosmonauts. Soviet Communist Party chief Leonid Brezhnev would state after the ceremony that, "Our science has come close to the creation of long-term orbital stations and laboratories, the decisive means of broad conquering of cosmic space" [12].

In an interview given the following month to the Communist Party newspaper *Sovietskaya Rossiya*, Gorbatko would also reveal that early on there had been some "minor differences" between the Soyuz 7 crewmembers, although no details were forthcoming. "Compatibility in all aspects can only be fully tested in flight," he observed, brushing away any questions. "Now, after the completion of the programme, I can say that we really did have compatibility. If, at the beginning of the flight there were some minor differences, then at the end we worked very harmoniously and our friendship grew in proportion to the hours we passed in space" [13].

More than a little misbehaviour

After demonstrating so much early promise as a cosmonaut, Soyuz 6 would prove to be the first and only spaceflight for Georgi Shonin. He would later serve in the Salyut 1 training group, and was assigned as the original commander of the Soyuz 10 mission, but it seems that he had earned something of a reputation for excessive partying and drinking. "We had come to the centre to try out our seats," he explained. "It was in December 1970. I arrived late ... and the engineer who was in charge of this try-out told my commander that I was late ... and then I was out. And so they took Shatalov.

"This was my fault," Shonin coyly admitted, "Only mine. I was ... they didn't allow me." When pressed, Shonin admitted that "we were very young men and my best friend was Rafikov. I was like Rafikov." Asked if that meant too many parties, he paused and smiled. "Yes ... maybe" [6]. Shonin was apparently hauled before Kamanin in March 1970 and told that his poor behaviour would not be tolerated, and if further incidents were reported he could face a five-year suspension.

Georgi Shonin in training.

Eleven months later, on 6 February 1971, Shonin was reported to Kamanin for allegedly turning up drunk to a training session, which had to be cancelled, and for consuming vodka during the session. By the time Shonin could be brought before Kamanin he was sober once again. For Kamanin it was a sad time; he had known Shonin for 11 years and had always liked the man, but action had to be taken. He ordered a medical evaluation for the following week, after which the physicians reported that Shonin had probably begun drinking in secret after his Soyuz 6 mission. He would be sent to a sanatorium for rehabilitation [14].

A stellar career comes to an end

Having eventually straightened himself out, Shonin was later assigned "about 1978" to the Almaz programme. "I was commander of [the Almaz space group] and responsible for their training," he mentioned with pride. As well as being the Almaz group's commander, Shonin also formed a provisional crew with Yuri Glazkov and Valeri Makrushin. During final training, however, the Almaz programme was cancelled and the crews dispersed to other projects. For Shonin, it would mean an end to his cosmonaut career after just the one space mission, Soyuz 6.

Georgi Shonin; selected in 1960, he would make only one flight into space.

"I retired from the cosmonaut detachment in October 1977. I was promoted to major-general, so our Air Force chief said, 'You're out; no generals fly in space. I have a lot of captains; they must fly, not generals.' And I went to the Air Force." He would later work on the Soviet shuttle programme. "I didn't work as a cosmonaut, but as a manager. I was commander of the division that was responsible for design and constructing and testing of Buran. Under my command there were a lot of people, especially engineers who built and ground-tested Buran. I was given my last award for my work on Buran; after the test flights in the atmosphere" [6].

Georgi Stepanovich Shonin would die of natural causes on 7 April 1997, aged 61 years.

Unlucky Zaikin

October 1969 would not only prove significant for the mission involving three Soyuz spacecraft flying simultaneously in orbit, but it would also mark the unfulfilled end of the cosmonaut career for yet another of the first group—Dmitri Zaikin.

In his first mission assignment Zaikin had served as replacement backup commander to Pavel Belyayev on the Voskhod 2 mission, after Viktor Gorbatko was removed for medical reasons. Reading from his diary notes, Zaikin gave more details: "The crews for Voskhod 2 were assigned for training on 15 August 1964 ... They were Belyayev–Leonov and Gorbatko–Khrunov. On 5 January 1965, Gorbatko was taken from the crew because he was ill and I was assigned in his place ... We stopped training on 7 March 1965. The flight took place on 18 March 1965 at 11:30 AM. It lasted one day and one night."

Asked in a 1993 interview what he did after the Voskhod 2 mission, Zaikin responded, "I trained like everybody and I was waiting to be assigned in a crew ... like many of us. And I studied at the Academy. Those who had flown already went abroad while I studied. And I gave them my notes so they could graduate as well!" [15].

Three years after serving on the Voskhod 2 backup crew, and having graduated from the Zhukovsky Air Force Engineering Academy, Zaikin was undergoing training for the command of an Almaz military space station mission when a medical examination revealed the devastating news that he was suffering from an ulcer. This immediately disqualified him from the flight, and he was replaced. Bitterly disappointed, he left the cosmonaut corps on 25 October 1969, realizing his dream of flying into space was over. But he would stay on at the training centre in a different role.

"In 1970, I was the deputy commander of the new cosmonaut group. You know; Dzhanibekov, Romanenko, Popov, Berezovoi, Isaulov, Kozlov, Dedkov, Illarionov ... I went with them everywhere, like on survival training. Volynov was the commander, but he was training at the time, so I was their 'nurse'."

"From 1972, I was the main engineer in the simulator department. And later, I became the main engineer of space technology and production on board stations [until 1987]. And now, I work in a laboratory to help the cosmonauts train for their work on board the station ... constructing electrical components."

Like every other cosmonaut, Zaikin (left) would always live in the shadow of the immensely popular Yuri Gagarin.

Asked by Bert Vis if he ever regretted becoming a cosmonaut, but if not, would he do it all again, Zaikin thoughtfully replied: "I think if I would get a chance again . . . I would want to be a cosmonaut again. But now I know all my mistakes and now I could avoid them all. I liked to tell the truth when I was young; now I think I wouldn't do it. I got some problems when telling the truth. I am sorry for the fact that I didn't get any practice as a pilot. I lost it during all those years. And I regret that."

Finally, Zaikin was asked about his opinion of his cosmonaut group, and how he felt about them on reflection. "Each of those who flew became another man. They all promised not to change, but they all did . . . because everyone praised them . . . they began to have some privileges. While we were training together we were friends, but after their flight, many wouldn't even say hello. Not only to me, but to many of us. It's some kind of illness in our society" [15].

NIKOLAYEV FLIES AGAIN

Had veteran cosmonaut Andrian Nikolayev not been onboard, it is unlikely that the flight of Soyuz 9 would have attracted as much media interest as it did. Making his second (and final) flight into space some eight years after his Vostok 3 mission, Nikolayev would be paired with a former spacecraft designer from Korolev's OKB-1 bureau, Vitaly Sevastyanov, who had actually given lectures on spacecraft

Sevastyanov and
Nikolayev would fly
together on Soyuz 9.

designs and systems to the fledgling cosmonauts—including Nikolayev—when they
first commenced their training back in 1960.

The other significant feature of the Soyuz 9 flight, which—as a science mission—
was not fitted with a docking system, lay in the fact that it would be the last mission
before the era of the Soviet space stations began. The following year, in April 1971,
Salyut 1 would be launched atop a Proton rocket, ushering in a whole new beginning
for the Soviet space programme.

An interested bystander to history

The Soyuz 9 crew, according to the Tass news agency, would undertake "medico-
biological research", as well as studying geographical and geological objects
including the snow and ice coverage of the planet, and meteorological research.
The launch took place on schedule at 10:00 PM Moscow time on 1 June 1970 and
would be shown on Moscow television just 35 minutes after it had taken place.

One interested viewer of the launch coverage was none other than Apollo
moonwalker Neil Armstrong, who was paying an official visit to Star City. To his
complete surprise he witnessed the launch on television at a casual function held in
the apartment of Major-General Georgi Beregovoi, who had managed to keep the

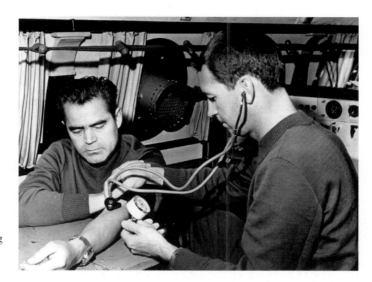

Nikolayev and
Sevastyanov in training
for their long-duration
mission.

launch a well-guarded secret from his guest until it had actually taken place. "This is especially in honour of your visit here," he would reveal as the men watched the lift-off, to which Armstrong responded that it was one of the nicest things they could do for him [16].

Further Tass announcements indicated that the spacecraft was not equipped for rendezvous or docking, and the solo flight was one purely devoted to science and medical studies of the crew on a long-duration flight. However, there was no indication as to how long the crew would remain in orbit.

One abnormality later reported was that both cosmonauts apparently suffered a slight deterioration in their vision; their eye muscles would not coordinate properly, and they had difficulty in perceiving colours. However these problems would not prevent them from some of their more crucial tasks. For instance, from their vantage point above the planet they took an optical fix on Lake Viedma, situated high in the Andes of southern Argentina. The flight also provided substantial benefits for ground controllers, who, using new radio navigational equipment, were able to track Soyuz 9 to within nearly a metre of its actual path in orbit.

The two cosmonauts were scheduled to spend two one-hour periods daily performing some of exercises designed to minimize any degeneration of their muscles, bones and circulatory system. With so little room available in the cramped spacecraft, these exercises were carried out using elastic torso straps, harnesses and simple chest expanders, while they kept check on their heart rates, blood pressure, respiration and other vital physiological functions. But they grew increasingly lax about continuing this regime, despite the urging of doctors. "We have no time to do all the exercises," Nikolayev finally complained, "because we are busy with all the experiments for the programme." The subsequent admonitions of the physicians would unfortunately have little effect on the crew's attitude. [17].

The flight continues

The sixth birthday of Nikolayev's daughter Elena fell on 8 June. That day his wife Valentina was given special permission to bring their daughter into the ground control centre so she could receive happy birthday greetings from her father by radio, and see him on a live television monitor.

On the tenth day of their mission the two men enjoyed the first of their prescribed days without duties, with no experiments scheduled and radio communications kept to a minimum. Both keen chess players, they had earlier challenged Nikolai Kamanin and CapCom Viktor Gorbatko to a game. After a lengthy game played over four communication periods and three orbits, the long-distance match ended in a draw.

A few days later it was noticed that the crew seemed to be quite fatigued and were making uncharacteristic and elementary errors, to the point where Sevastyanov accidentally engaged the craft's ASP automatic landing system. It was not a critical error, and one easily rectified, but to the doctors it indicated increasing inattentiveness by the crew.

The two men would eventually fly into their third week in space, breaking the previous endurance record of 13 days, 18 hours, 35 minutes and one second set by astronauts James Lovell and Frank Borman aboard their Gemini VII spacecraft in 1965, both of whom would graciously send congratulations to their cosmonaut counterparts.

Nikolayev and Sevastyanov would finally land in a triumphant televised touchdown after nearly 18 days in space, setting down in a field some 45 miles west of Karaganda. Rescue teams in four helicopters were already in the vicinity, landing soon after the spacecraft.

The two men had set a new space endurance record of 7 days, 16 hours and 59 minutes in flight, completing 286 orbits, and physicians were anxious to see how long-duration weightless spaceflight might have affected them. As they sat patiently in their couches the cosmonauts told recovery personnel they were feeling extraordinarily heavy, and it was decided to carry them once they had climbed out of the spacecraft and transport the cosmonauts to the medical facility at Star City for extensive check-ups and post-flight recuperation. On hand was veteran parachutist–physician Vitaly Volovich, who had aided in the recovery of several of the early cosmonauts, including Nikolayev after his Vostok 4 mission in 1962.

"Everything was alright. The cosmonauts were fine and had already begun to pack their space gear. Doctors leaned over the edge of the hatch, but their assistance was not needed.

Finally Andrian Nikolayev climbed out of the hatch. I carefully studied his face. He looked much the same as after his first flight, only he was a little pale and his face had a pinched look. We carefully helped him out of the compartment. Feeling the earth beneath his feet, he gestured away those supporting him. But then he began to sway.

"My legs don't seem to be supporting me too well. They feel cottony, like they aren't even mine. I'd better lie down." And Nikolayev lay down on a stretcher, breathing heavily.

I sat down next to Nikolayev to check his pulse. But Andrian sat up suddenly, moved my hand away and began to count the beats of his heart. His face grew even paler, but he counted till the end. Lying back on the cushion he said in a low voice: "One hundred twenty. Too fast." He paused, then added, "I feel dizzy." [18].

"Completely emaciated, sick people"

Nikolai Kamanin was also at the recovery site, and he was shocked at the men's appearance as they were carried aboard one of the helicopters. "When I entered the aircraft's cabin, Sevastyanov was sitting on the sofa, while Nikolayev was at a small table. I knew they were having a hard time enduring the return to the ground, but I had not counted on seeing them in such a sorry state. Pale, puffy, apathetic, without the spark of vitality in their eyes—they gave the impression of completely emaciated, sick people" [19]. Flying the cosmonauts over to Star City was postponed until further tests had been carried out at Karaganda. Following this, they would be carefully driven by limousine to the cosmonaut training centre in Moscow.

"For lack of muscle, we couldn't walk for seven days," Sevastyanov recalled in 2001. "Our thigh girths had shrunk by five-and-a-half centimetres. Cardiograms were suggestive of something very close to a severe attack. Actual lesions, however, could not be detected. The hearts had simply adapted to weightlessness and took normal gravity as [an] abnormality and a serious challenge. We had returned to Mother Earth as seasoned space dwellers who had gone a long way to adapt to the conditions of outer space. Normal health did not start to return before our second week back home" [20].

Nikolayev also had problems readapting to gravity. "Orbital stations were unheard of, and our mission was supposed to test uncharted waters ahead of the first such station, the Soviet Salyut. Cramped conditions aboard our Soyuz craft ruled out appropriate exercise, spelling quick physical degeneration in every respect. My heart lost twelve percent of its initial volume over our eighteen days in flight."

"The first day back home was quite an ordeal," Nikolayev added. "A few days later, however, Sevastyanov offered me to share a cigarette with him, in a washroom where he had hidden it during medical check-ups before going up. We had to lean against the wall crawling to our not-so-far-away destination. I smoked first, as mission commander, and then passed the half-smoked cigarette to Vitaly. At this moment our doctor came in. And instead of scolding us, which we certainly well deserved, he burst into jubilation. He told us he felt assured of our quick recovery seeing us smoking" [20].

Professor Anatoli Grigoryev, a full member of the Russian Academy of Sciences and Director of the Institute of Medical and Biological Problems in Moscow would later state, "I saw both men shortly after their flight. That 1970 mission was a bold venture into the unknown. The Soyuz craft offered too little room inside, severely restricting exercise. For bigger spaceships that followed, my Institute developed in-cabin training machines. It also came up with techniques for cosmonauts to keep fit through exercise" [20].

The debilitated condition of the cosmonauts, which came to be known as the Nikolayev Effect, clearly demonstrated for future space station crews a vital need to establish suitable exercise regimes.

Following two weeks spent in recuperation, the Soyuz 9 cosmonauts finally attended a formal reception at the Kremlin's Georgiyevsky Hall. Sevastyanov would receive his first gold star of Hero of the Soviet Union and Nikolayev his second. To his delight, Nikolayev was also promoted to the rank of major general.

Andrian Nikolayev never flew into space again, but Vitaly Sevastyanov would eventually gain a second mission when paired with Lt. Colonel Pyotr Klimuk on Soyuz 18.

PRISONERS OF THE EARTH?

A little over a decade after the pioneering flight of Yuri Gagarin, the Soviet Union hoisted the first of its space stations, Salyut 1, into orbit on 19 April 1971. Four days later a highly experienced crew of Vladimir Shatalov, Alexei Yeliseyev and Nikolai Rukavishnikov was launched on a mission to link up with and occupy the station. While the rendezvous and docking took place as planned, a mechanical fault in the docking tunnel meant that a hard dock and cosmonaut transfer could not be achieved. Despite Shatalov's best efforts the frustrated crew was eventually ordered to undock and return to Earth after less than two days in space.

Death in space

On the following flight the Soviet Union would suffer its worst ever disaster in space. The Soyuz 11 crew, launched on 6 June 1971, had finally achieved a successful link-up with Salyut 1 and carried out a record-breaking three-week occupancy of the station. Then, following a routine undocking from the station, the three cosmonauts perished during re-entry when all the air bled from their spacecraft due to a faulty valve.

On all Soyuz manned flights beyond Soyuz 1 the crews had not been provided with protective spacesuits and helmets as a simple but potentially hazardous means of saving precious space inside the crew compartments. However, because they were not provided with these lifesaving suits, the Soviet Union lost three cosmonauts—Georgi Dobrovolsky, Viktor Patsayev and Vladislav Volkov. The men were found dead inside their Descent Module after a planned automatic landing by parachute.

Of immediate concern to space analysts in the West, and particularly for NASA, was the fact that three men had perished in mysterious circumstances after nearly 24 days in space, and the Soviet space chiefs were keeping quiet about the cause of the tragedy. Immense political pressure was applied to the Soviet Union, as NASA was also planning long-duration space missions and had to ensure that the human physiology could stand up to the crushing weight of re-entry after prolonged weight-

The original crew of
Soyuz 11 was (from
left) Valery Kubasov,
Alexei Leonov and
Pyotr Kolodin.

lessness. Otherwise, it was feared that humans might become prisoners on their own planet for decades to come. Eventually, buckling under this extreme international pressure, the Soviet Union released its findings on the Soyuz 11 tragedy. It had been a simple mechanical failure, they stressed, and nothing to do with lazy hearts or other physical unknowns.

The deaths of the three cosmonauts caused an immediate hiatus in the Soviet space programme. It would be nearly two and a half years before Soyuz 12, incorporating new features and safety measures, was finally launched into orbit on 27 September 1973 carrying cosmonauts Vasili Lazarev and Oleg Makarov on a two-day test flight. Most notably, only two cosmonauts would now be accommodated on each mission, while spacesuits and helmets had become mandatory safety items, to be worn without exception during launch and re-entry.

Originally, Alexei Leonov had been appointed commander of the Soyuz 11 crew along with Valery Kubasov and Pyotr Kolodin, but Kubasov had developed a spot on his lungs as a result of insecticide sprayed on trees in Star City. Initially it was thought by doctors that this might be infectious tuberculosis, resulting in the crew being replaced in its entirety by Dobrovolsky, Patsayev and Volkov just 11 hours before the launch. Although Dobrovolsky's was a lesser trained crew, it was decided (despite Leonov's strong protests) to keep that training crew together, rather than Leonov's preferred option of replacing Kubasov with his positional backup, Volkov.

Leonov is adamant that his better trained prime crew would have known how to remedy the faulty valve once it occurred, and the disaster could have been prevented. "The loss of the Soyuz 11 cosmonauts was a terrible blow to the morale of the whole corps," he observed in his memoirs. "Everyone understood that we were in the business of testing spacecraft, and the deaths of these three men undoubtedly saved the lives of subsequent crews, because of the substantial modifications made, but their loss was a tragedy. Not only was I deeply saddened by what had happened, but I was

The crew who would fly the ill-fated Soyuz 11. From left, Vladislav Volkov, Georgi Dobrovolsky and Viktor Patsayev.

frustrated, too. Had I been allowed to fly in their place I am sure my crew would have survived" [21].

POPOVICH RIDES AGAIN

His one-time Vostok 3 partner in space Andrian Nikolayev had flown into space a second time in 1970, and the second of the so-called "Heavenly Twins", Pavel Popovich, would make his own second flight aboard Soyuz 14, some 12 years

The Soyuz 14 crew; Pavel Popovich and Yuri Artyukhin.

after he had ridden Vostok 4 into the skies and into the history books. Following that historic tandem flight, Popovich had taught at the Zhukovsky Air Force Engineering Academy, training future cosmonauts.

On 3 July 1974, Moscow radio announced that the two-man Soyuz 14 spacecraft had been launched at 9:51 PM, with a crew of Pavel Popovich and flight engineer Captain Yuri Artyukhin. For Artyukhin this would be his first flight into space, having joined the cosmonaut team a year after Popovich's 1963 mission. With Salyut 1 abandoned and deorbited after the Soyuz 11 tragedy, and the breaking up of the never-occupied Salyut 2 in orbit, Soyuz 14 would rendezvous and dock with the Salyut 3 space station, lofted into space atop a giant Proton booster just nine days earlier.

Jack Reilly, a spokesman for the nine American astronauts then in training for the ASTP mission in Moscow, said they had been given no advance notice of the launch, but were notified the following morning that a link-up with the station had been achieved. Meanwhile Tass announced that the crew of Soyuz 14 would conduct "joint experiments with the orbital scientific station Salyut 3" as well as making "a comprehensive check-up of the improved on-board systems of the Soyuz ship," which would be used in the joint Soviet–American spaceflight the following year [22]. As both cosmonauts were high-ranking officers in the Soviet Air Force, it was rumoured that some of their activities would involve covert reconnaissance. This would prove to be correct, as Salyut 3 was actually the second top-secret Almaz space station. Without knowing this, and because so little was reported about their activities, the latest flight did not attract much media attention in the West, which was now becoming inured to Soviet activities aboard their space stations.

Taking no chances

The flight of Soyuz 14 may not have attracted too much attention, but as *Newsweek* magazine reported later that week, the Russians were taking no chances of anything going wrong.

> "After failing at least three times in attempts to complete a Skylab-type orbital mission, the Russians were not about to take any unnecessary risks in their latest effort. As cosmonauts Pavel Popovich and Yuri Artyukhin, both 44, whirled around the earth aboard their Salyut 3 space station, ground control sternly refused to let them listen to the semi-final match between Poland and Brazil in the World Cup championship. The excitement, the controllers feared, might stir up the cosmonauts' pulse beats and blood pressure. But after a while, Soccer Nut Popovich could bear the suspense no longer. "How did they play? What's the score?" he demanded. Told that the Soviet Union's East bloc allies had eked out a 1–0 victory, Popovich exulted: "Bravo! Good guys those Poles!" [23].

Popovich and Artyukhin spent a total of 14 days aboard the Almaz station, at the end of which they would make a successful landing southeast of Dzhezkazgan in Kazakhstan after a total of 15 days and $17\frac{1}{2}$ hours in space. There was much jubilation

Soyuz 14 would be the last spaceflight for veteran cosmonaut Pavel Popovich.

for the Soviets, as it had proved to be the first wholly successful space station mission, coming almost three years after the loss of the Soyuz 11 crew. It had not been a lengthy mission, but the flight of Soyuz 14 would begin a period in which the Soviet Union began to take the lead in gaining long-duration space experience.

Soyuz 14 would also mark the final spaceflight for Pavel Popovich. He then undertook a number of senior positions at the cosmonaut training centre, before finally retiring in 1990. He now resides in Moscow.

HANDSHAKE IN SPACE

A month before the Apollo–Soyuz Test Programme (ASTP) joint spaceflight was to take place, leading U.S. Senator William Proxmire called for the cancellation of the entire project. A long-time critic of the $225 million price tag for the so-called mission of "détente," the senator said that he feared for the safety of the three American astronauts involved in the historic flight because, he said, "the Soviet record shows they have experienced failures in launch, rendezvous, docking and re-entry phases of spaceflight. The danger," he added, "is significant." Senator Proxmire's qualms would fortunately fall on deaf ears as training for the mission continued [24].

Training for togetherness

The ASTP mission had begun some years earlier in May 1972 when, climaxing more than 18 months of technical discussions between the space agencies of both nations, President Richard Nixon and Soviet leader Leonid Brezhnev had agreed at a Moscow summit to send American and Soviet spacemen on a historic joint Earth-orbiting flight in 1975.

Marking an important milestone in international cooperation in space exploration, Soviet cosmonauts had visited the Johnson Space Center in Houston for two weeks of initial briefings by space officials on the Apollo spacecraft and its systems. The two prime cosmonauts for the mission were both space veterans: Alexei Leonov, who, in 1965, had become the first person to walk in space, and flight engineer Valery Kubasov, who had earlier conducted welding experiments aboard Soyuz 6.

In a reciprocal move, nine U.S. astronauts later travelled to Star City, including the prime crew of Tom Stafford, Vance Brand and Mercury astronaut Donald "Deke" Slayton. A determined Slayton would be making his first spaceflight after being grounded with a minor heart problem in March 1962, only two months before he had been scheduled to fly the mission that would eventually be flown by his

Astronaut Tom Stafford (second from right in light suit) is shown around a Soyuz exhibition during a visit to Star City. An unknown Soviet official is at left, then Nikolayev, Shatalov, Stafford and Volynov.

Mercury colleague, Scott Carpenter. "I'd rather be a 50-year-old rookie than a 50-year-old has-been," he happily observed after being selected to fly the ASTP mission.

The Russians were initially reluctant to open up their secret base to foreigners, but their hand had been forced by Stafford, who had insisted he wouldn't fly unless he could inspect an operational Soyuz before lift-off. "I never fly in a spacecraft that I haven't been in on the ground," he insisted, and his argument won the day [25].

The two Russian and three U.S. crewmembers would each have to learn the others' language as well as training in both spacecraft, with Leonov laughingly admitting that learning English wasn't too bad in itself, but comprehending Tom Stafford's broad Oklahoma accent was something else again. Fortunately the five men got on exceedingly well, and after two years of teams of scientists, astronauts and cosmonauts shuttling between the United States and the Soviet Union, exchanging space technology and mutually developing new equipment, a true spirit of camaraderie had emerged.

Once both craft were in orbit they were scheduled to link up for two days, utilizing a common docking system that was the product of a joint design effort between the two nations. During this time the crew would transfer back and forth between their respective spacecraft, sharing meals and working on experiments. Also planned for their 9-day or 10-day mission, the U.S. astronauts hoped to conduct 27

NASA's crew for the ASTP mission comprised (from left) Deke Slayton, Tom Stafford and Vance Brand.

scientific experiments—5 in cooperation with the cosmonauts. For their part, the Soyuz 19 craft would be in orbit for 6 days.

In a mission aimed squarely at détente, several rituals would also be carried out for the benefit of the TV cameras. A symbolic "handshake in space" would take place once the hatches between the two craft had been opened, then flags and pine tree seeds were to be exchanged, halved medallions and plaques joined together, and certificates of achievement signed.

Meals would be shared, but at all times at least one crewmember from each nation would remain in his own spacecraft in the event an unexpected emergency separation became necessary. In extreme circumstances all five men could return to Earth aboard the much larger Apollo spacecraft.

Questions were certainly raised as to the outcome of the venture, and just how much the Soviet Union would benefit from technological exchanges. Apart from the common docking system, however, the two nations would use their own equipment, as the project was structured to keep any crucial exchanges of technical information and data to a minimum. Nevertheless, there was still an undeniable aura of distrust, secrecy and distortion going on being the scenes.

The Soviet ASTP
crew—Alexei Leonov
and Valery Kubasov—
prior to the launch of
Soyuz 19.

Détente in orbit

On 15 July 1975, Soyuz 19 lifted off from the Baikonur cosmodrome at 3:20 PM Moscow time, carrying Leonov and Kubasov into orbit. For the first time ever, a manned Soviet launch was shown live to television audiences around the world. Against a stark desert landscape, the red-banded rocket rose rapidly into a clear blue sky, building up thrust and accelerating as it hurtled toward orbit. Meanwhile, 10,000 miles away from Moscow, its Apollo counterpart was being prepared for a lift-off from the Kennedy Space Center in Florida just $7\frac{1}{2}$ hours later.

There was a great deal of contrasts in the two launches as the Saturn 1B lifted off in a spectacular and powerful display, easily overshadowing the Soyuz launch. Strapped down until it achieved full thrust, the Saturn V trembled and thundered on its pad before lifting slowly and majestically into the skies. Once both craft were successfully in orbit the Apollo crew would take the initiative, skilfully guiding their

Soyuz 19 lifts off to begin the much-heralded ASTP mission.

craft to a rendezvous with the passive Soviet Soyuz craft 140 miles above what was then West Germany.

Three hours after a perfect link-up between the two spacecraft, and precisely on schedule, astronaut Tom Stafford and cosmonaut Alexei Leonov greeted each other with handshakes, bear hugs and huge grins.

"Glad to see you," said Stafford.

"Very, very happy to see you," responded Leonov.

After the formality of receiving congratulatory messages from Leonid Brezhnev and Gerald Ford, and with television images beamed down to the ground, the men exchanged gifts of flags. The walls of the Soyuz craft were festooned with welcome signs and portraits of the five crewmembers drawn by Leonov, a talented artist. While Brand remained in the Apollo spacecraft, Stafford, Slayton, Leonov and Kubasov gathered around a green metal table in the Soyuz for a meal. The two cosmonauts, acting as hosts, unpacked a meal packaged in tubes, cans and plastic bottles, anchoring them to the table with rubber bands. The meal consisted of cans of cheese and turkey in jelly, tubes of cranberry dressing, green borscht, apple juice and strawberries, as well as plastic bags with what Stafford described as "little bitty . . .

Détente in orbit:
Stafford and
Slayton pose
with Leonov
aboard the
Apollo
spacecraft.

miniature loaves of Russian bread" [26]. To symbolically celebrate the occasion Leonov even produced squeezable tubes that were cheekily labelled "vodka", but which in reality held nothing more potent than borscht soup.

Over the two days of the link-up there would be several crew visits between the two spacecraft before final goodbyes and the undocking procedure.

Coming home

The Apollo crew would remain in orbit for another three days, but for the Soviet crew their mission was virtually at an end. Following re-entry, they would land gently and triumphantly amid an enormous billowing cloud of dust thrown up moments earlier by their retro-rockets, touching down in the central Asian wheat belt within six miles of the targeted landing site and just 310 miles from where they had been launched. Television cameras mounted on the recovery helicopters beamed live pictures of the landing and subsequent events around the world, as Leonov and Kubasov emerged smiling from their scorched, tilted spacecraft.

Obviously tired from their six-day flight, Leonov staggered slightly when he emerged, saying, "It was difficult, very difficult. We are a bit shaky due to tiredness and happiness" [27].

The Apollo crewmembers were sound asleep when Leonov and Kubasov touched down, but they later transmitted congratulations to their cosmonaut colleagues. The three astronauts would return home at the end of their 138th orbit, although some nitrogen tetroxide rocket propellant escaped into their cabin through a pressure relief valve as the spacecraft plunged to a Pacific splashdown. The highly toxic, corrosive fumes made the crew cough and gag, as well as causing burning and irritation to their eyes.

The Apollo craft slammed down hard into the water, "like a ton of bricks," as Stafford would later describe the impact [28]. By this time Brand had actually passed out from the effects of the toxic fumes. The three men would eventually make a full recovery, although tests would reveal some damage to their lungs from exposure to the poisonous gas.

The Soyuz 19 mission would bring to an illustrious end the spaceflight career of Alexei Leonov. The much-revered cosmonaut had not only been the first person to conduct an EVA in space, but if fate had decreed otherwise he might even have flown as commander of the ill-fated Soyuz 11 mission. Had the Soviet L-1 programme gone ahead, Leonov would likely have been in charge of the first lunar crew, flying in the compact LK landing craft. However, he would cap his magnificent career with the highly successful ASTP mission as commander of Soyuz 19.

Promoted to the rank of major-general, Leonov later became Deputy Director at the cosmonaut training centre and the commander of the cosmonaut team. He stood down from the cosmonaut team on 26 January 1982, and then became the First Deputy Director at the training centre, retiring from the position on 12 September 1991. Retiring from the Air Force the following year he decided to take up a position with Alfa Bank, now the largest private commercial bank in Russia.

"I might have continued my military service, but the political situation was so difficult that I didn't know what to do," Leonov recalled in 2005. "I was a fighter

pilot, but it was too late for me to go to the civilian airlines. There were no places for me to teach, and the pay was so low anyway. I met with a friend who worked with Sergei Korolev ... He knew about my retirement and that I had been a member of the State Commission of the Military–Industrial Complex, so he knew that I understood well industry and economics. Russia's new economy was beginning to form at that time. Privatization had started. He proposed that I join the Alfa Group and lead the Alfa Capital Fund as its president. I worked there until 1998.

"Alfa Bank became a commercial bank and there were reorganisations; a number of highly talented and clever staff that worked with me were let go. I told them that I would not leave them; we would go on together. We created Baring Vostok Capital Partners with some foreign partners, which has brought together very clever and intelligent people, an absolutely wonderful

A recent photo portrait of Alexei Leonov. (Courtesy Francis French)

A mock-up of the LK (*Lunniy Korabl*) spacecraft, designed to land a single cosmonaut on the Moon before the Americans.

group, to create Russia's leading private equity organization. I became executive director from the Russian side. We continue to work with Alfa Bank, where I remain the vice-president and do various projects ... The only problem is that I live in Star City and spend four hours each day for my trip to work and back" [29].

To his delight, Leonov would also receive vicarious recognition in the field of science fiction. In Arthur C. Clarke's 1982 book *2010: Odyssey Two* (released theatrically two years later as the movie *2010: The Year We Make Contact*) one of the spaceships was named *Cosmonaut Alexei Leonov*. Similarly, in the 1998 novel *Star Trek: Avenger*, written by William Shatner, the USS *Leonov* (registry NCC-648) was a Gagarin-class science scout affiliated with the Federation Starfleet during the last quarter of the 23rd century.

In a 1999 interview conducted with Reuters news agency, a pensive Leonov would reflect on his nation's failed top-secret plans to fly cosmonauts to the Moon ahead of the United States. Soviet rocket scientists and designers have long and reluctantly agreed they were not even close to landing a cosmonaut on the Moon, as they were seriously lacking in both finances and a sure-fire method for getting a man from lunar orbit onto the surface and back again. They had been close to flying a man around the Moon, but that race was lost when an audacious move by NASA saw three astronauts sweep around the Moon on the Apollo 8 mission in December 1968. But Leonov is adamant the Soviet Moon effort never recovered from the loss of Sergei Korolev in 1966.

"Some people today say there wasn't enough money," he stated. "Nothing of the kind. We had the money but we only needed to spend it properly. Mishin says the Defence Ministry didn't give us money. This is not true. We did not properly analyse things ... This was his mistake."

Mishin's response to Leonov's statement was dismissively terse: "Leonov is a mouse. He doesn't understand anything."

One of Leonov's colleagues who also trained for a lunar flight was Vitaly Sevastyanov, and like Leonov he is often wistful when looking up at the Moon. "I don't allow myself to say perhaps I could have landed on the moon," the former cosmonaut observed. "That couldn't have happened, but perhaps I could have flown around the moon. But it didn't work out. Of course there is certain regret" [30].

While Leonov and Sevastyanov were sadly reminiscent about the termination of the once-viable lunar programme, another of the pilots involved, Pavel Popovich, was typically jocular in describing an incident associated with one particular unmanned mission:

"In the late 1960s we were getting ready for a flight around the moon. At the time we sent to the moon the so-called probes, the very same Soyuz spacecraft, but with no crews in them. Each one of such probes was to fly around the moon and return to Mother Earth. A major problem was for the probes to land. Of all probes only one landed safely. When we realised we would never make it to the moon, we decided to engage in a little bit of hooliganism. We asked our engineers to link the on-the-probe receiver to the transmitter with a jumper wire. Moon flight missions were then controlled from a command centre in Yevpatoria, in the Crimea.

When the probe [designated Zond 5, September 1969] was on its path round the moon, I was at the centre. So I took the mike and said: "The flight is proceeding according to normal; we're approaching the surface ...." Seconds later my report—as if from outer space—was received on Earth, including the Americans. The U.S. space advisor Frank Borman got a phone call from President Nixon, who asked: "Why is Popovich reporting from the moon?" My joke caused a real turmoil. In about a month's time Frank came to the USSR, and I was instructed to meet him at the airport. Hardly had he walked out of his plane when he shook his fist at me and said: "Hey, you, space hooligan!" [31].

VOLYNOV AND A WHOLE NEW DANGER

Soyuz 20 was launched into space on 17 November 1975, but on this occasion the spacecraft was unmanned. It was designed to conduct a long-duration test of the Soyuz vehicle and improved onboard systems, as well as carrying a biological payload. The spacecraft would remain in orbit until it was safely recovered on 16 February 1976, having completed a total of 1,470 orbits.

Soyuz 21 in space

It would be nearly eight months before the launch of Soyuz 21, carrying a two-man crew of Boris Volynov and flight engineer Vitaly Zholobov—the first Soyuz crew since the highly-successful ASTP mission almost a year earlier. On this flight, launched 6 July 1976, the crew was headed for a series of what Tass referred to as "joint exercises" with the Salyut 5 space station, which had been placed in orbit just two weeks earlier. It was, in fact, yet another name-disguised Almaz military station that they would man for a period later revealed by Volynov to be in the vicinity of 60 days.

Unaware of the military application of what would prove to be the last dedicated Salyut/Almaz station, neither Soviet commentators nor Western space analysts saw this latest manned mission as anything more than a relatively routine continuation of Soviet space experiments. "We have not set any basically new engineering problems for the spaceship," said veteran cosmonaut Lt. General Vladimir Shatalov of the flight. "It is just another working start under our orbital programme" [32].

Mission commander Boris Volynov was making his second and final flight into space after being selected into the first cosmonaut team back in 1960. Although he had eventually flown aboard Soyuz 5, his 16 years as a cosmonaut had seen him relegated to the backup role for several missions, so it was certain he enjoyed being part of a prime crew for a change. Meanwhile for Vitaly Zholobov, who had been selected as a cosmonaut candidate in 1963, Soyuz 21 would prove to be his first and,

The Soyuz 21 crew of Boris Volynov and Vitaly Zholobov.

Volynov and
Zholobov in
training for their
mission, which
would end under
dire circumstances.

like Volynov, last space mission. At the time of the flight, Alexei Leonov would
describe the Soyuz 21 crew as "perfectly trained for the mission, and furthermore very
good athletes. Both cosmonauts are on volleyball and basketball teams and are very
good soccer players" [32].

A home in orbit

After docking with Salyut 5, the two cosmonauts spent a further five hours
checking systems and preparing to board the space station for what was expected
to be a lengthy stay. After they had successfully transferred, fellow cosmonaut
Alexei Yeliseyev told a television reporter the two men had initially carried out a
visual check of Salyut and reported that everything onboard was working normally.
"Now the basic work of the crew will be aboard the station," he added [33].

As well as using special cameras to record strategically important sites and military exercises being conducted in Siberia, the two cosmonauts would perform a number of medical, biological and technical assignments. It was, despite the proposed length of the mission, an onerous timetable to which the cosmonauts had to work, and unmistakable signs of stress began to impact on their health after weeks of working 16-hour days. This was close to twice the workload they should have been given, resulting in the men neglecting their planned exercise regime. It seemed the earlier lessons of Soyuz 9 were still being neglected. Worse still, there was a distinct personality clash, and the two men were constantly bickering and using hostile language toward each other.

Adding to the problems, Zholobov had developed worrying signs of claustrophobia and seemed clearly disorientated, and onboard medication did little to calm him down. Things got so bad for the steadily weakening Zholobov that Volynov finally had to do the unthinkable—he requested that the mission be cut short in order to get his companion back to the ground for treatment. After several meetings by medical staff analysing the situation, permission was finally, albeit reluctantly, given.

"There was great stress," Volynov said during an interview at Rex Hall's home during a visit to London in 2001. "After that stress, from a medical point of view, Vitaly had medical disorders. That's why there were a few medical checks. Doctors were sitting on Earth. We put sensors on ourselves and sent telemetry on ourselves. There were a few meetings by the doctors , after analysing the data through the radio connection. And there was the decision not to complete the flight, because that would be dangerous to Zholobov's health. It doesn't depend on whether or not you wish to land. On the transportation ship there is a braking when you go through the dense layers of the atmosphere. There are overloads ... 3 or 4 Gs. A man must have a healthy heart. During the overloads, the pulse goes up and the blood pressure increases. So there was nothing we could do than return" [2].

Zholobov was by now in such a weakened state that he was of very little assistance to Volynov as he prepared for the undocking and return to the ground. "I was working alone; he wouldn't help me," Volynov would recall. "I was to load the transportation; I was to prepare to fly in automatic mode. There were lots of jobs to be done; gather all the results of the missions, place them on board the transportation vehicle, and make everything ready for the return. That's why before landing I practically didn't have any sleep" [2]. By now Zholobov was so confused that Volynov had to strap him into his couch before climbing aboard for the undocking procedure.

Hurried return

The undocking itself would not go well. As Volynov attempted to disengage from Salyut 5 a latch failed to release correctly, which meant that as he fired the manoeuvring jets to back away from the station the docking gear jammed, resulting in the Soyuz craft being undocked but still attached to Salyut 5. A second attempt managed to loosen the latches slightly, but they were still in dire trouble. It took another full orbit and some instructions from the ground before the latches finally released, allowing Volynov to back away from the station.

Soyuz 21 would successfully re-enter, but as it descended for a parachute landing near the town of Kokchetav on the Asian steppes of Soviet Kazakhstan, strong, gusting winds caused an uneven firing of the soft-landing rockets. As a result the spacecraft impacted hard with the ground, literally bouncing across the ground until it came to a standstill, tipped on its side.

In his weakened state Volynov found it difficult to unbuckle his harness and open the spacecraft hatch, and when he had accomplished this he crawled out onto the soft ground, where he found he could not stand unaided. After considerable effort he managed to release Zholobov, whose helmet had jammed on an obstruction inside the craft, and dragged him out onto the ground.

Landings in darkness were avoided if at all possible by Soviet space controllers, as they tended to hamper the recovery operations and rescue crews, but the emergency return situation meant that it would take 40 minutes after touchdown before helicopters could reach the two weakened cosmonauts.

"For three days and nights we practically didn't walk at all," Volynov said of their post-flight problems [2]. He had lost seven kilograms in weight, and said that they had both suffered from lack of sleep and exercise during the aborted space station mission.

As expected, the Tass news agency declared that the entire programme of the mission had been successfully carried out and the general condition of the cosmonauts was "satisfactory", but other reports described the two participants as suffering from "sensory deprivation". It had also been noticed in an article appearing in the Russian daily newspaper *Izvestia* just a week before the mission was abruptly terminated that psychologists monitoring the well-being of the crew had urged ground control to allow music to be played to the crew to ease the effects of prolonged isolation. It was quite evident to Western scrutineers that a potentially serious problem involving the crew had caused a premature end to what was expected to be a longer duration mission.

In October 1999 author Rex Hall and Bert Vis interviewed Dr. Rostislav Bogdashevsky, who had been the head of psychological training of cosmonaut crews. He stated that the psychological aspect of this troubled flight was actually secondary to what he termed "something else". When pressed, Bogdashevsky said:

"Primarily, there were still medical problems. It was a mistake made during the selection. It turned out that one of the crewmembers [obviously referring to Zholobov] had some unpleasant experiences here on Earth. He had been involved in a car accident and even went unconscious for a while. Nobody knew that. And there were some technical reasons. Besides all that, they had a very bad work regime. They were to work at night and to sleep during the day. They were told to do a task and then to do the opposite thing. They were desynchronised.

On top of that, there were a number of technical malfunctions. The crew was exhausted, nervous. Psychologically as well. Finally, we decided that nobody could be responsible for their actions during the flight and at that time . . . I talked personally to the head of the State Commission, [Mikhail] Grigoryev. He was the first deputy of the Rocket Forces. And I told him that I could not give any guarantees with respect to the actions of the crew. Anything could happen at any time and I suggested an immediate landing. I don't think anyone could stand such a flight. It was a terrible experiment . . . The programme was very complex and the techniques were new.

It was not a humane flight . . . it was an inhumane mission. Being a physician, I could imagine what it must have been like. I called that flight an experiment of the nerves. We have a famous doctor, Ivan Petrovich Pavlov, and he had [an assistant] Petrova. She was carrying out experiments with dogs, and she created similar conditions for the dogs. Every time the results were the same: the nervous systems were ruined completely. She took different kinds of dogs, with different nervous systems. Strong systems, weak systems . . . all of them had nervous breakdowns" [34].

Whether as a result of their condition or a wrongly perceived laxity in maintaining their exercise regime aboard the station, neither man would fly into space again. Volynov would later work as a controller for Salyut 6 and Salyut 7 and helped train two Mongolian cosmonauts for a 1981 Interkosmos mission. In November 1982 he was given the command of the cosmonaut detachment, a post he would hold until his retirement on 17 March 1990, as a colonel in the air force reserves.

THIRTEEN YEARS BETWEEN FLIGHTS

By mid-1976, all 12 of the first cosmonaut group destined to make flights had done so—several of them twice. The only cosmonauts from that group who would fly into space again were Bykovsky and Gorbatko, and both of them would make two more trips into the cosmos over the next four years.

First in a series

Soyuz 22 would mark the return of Valery Bykovsky to space, after an absence of 13 years. Then aged 42, he and civilian flight engineer Vladimir Aksyonov, 41, lifted off from the Baikonur launch site on 15 September 1976 in a spacecraft originally built as a backup vehicle for the ASTP mission the year before. The government newspaper *Izvestia* reported at the time that its androgynous docking collar would play no part in this mission and had been removed. The flight was also reported to be the first of a series of space projects that would one day involve crewmembers from nine socialist countries.

First indications were that the flight would be of short duration, and that the cosmonauts would not attempt to link up with the orbiting space station Salyut 5, hurriedly vacated only 23 days earlier by the Soyuz 21 crew of Volynov and Zholobov. Instead, they were on a double-barrelled mission as a precursor to longer duration missions, as well as conducting the Soviet Union's first joint space

Valery Bykovsky (top) would return to space on Soyuz 22, along with Vladimir Aksyonov.

experiment with East Germany. They carried onboard what was described as "multi-zonal photographic equipment" (actually a Karl Zeiss MKF-6 multispectral camera) that had been manufactured in East Germany, and which was designed to enable them to photograph parts of East Germany and Russian territory in "six spectral ranges", as Tass described the equipment's function. The photographs produced could then be used for geological and geographical research that would have economic benefits, Tass reported. Bykovsky would state that by using this equipment they could accomplish in five minutes what would take around two years with conventional photographic survey flights in aircraft.

As well as conducting 24 photographic survey sessions, in which they took 2,400 images covering 30 specific regions, the two cosmonauts also carried out a number of experiments to further research into the conditions of life in space, studying the effects of weightlessness on fish eggs, duckweed and maize seedlings. *Trud*, the Labor newspaper, reported a variation in the seedling experiment aimed at discovering whether the plants' growth in space would be affected by short bursts of artificial gravity. One box containing the seedlings was attached firmly to the spacecraft, thus experiencing all the movements of the Soyuz craft as it was subjected to the normal thumps, bumps, vibrations and rotation of the vehicle. A second box of seedlings was suspended on springs to absorb these shocks, however minor, making the plants subject to truer and more constant absence of gravity. Meanwhile, for the sake of post-mission comparison, a third box of the same seedlings was grown on the ground.

After eight days in orbit the two men returned safely to Earth, parachuting to a soft landing at 10:42 AM Moscow time, in the vast steppes of northern Kazakhstan, at a point 93 miles northwest of the town of Tselinograd. Both men were well and in good condition following their flight.

A SECOND MISSION FOR GORBATKO

On the following mission, Soyuz 23, cosmonauts Vyacheslav Zudov and Valeri Rozhdestvensky were launched toward a rendezvous and link-up with the unoccupied Salyut 5, but a failure in the rendezvous approach electronics saw a premature end to the mission. Unable to dock with the space station, the crew had to make an emergency return to Earth in the Soyuz spacecraft once its orbit had brought it over the recovery zone. Unhappily for the crew, their night landing brought them down squarely into the blizzard-swept Lake Tengiz, in the midst of howling winds. It would make for an incredibly difficult and risky recovery, and the crew would have to wait inside their spacecraft for more than ten hours to be rescued. Zudov was asked to describe that night.

> "It was 23 below zero on that night ... it was foggy, a dense fog. The water was salty enough for it not to freeze, and it was mushy ... with ice crystals in it, and the helicopter pilot couldn't tell where the fog stopped and the water started ... he couldn't land. The amphibious type of rescue ... also. They looked for (us) but they got lost and went in the wrong direction. We were in a very difficult condition. We turned upside-down. The reserve parachute went out into the water; it fell down to the bottom and started pulling us with it. It was very difficult for us. We were sitting there ... we were freezing ... we couldn't get out because the hatch was upside-down. Temperature was very low. If we had come out of the craft and gone to the surface of the water we would have died because it was so cold there: 23 below zero" [35].

Eventually a rescue helicopter crew managed to secure a cable and slowly towed the craft ashore, while other helicopters shone searchlights on the spacecraft to ensure it did not come loose during the dramatic operation.

A repeat mission

It would be up to the backup Soyuz 23 crew of Viktor Gorbatko and flight engineer Yuri Glazkov, both of the Soviet Air Force, to take on a mission the original crew, through no fault of their own, had failed to complete.

The crew of Soyuz 23: Viktor Gorbatko and Yuri Glazkov.

Lift-off of Soyuz 24 took place on 7 February 1977, and this time all went as planned and a successful hard dock with Salyut 5 was achieved on their 18th orbit. "After the approach of the Soyuz 24 ship to the Salyut 5 station, mechanical docking of the vehicles and the link-up of their electrical communications was effected," Tass reported. It added that Gorbatko and Glazkov "continue their programme" and "feel fine" [36].

It would prove to be a comparatively short flight. After 17 days aboard the space station the two cosmonauts began making final preparations for their return to Earth, leaving Salyut 5 to continue in orbit under automatic control. It was said that they had continued the work begun by Soyuz 21 cosmonauts Volynov and Zholobov the year before. They had photographed the Sun and carried out Earth observation and resources photography as well as continuing a study of glacial precipitation begun by the Soyuz 21 crew, cloud tests, and agricultural surveys of the Soviet Union. They also carried out biological and medical studies, successfully conducted some experiments in soldering techniques, but less successfully attempted to cast metals.

Once the two cosmonauts were safely strapped into their Soyuz 24 spacecraft, Gorbatko skilfully undocked from Salyut 5, easing back from the orbiting laboratory. Then, as Glazkov recorded in his co-authored book *A Spaceship in Orbit*, they prepared for their return through the atmosphere.

"We are getting ready for the landing. Soyuz 24 has left the Salyut 5 station and is circling our planet on its own. To set it on its flight back to Earth, it is necessary to diminish the orbital velocity of the spaceship, to make it less than the orbital velocity. For this purpose it becomes necessary to orientate the spaceship for retardation. We watch the Earth's surface through the sighting device, and slowly turn the spaceship. It takes some time for the command from the control handle to reach the engines. It is first taken up by the gyroscopes, and then passed on to the logical blocks, determining the duration of work for the engines and the sequence of the ignition. Only after that the command is given to the valves which feed the fuel to the engines.

I press the key on the panel and there appears the green light signalling that the gyroscopes are ready for work. I give the next command, and the display flashes: MANUAL ORIENTATION. Now, when the control handle is shifted, the signal will be directed to the gyroscopes and then to the engines. The orientation engines on board Soyuz are different. Some of them are lower and others are of higher thrust. The new command is given and the panel flashes: ORIENTATION ENGINES. This means that the low-thrust engines have been selected.

The commander [Gorbatko] shifts the control handle and the panel starts to flash, which is a sign that the orientation engines cut in and off at regular intervals, and the spaceship changes its position. The Earth's "run" in the sighting device changes its direction, approaching the one that is required. In order to brake the spaceship, the nozzle of its correcting-and-braking engine should face onward in the flight direction. It is important for the Earth to "run" in the sighting device not from top to bottom, as in the case of speed-up, but the other way round.

The panel does not flash anymore. The orientation engines are cut off, and now the spaceship rotates by inertia alone. The commander sets the handle to its initial position, the panel starts to flash again, the engines resume their work, now lessening the angular velocity. The spaceship's rotation is being retarded and will soon stop altogether. The required position has been achieved. We can fire the main engine for braking. This will be done exactly at a given time, so that the spaceship could touch down at a place selected beforehand" [37].

Following the braking procedure, the Soyuz 24 spacecraft would separate into separate modules. The Instrument Module and Orbital Module were discarded and would burn up in the atmosphere, while the Descent Module containing the cosmonauts experienced a normal ballistic re-entry before parachuting to the ground.

Journey's end

The return process went smoothly, and the Soyuz 24 Descent Module landed amid high winds and low clouds, frost and sub-zero temperatures some 22 miles northeast

Viktor Gorbatko would eventually make three spaceflights.

of Arkalyk in northern Kazakhstan, where the crew was recovered without incident. "The cosmonauts feel fine after the landing," Tass reported soon after. "The planned programme of exploration of two expeditions in the Salyut 5 orbital scientific station has been completed successfully," it added, indicating (correctly, as it would turn out) that the Soyuz 24 mission would be the last involving that particular space station [38].

Two days after the crew had landed, a small capsule containing what was thought to be some military reconnaissance photography was automatically ejected on command from Salyut 5 and recovered after re-entry. The space station would then continue in orbit until 8 August, when it was brought down as planned by remote control and burned up in the atmosphere.

Yuri Glazkov would never fly into space again, but in 1989 he would become Deputy Chief of the Gagarin Cosmonaut Training Centre, retiring in 2000. He would also achieve modest fame as the author of some science fiction books.

Viktor Gorbatko had now flown twice, but he would still have another flight to come three years later, in July 1980. In accomplishing this third mission, he would become the last of the original cosmonaut group to fly into space.

REFERENCES

[1] Francis French and Colin Burgess, *In the Shadow of the Moon: A Challenging Journey to Tranquility 1965–1969*, University of Nebraska Press, Lincoln, NE, 2007.
[2] Boris Volynov interview with Bert Vis, London, 16 March 2001.
[3] James Oberg, "Soyuz 5's flaming return," *Flight Journal*, **7**(3), 56–60, June 2002.
[4] Alexander Zheleznyakov, "Yevgeny Khrunov remembered," *Orbit* magazine, No. 47, October 2000.

[5] L. Lebedev, B. Luk'yanov and A. Romanov, *Sons of the Blue Planet*, Amerind, New Delhi, 1973.

[6] Georgi Shonin interview with Bert Vis, Star City, Moscow, 9 August 1993.

[7] Evgeny Riabchikov, *Russians in Space*, Doubleday & Co., New York, 1971.

[8] *Time* magazine, "Orbital Troika," Friday, 24 October 1969.

[9] Viktor Gorbatko interview with Bert Vis, onboard MV *Akademik Kurchatov*, Rotterdam, 3 August 1994.

[10] Rex D. Hall and David J. Shayler, *Soyuz: A Universal Spacecraft*, Springer/Praxis, Chichester, U.K., 2003.

[11] Sydney *Daily Telegraph*, "Red space mission 'successful'," 19 October 1969.

[12] Sydney *Sun*, "Cosmic celebration," 24 October 1969.

[13] Auckland (NZ) *Herald*, "Cosmonaut tells of friction," 18 November 1969.

[14] Encyclopedia Astronautica, "Shonin: Incident with Shonin during DOS training, 6 February 1971." Website: *http://www.astronautix.com/astros/shonin.htm*

[15] Dmitri Zaikin interview with Bert Vis, Star City, Moscow, 13 August 1993.

[16] Sydney *Daily Mirror*, "U.S. Moon-man sees blast-off," 3 June 1970.

[17] Sydney *Sun*, "Reds set for long space trip," 5 June 1970.

[18] Vitaly Volovich, *Experiment: Risk*, Progress, Moscow, 1986.

[19] Bart Hendrickx, "The Kamanin diaries, 1960–1963," *Journal of the British Interplanetary Society*, **50**(1), 33-40, January 1997.

[20] The Voice of Russia, "Pioneering experience of long-term weightlessness, 2001." Website: *http://www.vor.ru/Space_now/Cosmonauts/Cosmonauts_3.html*

[21] David Scott and Alexei Leonov, *Two Sides of the Moon*, Simon & Schuster, London, 2004.

[22] San Francisco *Chronicle*, "Successful Russian space maneuver," 4 July 1974.

[23] *Time* magazine, "Détente in space," 22 July 1974.

[24] Sydney *Sun*, "Russian gear danger to spacemen," 11 June 1975.

[25] *Newsweek* magazine, "A visit to Baikonur," 5 May 1975.

[26] Paul Recer, "Yanks, Russ link up in space in triumph of science, amity," San Francisco *Chronicle*, 18 June 1975.

[27] Sydney *Daily Telegraph*, "2 cosmonauts land in Russia," 22 July 1975.

[28] Tim Furniss, *Manned Spaceflight Log*, Jane's, London, 1983.

[29] Charles W. Borden interview with Alexei Leonov for "Aeroflot" in-flight magazine article, *Across Space and Time*, No. 2, 2005.

[30] Adam Tanner, "Space Race lost: Russians remember," 12 July 1999. Website: *http://www.space.com/sciencefiction/russia_moon_wg.html*

[31] Voice of Russia, "Space exploration is a lifelong job," 2007. Website: *http://www.vor.ru/English/People/programm.phtml?act = 29*

[32] *International Herald Tribune*, "Russians orbited, expected to link to Space Station," 7 July 1976.

[33] Singapore *Times*, "Successful link-up as skymen dock Salyut-5," 8 July 1976.

[34] Dr. Rostislav Bogdashevsky interview with Rex Hall and Bert Vis, Star City, Moscow, 10 October 1999.

[35] Vyacheslav Zudov interview with Bert Vis, 7th Planetary Congress of the Association of Space Explorers, Berlin, 4 October 1991.

[36] *International Herald Tribune*, "Russia reports linkup in space," 9 February 1977.

[37] Yuri Kolesnikov and Yuri Glazkov, *A Spaceship in Orbit*, Mir, Moscow, 1984.

[38] Sydney *Daily Telegraph*, "Soviet spacemen return safely," 26 February 1977.

12

Orbits of co-operation and the end of an era

In April 1967, the military forces of socialist countries allied to the Warsaw Pact—namely Bulgaria, Cuba, Czechoslovakia, the German Democratic Republic, Hungary, Mongolia, Poland and Romania—united their efforts in space exploration and research under the leadership and administration of the Soviet Union. The programme would become known as Intercosmos.

THE INTERCOSMOS PROGRAMME

Unmanned research satellite flights subsequently began, launched on Soviet rockets. Nine years later, on 16 July 1976, the Soviet Union agreed to fly "guest cosmonauts" from Intercosmos member nations aboard its Soyuz and Salyut spacecraft. Two carefully selected candidates from each participating nation—all jet pilots but with little or no background in science or engineering—would begin arriving at Star City in December 1976. They were from Czechoslovakia, Poland and East Germany, while a second group from Bulgaria, Cuba, Hungary, the Mongolian People's Republic and Romania would report for training in March 1978. They would be joined early in 1979 by two candidates from Vietnam.

The "red-handed" cosmonaut

By means of evaluation, the two candidates from each nation would either be selected as a prime or backup crewmember, and they would each be assigned to a senior Soviet cosmonaut for specific mission training. While this training may have been somewhat rushed, their spaceflights were mostly vehicles for propaganda, so they were effectively just along for the ride, and to conduct experiments aboard the Salyut stations.

The Intercosmos crew
of Soyuz 28: Alexei
Gubarev (left) and
Czech researcher
Vladimir Remek.

The first Intercosmos mission was Soyuz 28, a highly political flight under the command of Colonel Alexei Gubarev. Travelling with him on a seven-day mission to the Salyut 6 space station was the first non-Russian, non-American space traveller, 29-year-old Captain Vladimir Remek of the Czech Army Air Force. Their flight was launched on 2 March 1978 and would link up successfully with the orbiting space station, then tended by cosmonauts Yuri Romanenko and Georgi Grechko.

The fact that the Intercosmos space explorers were actually little more than passengers aboard any Soviet spacecraft is best exemplified by a humorous story that made the rounds after Vladimir Remek's return from space on 10 March 1978. As space historian James Oberg notes in his book *Red Star in Orbit*, Remek's hands were rumoured to have been bright red at his post-flight medical examination. "How could Remek have gotten red hands in space?" asked Oberg.

> "So the doctors asked him when he arrived in Moscow as he was preparing to leave his brief space career for a new profession of public appearances and parades all over Czechoslovakia. Remek looked at his hands, which were still sore, and laughed. 'Oh, that's easy,' he replied. 'On Salyut, whenever I reached for a switch or a dial or something, the Russians shouted, 'Don't touch that!' and slapped my hands.' And they reddened.
>
> As it turns out, the red-hands story is apocryphal; it never really happened. But within days of Remek's triumphant return to earth, it was gleefully passed from friend to friend in Prague, for the amusement but also to satirise the political aspects of the flight" [1].

The flight of Soyuz 28 took place a decade after Warsaw Pact forces had invaded Czechoslovakia, with Soviet tanks rumbling menacingly into the streets in order to suppress a thriving, popular culture of freedom known as the Prague Spring. It would be suggested that the Soviets had selected a Czechoslovak cosmonaut to help soften a long-prevailing, anti-Russian feeling in that country. But not according to Remek, who today lives in Russia and has maintained his left-wing ideals.

"I don't think so," he told Radio Prague. "If the Soviet leaders had any problems at that time it wasn't a sense of guilt for entering Czechoslovakia. It could've been partly political, but what was really important was that we were among the strongest partners in the Intercosmos programme, and our people were also on the U.N. space committee. And maybe we weren't the worst among those who prepared for the flight!" [2].

Bykovsky calls it a day

The second manned launch of an Intercosmos flight, Soyuz 30, took place on 27 June 1978 under the command of Colonel Pyotr Klimuk. On this occasion the cosmonaut–researcher was Polish Air Force Lt. Colonel Miroslaw Hermaszewski. They would occupy the Salyut station for nearly seven days in company with resident

Valery Bykovsky and Sigmund Jähn.

crewmembers Vladimir Kovalyonok and Alexander Ivanchenkov before returning to Earth at the end of their mission.

On the following flight, Soyuz 31, Valery Bykovsky would make a triumphant return to space along with Lt. Colonel Sigmund Jähn of the East German Air Force. Their voyage began with a launch from the Baikonur cosmodrome on 26 August 1978, ferrying supplies to the orbiting Salyut 6. They would be the second Intercosmos crew to visit Kovalyonok and Ivanchenkov, who had already been in space for 71 days.

Married, and the father of two children, Jähn had graduated from East Germany's military flying school and in 1966 was sent to the Soviet Union to study at the Gagarin Air Force Academy. Prior to his flight, he had almost two years of classes and joint training at the cosmonaut training centre.

On 3 September, Bykovsky and Jähn made a safe landing in the Soyuz 29 capsule, thumping down in a cloud of dust some 87 miles southeast of Dzhezkazgan, a town (and former gulag site) in the Central Asian Republic of Kazakhstan. As planned, they had left their own Soyuz 31 ferry craft attached to Salyut 6 for a later crew return, returning in the spacecraft Kovalyonok and Ivanchenkov had originally flown to the station. Their departure from the station had been shown on Russian television.

The two men spent a total of seven days aboard the Salyut 6 station along with the two-man resident crew, successfully carrying out a number of East German experiments, as well as cardiovascular tests on Kovalyonok and Ivanchenkov. The experiments they worked on included multi-spectral photography of Europe, studying the effects of weightlessness, observing the growth of semiconductor crystals, and the formation of new metallic and non-metallic materials.

Soyuz 31 would mark the final spaceflight for Valery Bykovsky. He was later assigned as backup commander for another Intercosmos flight in September 1979 along with Vietnamese pilot–researcher Bui Thanh Liem, but the prime crew would fly that mission. During this time he also worked as a training official at the cosmonaut centre, eventually leaving the cosmonaut team on 2 April 1988 to accept a position as director of the House of Soviet Science and Culture in East Berlin [3].

Failure of a mission

The Intercosmos Programme continued on 10 April 1979 when mission commander Nikolai Rukavishnikov (the first non-military commander of a Soviet spacecraft) teamed up with Bulgarian Georgi Ivanov on the Soyuz 33 mission. It would be the third flight into space for Rukavishnikov, having previously flown on Soyuz 10 and later on Soyuz 16—the ASTP "dress rehearsal" mission.

For Georgi Ivanov, it would be his first and only spaceflight, but it was still a dream come true for the air force major from Lovech District in Bulgaria. In 1961, then aged 21, he had wistfully told his fiancée: "I was born too early, Natasha. It's hardly likely that any Bulgarian will fly to outer space before the year 2000" [4].

The Soviet–Bulgarian flight would be marred from the outset by what Tass later described as the shakiest blast-off in Soviet space history, with winds of up to nearly

30 miles an hour buffeting the spacecraft during lift-off from Kazakhstan's Baikonur space centre. "Not one manned spacecraft has blasted off in such bad weather [as that] in which the Soyuz 33 lifted off," Tass said in reporting the flight. Meanwhile, aboard Salyut 6, Vladimir Lyakhov and Valery Ryumin, who had taken up residence in the station some six weeks earlier, were preparing to receive their visitors after watching the launch on a newly installed television hook-up.

Unfortunately the docking between Soyuz 33 and the Salyut 6 station had to be abandoned when the approach correction power unit aboard the Soyuz craft malfunctioned, causing the main engine to fire erratically for just three seconds before shutting down. Unable to complete the docking, Rukavishnikov was finally ordered to retreat from the space station and prepare for an early re-entry. It was the second such blow for the veteran cosmonaut who had also returned to Earth early when his Soyuz 10 craft had failed to dock with Salyut 1 eight years earlier.

The bitterly disappointed crew would land in darkness almost 200 miles southeast of Dzhezkazgan at 7:35 PM Moscow time on 12 April 1979 (ironically the 18th anniversary of Gagarin's flight) after a two-day mission. Some thought was given to launching a second mission, this time with the backup crew of Yuri Romanenko and Alexander Alexandrov, but it might have created a political dilemma with Bulgaria being able to claim that two of its citizens had gone into space, so any plans for a second mission were discarded.

The next Intercosmos visitor to the Salyut 6 station, now manned by the resident crew of Leonid Popov and Valeri Ryumin, was Lt. Colonel Bertalan Farkas of the Hungarian Air Force. In company with mission commander Valeri Kubasov, they would be launched on 26 May 1980 aboard the Soyuz 36 spacecraft, heading for a rendezvous and docking with their orbiting colleagues. This time a docking was achieved, and the eight-day mission was carried out with routine competence. At mission's end, returning in the Soyuz 35 spacecraft, they landed without incident 87 miles southeast of Dzhezkazgan.

Final flight of a "Little Eagle"

In May 1979 Vietnam had been officially approved as one of a number of Soviet bloc countries to participate in the Intercosmos Programme, and two Vietnamese military pilots were subsequently selected as candidates for a joint orbital mission. Lt. Colonel Pham Tuan, from the Vietnamese Air Force, and Senior Captain Bui Thanh Liem of the Red Star Regiment, had already begun their training a month earlier at the Star City training centre.

In September of that year, Pham was teamed with veteran cosmonaut Viktor Gorbatko, while Bui began further training with another member of the original cosmonaut team, Valery Bykovsky. It would later be announced that Pham and Gorbatko would be the prime crew, with Bui and Bykovsky (who had already flown as commander of the East German Intercosmos mission) acting as their backup crew on a mission to the orbiting Salyut 6 space station, scheduled for the following year.

The official Vietnamese News Agency (VNA) reported at the time that Colonel Pham, married with a daughter aged four and born into a rural peasant family in

The backup crew for Soyuz 37: Valery Bykovsky and Vietnamese fighter pilot Bui Thanh Liem.

Quoc Tuan, North Vietnam, had joined the army in 1963 but was assigned to the Air Force and sent to the Soviet Union for training, graduating in 1968. Returning to Vietnam, he became a jet fighter pilot and, according to the VNA: "In a fight against a massive raid by United States strategic aircraft on Hanoi, Haiphong and other urban centres towards the end of 1972, he shot down one B-52 bomber." This aircraft loss, however, has never been confirmed and is strenuously denied by the United States.

"From the very start, both candidates had to overcome a formidable number of difficulties," according to Alexei Leonov. "There wasn't much time for training. True, we already had quite a lot of experience in conducting such training and that made the task somewhat easier. Another thing, I think, that helped was that they had the guidance of veterans like Viktor Gorbatko and Valery Bykovsky ... By the time the Vietnamese began their training, Gorbatko had already made two spaceflights, and Bykovsky had made three" [2].

As an essential part of their training, Pham Tuan and Bui Thanh Liem had to learn theoretical subjects that were entirely new to them, such as space navigation and celestial mechanics. They also studied the construction of the Soyuz spacecraft and its carrier rocket, while at the same time undertaking lessons to improve their knowledge of the Russian language. While the men found it extremely difficult to master space-flight terminology, both managed to cope and completed their pre-flight training with high marks.

The prime crew for Soyuz 37, Pham Tuan and Viktor Gorbatko.

"We were always given special assistance by the methodologists, the scientific and technical personnel and, of course, our instructors Viktor Gorbatko and Valery Bykovsky," Pham would later reflect.

Gorbatko would say that he felt fortunate to have such a good partner. "Pham Tuan proved to be a very capable student. His diligence and persistence enabled him to get ready to carry out not only his own flight, but also a wide-scale scientific programme" [4].

Pham Tuan and Viktor Gorbatko would launch aboard the Soyuz 37 spacecraft on 23 July 1980, on the sixth Intercosmos mission. Pham, it was noted at the time, was not only the first Vietnamese, but the first person from a Third World country to fly in space. They docked successfully with the aft port of Salyut 6 the following day, and would be offered a traditional Russian greeting of bread and salt by the resident crew of Leonid Popov and Valery Ryumin, who had occupied the station on 10 April. At the outset of the mission Pham is said to have suffered badly from a headache and loss of appetite, but he would quickly recover.

During their week-long mission, Pham and Gorbatko would carry out around 30 scientific, medical and biological experiments. Many of these were concerned with Vietnam's economy, and were designed to aid in the study of the nation's timber areas, particularly after the massive amounts of defoliants and fire bombs dropped during the war, erosion, silting and irrigation. Photographs taken from

Salyut 6 would later be used to determine the exact boundaries of flooded areas and to evaluate sea pollution by river discharges—vitally important for the fishing industry. The crew would also observe the growth in weightlessness of a fern—a plant that is widespread in South East Asia and is a little like a small nitrogen factory. Like clover or leguminous plants it improves the yield of other crops and fertilizes the soil.

Pham and Gorbatko returned to Earth in the Soyuz 36 spacecraft, leaving a comparatively fresh craft for Popov and Ryumin, who would move it from the aft to the forward docking port on 1 August, freeing the aft port to receive a Progress automatic cargo craft.

The two men would make a safe touchdown 112 miles southeast of Dzhezkazgan on the evening of 31 July. For 33-year-old Lt. Colonel Pham Tuan it would prove to be his first and only journey into space, while for Gorbatko this would be his third—but also final—spaceflight. Backup researcher Bui Thanh Liem would return to Vietnam, but sadly on 1 September the following year he was killed in the crash of his MiG-21 fighter jet, aged 32 years.

The backup crew for Soyuz 38 comprised Cuban candidate José Armando López Falcón and Yevgeny Khrunov. The mission would eventually be flown by the prime crew of Romanenko and Tamayo Méndez.

The first "black" cosmonaut

On 18 September 1980, a new international crew was launched to link up with Salyut 6. A Cuban cosmonaut was onboard the Soyuz 38 spacecraft, together with mission commander Yuri Romanenko. This was 38-year-old Arnaldo Tamayo Méndez who not only became the first Latin American in space, but was widely proclaimed at the time as the first black cosmonaut.

On this, the seventh Intercosmos flight, the two-man crew would be greeted in space by Leonid Popov and Valeri Ryumin, who by that time had been working in orbit for a record 160 days. The seven-day flight programme would include about 15 experiments prepared jointly by Soviet and Cuban specialists, and some started by previous international expeditions. Among the medical–biological research, of particular interest was a study of the peculiarities of sucrose crystallization in weightless conditions, intra-cellular processes and the central nervous system.

Romanenko and Tamayo Méndez returned safely on 26 September, re-entering in the Soyuz 38 spacecraft in which they had been launched and landing just two miles from their designated landing zone. It was not clear why the two men returned in Soyuz 38, as all previous Intercosmos crews had re-entered in an earlier spacecraft, leaving a fresh Soyuz craft for the long-duration occupants of Salyut 6.

Mongolia and Romania join the space club

Just three weeks before the maiden flight of the first U.S. space shuttle, *Columbia*, a researcher from the Mongolian People's Army with a true tongue-twister of a name became the eighth Intercosmos guest cosmonaut. Captain Jugderdemidyin Gurragcha, then 33 years old, would join veteran cosmonaut Colonel Vladimir Dzhanibekov on a flight to Salyut 6 aboard the Soyuz 39 spacecraft.

Launched on 22 March 1981, Soyuz 39 docked with the space station the following day. Now onboard Salyut 6 were Colonel Vladimir Kovalyonok and Viktor Savinykh. They had been launched aboard the new-generation Soyuz T-4 spacecraft on 12 March 1981, with Savinykh becoming the 100th person to fly into space. Gurragcha and Dzhanibekov conducted more than 20 experiments during their stay aboard Salyut 6, including experiments which would lead to a better tectonic map of Mongolian territory. The two men landed aboard Soyuz 39 on 30 March, and like the Intercosmos crew before them, they landed in the spacecraft in which they had been launched eight days earlier.

The Soyuz 40 spacecraft was launched on 14 May 1981, carrying a crew of Colonel Leonid Popov and Dumitru Prunariu, a senior lieutenant in the Romanian Army Air Force. The flight programme included docking with the Salyut 6/Soyuz T-4 complex in orbit, and conducting experiments onboard with cosmonauts Kovalyonok and Savinykh. As on earlier Intercosmos missions the seven-day flight programme was carried out successfully and the two men would undock their Soyuz 40 spacecraft from Salyut 6 on 22 May, making a safe landing soon after some 78 miles east of Dzhezkazgan.

Prunariu was the ninth Intercosmos guest cosmonaut–researcher from the Soviet bloc countries aboard a standard Soyuz flight. Interestingly, the Soviet Union did not have any North Koreans in cosmonaut training; as the tenth socialist country it was the only one not involved in the Intercosmos space programme. Subsequent missions, beginning in 1982 with French *spacionaute* Jean-Loup Chrétien, would carry invited guest cosmonauts from a variety of countries, willing to pay to have one of their nationals conduct research in orbit on its behalf.

The last of the first

Significantly, the July 1980 Intercosmos flight of Viktor Gorbatko would mark the very last flight of any of the first cosmonaut group, formed 20 years earlier. Twelve of that original group of 20 cosmonaut candidates would eventually fly into space, with 5 making a single flight (Gagarin, Titov, Belyayev, Khrunov and Shonin) another 5 flying twice (Nikolayev, Popovich, Komarov, Leonov and Volynov) and 2 (Bykovsky and Gorbatko) entering space three times.

Of the remaining eight, seven would never fly for medical or disciplinary reasons, while Valentin Bondarenko had lost his life in a terrible accident.

By the time Viktor Gorbatko flew his (and his group's) final space mission in July 1980, five of Korolev's "Little Eagles" had died: Belyayev, Bondarenko, Gagarin, Komarov, and Nelyubov. The group would also lose Varlamov later that year.

Viktor Gorbatko at Sydney's Powerhouse Museum during a visit to Australia in April 2005. (Photo: Jon Reid, Sydney *Morning Herald*)

Reduced to only 11 members, this group photograph was taken of the surviving members of the first cosmonaut group in March 2000. Back row (from left): Khrunov, Volynov, Leonov, Titov and Zaikin. Front row: Gorbatko, Bykovsky, Kartashov, Nikolayev, Rafikov and Popovich. Three more of the group would be lost that same year.

At the time of writing this book, sadly only six men survive from that first cosmonaut group: Bykovsky, Gorbatko, Leonov, Popovich, Volynov and Zaikin.

LOSSES IN THE MILLENNIUM YEAR

The year 2000 was celebrated worldwide, significantly ushering in a brand new millennium (although many pedants argued that this would not occur until 2001). It would also prove to be a particularly sad year for the first Soviet cosmonaut detachment: in the space of just four months three group members—all men in their mid-sixties—would pass away. Mars Rafikov, his place in the first team unrecognized for more than two decades, had been dismissed quite early from cosmonaut training, but Yevgeny Khrunov and Gherman Titov will always be favourably regarded and remembered for their achievements in Soviet spaceflight history.

Challengers of the space era

To mark the 40th anniversary of the selection and arrival of the first cosmonauts at the Russian cosmonaut training centre, celebrations were held in Star City in March 2000, attended by all living members of the first cosmonaut detachment, including

Yevgeny Khrunov photographed with researcher Alex Panchenko in 1999. (Photo courtesy Alex Panchenko)

Yevgeny Khrunov, then a comparatively young 67-year-old. However, he was noticeably absent from far more formal celebrations held on 12 April to mark Cosmonauts Day.

Just five weeks later, on 19 May, Khrunov died from a heart attack and was buried in Moscow's Ostankino cemetery. A prolific collector of books, his home library was crammed with several thousand publications at the time of his death, including several that he had written, such as the 1974 book *Man as an Operator in Open Space*, *Conquering Zero Gravity* (1977) and *Outside Space Ship on Orbit* (co-written with Yuri Glazkov and Dr. L.V. Khachaturyants) as well as the fictional *The Road to Mars*, co-authored by Khachaturyants. He was also the principal subject of the 1970 book *Cosmonaut: Son of the Land of Tula*.

Mars Rafikov had also authored a book, an autobiography called *My Parni iz Otryada X* (loosely translated as "We Members of Group X"), for which he spent a lot of time unsuccessfully seeking a publisher. On 23 July 2000, just two months after the death of Khrunov, 66-year-old Rafikov passed away from natural causes at his home in Almaty (prior to 1991 known as Alma Ata) in the Republic of Kazakhstan. He was laid to rest in a local cemetery.

Then, on 21 September 2000, the body of Gherman Titov was discovered in the sauna of his Moscow apartment on Khovanskaya Street. He had apparently died the day before. Initial media reports on the famed cosmonaut's death centred on speculation that he had died from accidental carbon monoxide poisoning; however, this

An undated photograph of
Mars Rafikov.

Gherman Titov: first man to
spend a day in space.

Titov's grave in the
Novodevichye cemetery,
Moscow. (Photo courtesy
Bert Vis)

notion was quickly dismissed when it was pointed out that the sauna was electrically
operated. The 65-year-old was known to have been suffering from recent heart
problems, and it was later determined at an official post mortem that this condition
had led to his death by "cardiac insufficiency"—a heart attack.

Four days later, on 25 September, Gherman Titov's funeral was held at
Moscow's elite Novodevichye cemetery, where he was buried with full military
honours alongside such notables as former premier Nikita Khrushchev, aircraft
designers Sergei Ilyushin and Andrei Tupolev, as well as assorted generals,
writers and composers and three spaceflight colleagues: Pavel Belyayev and Georgi
Beregovoi from the Air Force detachment, and Boris Yegorov, the medical doctor
who flew on Voskhod 1.

Times had changed; in earlier days his remains might have been given an
honoured place in the Kremlin wall alongside Yuri Gagarin and Vladimir Komarov.
Many of his colleagues in attendance openly wept as Titov's body was laid to rest,

and Pavel Popovich later reflected on behalf of the cosmonaut detachment that "We have not just lost a comrade in arms, but a true friend."

Death of a space pioneer, and controversy

Following the millennium year of 2000 (and up to the time of publication of this book), the world would lose yet another two of the first cosmonaut group—Andrian Nikolayev and Anatoly Kartashov. Kartashov, who saw his chance to fly into space evaporate after being ruled medically unfit, died in Kiev on 11 December 2005 following what was described as "a severe and prolonged illness". He was survived by his wife Yuliya and two daughters, Lyudmila and Svetlana.

With his active cosmonaut days at an end, Andrian Nikolayev had subsequently served as first deputy head of the Yuri Gagarin Cosmonaut Training Centre, retiring from the armed forces as a major-general in 1982. Nikolayev, the third Soviet cosmonaut to fly into space who later set a second space endurance record on the 1970 Soyuz 9 mission, was lost on 3 July 2004, aged 74. He suffered a heart attack that day while officiating at the All-Russian rural sports game at Cheboksary, the capital of his native Chuvash Autonomous Republic.

Although he was rushed to hospital, Nikolayev could not be revived and was pronounced dead. Three days later his funeral was held at the Space Exploration Museum in his home village of Shorshely, attended by friends, relatives, former Russian president Boris Yeltsin and a number of fellow cosmonauts, including Nikolayev's "Heavenly Twin" Pavel Popovich.

However, as reported by space analyst James Oberg, Nikolayev's funeral was marred by controversy and a bitter dispute. His family and friends had wanted his remains interred in Moscow, near the cosmonaut training centre, but according to Oberg, "he died while visiting his home province, and the local leader decreed that he be buried there, at a space museum in his honour." Yelena Mayorova, Nikolayev's daughter with Valentina Tereshkova, had told local authorities to ship her father's remains to Moscow. Nikolai Federov, president of the Chuvash Autonomous Region, promised that her wishes would be "considered" [5]. He then promptly organized a funeral for the

Veteran Soviet cosmonaut Andrian Nikolayev.

Andrian Nikolayev's funeral in Shorshely. Among the mourners in this photograph are three of his first-group colleagues: Boris Volynov, Viktor Gorbatko and Alexei Leonov.

Monday, with interment in Shorshely. Mayorova, supported by her mother and several of Nikolayev's cosmonaut colleagues, objected strongly to these plans, claiming that her father (who did not leave a will) had wanted to be buried in Moscow.

Despite their objections the interment took place as scheduled. Mayorova (who refused to attend the funeral) subsequently demanded that her father's remains be dug up and flown to Star City for reburial, but Federov refused. According to *The Times'* Moscow reporter, Jeremy Page, Federov accused Mayorova of neglecting her father after his divorce from Tereshkova, saying the former cosmonaut had lived a lonely existence in Moscow, neglected by his former wife and daughter. "Those women, they did nothing for him," Federov is reported as saying. "He even had to wash his own socks" [6]. It was a sad postscript to such a remarkable life for a man once so revered throughout the Soviet Union and elsewhere.

The world's first cosmonaut reflects

It is probably fitting that as the first person to fly into space, Yuri Gagarin is given the last word on space exploration and our future.

The year before he died, Gagarin told a Russian news reporter that: "Spaceflights cannot be stopped. This is not the work of any one man or even a group of men. It is a

The world's first spaceman,
Yuri Alekseievich Gagarin.

historical process which mankind is carrying out in accordance with the natural laws of human development."

REFERENCES

[1] James E. Oberg, *Red Star in Orbit: The Inside Story of Soviet Failures and Triumphs in Space*, Random House, New York, 1981.
[2] Radio Prague, "Czech Republic 2004." Website: *http://incentraleurope.radio.cz/ice/article/101738*
[3] Douglas B. Hawthorne, *Men and Women of Space*, Univelt, San Diego, CA, 1992.
[4] E. Knorre, "Bulgaria's First Spaceman," *Soviet Weekly*, Moscow, 23 April 1979.
[5] MSNBC Technology and Science, "Cosmonaut controversy," 7 July 2004. Online at: *http://www.msnbc.msn.com/id/5379793*
[6] Jeremy Page, "Unearthly row over cosmonaut's burial," TimesOnline, 8 July 2004. Website: *http://www.timesonline.co.uk/tol/news/world/article454815.ece*

Appendix A

Biographies in brief

Although the lives of the 20 individuals in the first Soviet cosmonaut detachment are examined within the book, it was felt that a brief summary of each man's life should also be included as a quick reference.

Anikeyev, Ivan Nikolayevich   The son of Nikolai Nikolayevich and Natalya Ivanova, Ivan Anikeyev was born on 12 February 1933 in the town of Novaya Pokrovka (now Liski), located 250 miles south of Moscow in Voronezh Region, West Central Russia. He would complete ten years of schooling and in March 1952 became a student at the Stalin (later renamed the 12th) HAFP School for navy pilots, graduating on 30 July 1956 with the rank of lieutenant. Anikeyev then served in the 255th Regiment, 91st Fighter aviation division of the Northern Naval Fleet from August to October 1956, and later as a senior pilot with the 524th Regiment of the 107th Fighter aviation division. He was promoted to senior lieutenant on 17 August 1958. Following his selection as a cosmonaut he reported for training on 7 March 1960. On 30 August that year he received a further rise in rank to that of captain. He completed his training on 5 April 1961 and was confirmed as a Cosmonaut of the Air Force. Anikeyev would serve as CapCom for the Vostok 3/4 tandem flight in August 1962, but would never receive an actual mission assignment. He is said to have been placed on report earlier that year for being drunk in uniform along with Mars Rafikov. Together with Grigori Nelyubov and Valentin Filatyev, he was dismissed from the cosmonaut team for disciplinary reasons following an unruly evening confrontation with military guards at a railway platform near the TsPK training centre early in 1963. All three men were reassigned to their former military unit. Following his dismissal, Anikeyev left the training centre on 17 April 1963 and returned to his navy unit in the north, where he remained on active service as a pilot until 1965. For the next 10 years he served as a ground controller/navigator in the Air Force, finally retiring from active military service on 12 September 1975. Having transferred to reserve status, he and his wife

Galina Pavlovna Anikeyeva moved to the city of Bezhetsk in the Kalinin district. He died there of cancer on 20 August 1992.

Belyayev, Pavel Ivanovich   The future commander of Voskhod 2 was born in the Chelischevo settlement of the Rostlyansy district in the Vologda Region on 26 June 1925, the first son of Ivan Parmenovich and Agrafina Mikhailovna Belyayev. From 1932 he spent six years studying at the Minkovsky secondary school and completed his secondary education in the Sverdlovsk Region in 1942. He then took on work as a turner at a plant in Kamensk-Uralsk before entering the Soviet army at the age of 18, turning to aviation. Graduating from the Yeisk HAFP School in 1945 he flew as a fighter pilot against the Japanese in the last phase of World War II and then served with a number of fighter units in the Far East and Pacific before being assigned to the Red Banner Air Force Academy in Moscow in 1956. Following his graduation late in 1959 he became a squadron commander with the Black Sea Fleet Air Force before his selection as a cosmonaut. At 35 years old and the oldest member of the first cosmonaut team, he was chosen as their first commander. He was awarded the title Cosmonaut of the Air Force on 5 April 1961. In April 1964 he was selected to command the Voskhod 2 mission, during which Alexei Leonov became the first human to walk in space. It would be Belyayev's only space mission and he was subsequently given administrative duties, involved in training new cosmonauts and as head of the Almaz cosmonaut training group, and for a couple of months was involved in the Zond programme. In November 1969 he was assigned as commander of the first Salyut–Almaz space station, but he had been concealing severe stomach ulcer problems, resulting in his being hospitalized in Moscow the following month. Several weeks later he developed purulent peritonitis during an operation on his ulcer and died on 10 January 1970. He was buried in the Novodevichy cemetery, Moscow. Colonel Belyayev was survived by his wife, the former Tatyana Prikatshikova and their two daughters; Irina, born 27 October 1949, and Lyudmila, born 20 March 1955.

Bondarenko, Valentin Vasilyevich   Having turned 23 years of age only nine days before he was selected for training, Senior Lieutenant Valentin Bondarenko was the youngest member of the first detachment of cosmonauts. Born into a working family in the Ukrainian city of Kharkov on 16 February 1937, he was the second son of Vasily Grigoryevich and Olga Ivanovna Bondarenko. He completed his seven years of schooling in Kharkov in 1954 and subsequently volunteered for the Soviet army. Bondarenko later attended the Armavir HAFP School, graduating with the rank of lieutenant on 2 November 1957. He then briefly served as a fighter pilot with the Soviet Air Force in the Baltics. Married to Galina Semenyovna (known as Anya), they had a son, Aleksandr, born in 1956, who now works as an engineer in the training centre at Star City. He would receive a promotion to senior lieutenant on 2 December 1959, around the same time he was interviewed for possible selection as a cosmonaut. Bondarenko reported to Moscow for training on 25 March 1960, together with Pavel Belyayev, Dmitri Zaikin and Valentin Filatyev. On 23 March 1961, less than three weeks before Yuri Gagarin's Vostok flight into space, Bondarenko was shockingly injured when an oxygen-fuelled fire erupted inside an isolation test chamber in

Moscow. Literally burned from head to foot, he was rushed to the nearby Botkin Hospital, where he died several hours later. With his widow and five-year-old son in attendance, Valentin Bondarenko was buried in Kharkov in complete anonymity and amid strict State secrecy. Although persistent rumours mentioned the death of a young cosmonaut in a training fire, the truth behind this story (and Bondarenko's name) would not be revealed until 1986, when a series of articles about the first Soviet cosmonauts was published in the Soviet government's *Izvestia* newspaper. Some details of his death were first revealed in the 1984 book *Russian Doctor*, by Vladimir Golyakhovsky.

Bykovsky, Valery Fyodorovich   Pavel Bykovsky was born on 2 August 1934 in the Pavlovsky-Posad settlement of Moscow Region to Fyodor Fedotovich and Klavdia Ivanova Bykovsky. In 1940 the family moved to Kuybyshev (now Samara), then to Syzran and later Teheran, Iran. In 1948 they returned to Moscow, and while completing his secondary schooling Bykovsky enrolled at the Moscow City Aviation Club. In fulfilling military service he undertook initial pilot training at the 6th military aviation school in Kamenka, graduating in 1953, and for the next two years was given advanced flight instruction at the Kachinskoye HAFP School, later serving with various fighter regiments of the Soviet Air Force. Following his selection in 1960, Bykovsky was assigned to the advanced training group of six cosmonauts on 11 October, and was awarded the title Cosmonaut of the Air Force on 25 January 1961. He would serve as backup pilot to Andrian Nikolayev for Vostok 3 and complete his first mission as pilot of Vostok 5, launched on 14 June 1963 and flown in tandem with Valentina Tereshkova on Vostok 6. To this day his flight remains the longest solo space mission. He was next given command of Soyuz 2, cancelled following the loss of Soyuz 1, and was then reassigned to the group training for lunar Project L missions. He completed a second flight in September 1976 as commander of Soyuz 22, and then Soyuz 31 in 1978, an Intercosmos flight to Salyut 6 with German researcher Sigmund Jähn. His last assignment was backup commander for Soyuz 37. From July 1968 he also served as commander of the cosmonaut detachment. Promoted to colonel, he resigned from active military service and the cosmonaut team on 2 April 1988, and then served as the director of the Soviet House of Science and Culture in Berlin, returning to Moscow in 1990. Married to Valentina Mikhailovich Sukhova, they have two children: Valery, born 1963 (killed in an aircraft crash in Afghanistan in 1986), and Sergei, born 12 April 1965.

Filatyev, Valentin Ignatyevich   Valentin Filatyev was born in the village of Malinkova in the Ishimsk district of Russia's Tyumen Region on 21 January 1930. His father Ignat was killed during the course of World War II. In 1945 Valentin completed his seventh-grade education and enrolled at the Ishimsk Pedagogical School to study teaching, graduating in 1951. In August that year he became a student pilot at the Stalingrad HAFP School. Following his graduation in October 1955 he served as a fighter pilot in the 472nd regiment of the 15th Fighter aviation division in the Anti-Aircraft Defence Force until December 1956 when he transferred as a senior pilot to the 3rd Fighter Regiment. Promoted in rank to senior lieutenant on 24 December 1957, he joined the cosmonaut team on 25 March 1960. Filatyev received the rank of captain on 9 May 1960, and in September 1961 he undertook studies at the Zhukovsky Air Force Engineering Academy. He was awarded the title Cosmonaut of the Air Force three months later, on 16 December 1961. Unfortunately, his engineering studies were terminated in April 1963 after he and fellow cosmonaut trainees Ivan Anikeyev and Grigori Nelyubov were confronted by a military patrol at the Chkalovskaya railway station and reported for their failure to comply with regulations, as well as Nelyubov's arrogant behaviour. All three would be dismissed from the cosmonaut team. Filatyev departed on 17 April 1963, and over the next six years served as a pilot in several units of the Soviet Air Defence Force, promoted to the rank of major on 24 February 1967. He retired from active service on 29 November 1969, moved to the city of Orel, and until 1977 was employed in a State Institute known as Gripropribor. He later became a teacher of civil defence courses in Orel, retiring after 10 years in 1987 to become a pensioner. Filatyev was married twice, to Larisa (by whom he had a son, Viktor) and then Valentina Prokofyevna. He died of lung cancer in Orel on 15 September 1990, aged 60.

Gagarin, Yuri Alekseievich   The first person ever to fly in space was born on 9 March 1934 in the village of Klushino, Smolensk Region, to Alexei Ivanovich and Anna Timofeyevna Gagarin. Following World War II the family moved to Gzhatsk where Gagarin resumed his interrupted education until 1949. At the age of 16 he undertook studies in metallurgy as an apprentice foundryman at the Lyubertsy steel mill in Moscow, graduating with honours two years later. Following this, he attended a higher technical school in Saratov and joined the Saratov Aviation Club. An instructor persuaded him to apply for the Air Force, and in 1955 he entered the Orenburg HAFP School, graduating in 1957. An assignment to the Northern Fleet followed, based at Zapolyarny, north of the Arctic Circle. He applied for cosmonaut training and was interviewed at his base in October 1959. After his selection on 7 March 1960, Gagarin quickly established himself as a leading flight candidate. In the summer of 1960 he was assigned to the advanced group of six cosmonauts training for the first manned Vostok flight. He was awarded the title of Cosmonaut of the Air Force on 25 January 1961. Three months later, on 12 April 1961, he became the world's first spaceman when he completed a single orbit of the Earth, landing by parachute after a mission lasting 108 minutes. On 25 May, he was appointed commander of the cosmonaut group, a position he held for five years. In September 1965 he started Soyuz training, and was assigned to Soyuz 1 a year later as backup pilot to Vladimir Komarov. Five days after the fatal crash of Soyuz 1 he was officially removed from flight status. With the rank of colonel he continued to train, graduating from the Zhukovsky Air Force Engineering Academy in February 1968. The following month, on 27 March, the 34-year-old cosmonaut was killed in the crash of a MiG-15 jet trainer along with his instructor, Vladimir Seryogin. He is survived by his wife, the former Valentina Ivanovna Goryacheva, and daughters Yelena, born 10 April 1959, and Galina, born 7 March 1961.

Gorbatko, Viktor Vasilyevich   Born to Vasili Pavlovich and Matrena Aleksandrova Gorbatko on a collective farm in the Northern Caucasas settlement of Ventsy-Zarya on 3 December 1934, Viktor Gorbatko finished his secondary schooling in 1952. Called for military service, he completed initial flight instruction at the Pavlograd military aviation school, and from 1953 attended the Bataisk HAFP School. Graduating on 23 June 1956 along with fellow graduate and future cosmonaut Yevgeny Khrunov, they served as fighter pilots and senior lieutenants with the Air Force until their selection in 1960. Further promoted to the rank of captain on 30 August 1960, he was awarded the title of Cosmonaut of the Air Force on 5 April 1961. Gorbatko's first assignment was backup commander to Pavel Belyayev on Voskhod 2. He was then a backup crewmember for the (later cancelled) Soyuz 2 mission, after which he was assigned to the manned lunar landing programme, learning to fly helicopters. He next served as backup to Khrunov on Soyuz 4, and graduated from the Zhukovsky Air Force Engineering Academy in 1968. His first flight came on 12 October 1969 as research engineer aboard Soyuz 7. From 1971 he was assigned to the Almaz programme and also backed up the Salyut 3/5 missions before his second spaceflight on 7 February 1977, launched as commander of Soyuz 24 which linked up with the Salyut 5 space station on a 17-day mission. Following backup commander duties for Soyuz 31, Gorbatko's third and final spaceflight began on 23 July 1980 as commander of the Soyuz 37 flight to Salyut 6 with Vietnamese Intercosmos researcher Pham Tuan. From 1978 to 1982 Gorbatko was a training instructor, then commander of the cosmonaut detachment. He left the cosmonaut team on 28 August 1982 to become head of faculty at the Zhukovsky Air Force Engineering Academy, where he was promoted to major-general in December 1982. He was later elected to the Congress of People's Deputies, in April 1989. Pursuing an interest that began with collecting stamps about the launch of Sputnik in 1957, he is currently president of the Union of Philatelists of Russia. He and first wife, the former Valentina Pavlovna Ordynskaya (who died in 1997), had two daughters: Irina (born 9 October 1957) and Marina (born 4 April 1960). His current wife is Alla Viktorovna.

Kartashov, Anatoly Yakovlevich   Perhaps the most hapless member of the first cosmonaut group, it was a minor physical ailment that would disqualify Anatoly Kartashov from flying into space, but would not prevent him from a later and lengthy career as a test pilot. Born 25 August 1932 in the settlement of Pervoye Sadovoye in Russia's Voronezh Region to Yakov Prokoyevich and Eyrosinya Timofeyevna Kartashov, his father was killed in combat against the Nazi invaders in 1942. From 1948 to 1952 Anatoly was a student of the Voronezh aviation college, learning to fly at the same time at the Voronezh flying club. From October 1952 to 1954 he undertook further studies at the Chuguyev HAFP School. From then until his selection as a cosmonaut he served as a pilot and senior pilot in a fighter regiment of the Soviet Air Force, based near the city of Petrozavodsk. With the rank of captain, Kartashov joined the first cosmonaut team for training on 28 April 1960 and was an original member of the advanced team of six cosmonauts training for an early Vostok mission. On 16 July he suffered pinpoint haemorrhaging along his spine during a centrifuge test and was subsequently grounded by medical staff. Removed from the advanced training team, he was later forced to stand down from the cosmonaut group. On 7 April 1962, with no prospect of a spaceflight, he resigned and returned to service as a military pilot in the Far East, and later near the city of Saratov. He would also fly as a test pilot in the State scientific research institute of the Soviet Air Force, finally retiring from active duties in the Air Force as a test pilot, first class, with the rank of colonel on 5 May 1985. Kartashov then took on duties as a civilian test pilot with the Antonov design bureau in the city of Kiev. He died there on 11 December 2005 after what was described as "a severe and prolonged illness", and is survived by his wife Yuliya Sergeyevna and two daughters: Lyudmila, born in 1960, and Svetlana, born in 1967.

Khrunov, Yevgeny Vasilyevich   Yevgeny Khrunov was born 10 September 1943 in the village of Prudy, in the Volovsky district of Tula Region to parents Vasily Yegorevich and Agrafena Nikolayevna Khrunov. Graduating from seventh grade of middle school in Nepryadovo village, he enrolled at the Kashira Agricultural Secondary School to study farm machinery, finishing his studies in 1952. Drafted into the Soviet army the following year, his interest in flying resulted in an application for admission to the Pavlodar military aviation school, and he would graduate from the Bataisk (Serov) HAFP School in 1956. He then served as a senior fighter pilot in a regiment of the Odessa Military Command in Moldavia until his selection as a cosmonaut on 9 March 1960. Khrunov was awarded the title of Cosmonaut of the Air Force on 6 March 1961, and would assist as CapCom during Yuri Gagarin's Vostok mission. He later undertook EVA training as Alexei Leonov's backup for Voskhod 2, and graduated from the Zhukovsky Air Force Engineering Academy in 1968. A crew-member for the cancelled Soyuz 2 mission, he would subsequently become the first person to transfer to another manned spacecraft during the Soyuz 4/5 mission in January 1969. It would be his only spaceflight. He subsequently trained with several crews which did not fly, and then trained as backup commander for the Soyuz 38 Soviet–Cuban mission, flown in 1980. Khrunov would have flown as commander of Soyuz 40 but stood down from the cosmonaut team in December 1980 and became Chief of the U.S.S.R. State Committee for Foreign Economic Relations. After resigning from active military service in October 1989 with the rank of colonel he participated in the Chernobyl nuclear power station accident clean-up and then took up work as an engineer in the transport industry, becoming a managing director of several companies. Aged 66, Khrunov died of natural causes in Moscow on 19 May 2000, leaving behind his wife, the former Svetlana Anatolyevna Sokolyuk, and two children: Valeri, born 13 July 1959, and Aleksandr, born 19 October 1976. He was laid to rest in Moscow's Ostankino cemetery, near where he lived.

Komarov, Vladimir Mikhailovich   Vladimir Komarov was born in Moscow on 16 March 1927, the son of Mikhail Yakovlovich and Ksenya Egnatyevna Komarov. From 1935 to 1941 he studied at secondary school No. 235 in Moscow, and in 1945, aged 15, he enlisted in the army, studying at the Air Force Moscow specialized school No. 1. He received pilot training at the 3rd Sasov Air Force School, and completed a first-year course at the Borisoglebsk HAFP School. In 1949 he would also graduate from the Bataisk HAFP School. From 31 December 1949 until 28 November 1951 he flew as a senior pilot with No. 328 Regiment of the North Caucasian air defence forces, based in Grosny, later serving with No. 486 Regiment. In 1959 he graduated from the Zhukovsky Air Force Engineering Academy and became a test pilot–engineer attached to the Ministry of Defence at Chkalovsky. Enrolled as a cosmonaut trainee on 7 March 1960, he was awarded the title of Cosmonaut of the Air Force on 5 April 1961. His first mission assignment was as backup pilot to Pavel Popovich on Vostok 4, and then backup to Valery Bykovsky on Vostok 5, but was removed from training temporarily due to ill health. On 12 October 1964 Komarov made his first spaceflight as the commander of the three-man Voskhod 1. Following the success of this mission, he was promoted to the rank of colonel and was next assigned as commander of the first manned Soyuz mission, then planned as a tandem mission with Soyuz 2. From the outset his flight was beset with technical problems, causing the cancellation of Soyuz 2 and forcing an early return to Earth. Just when a safe landing seemed assured a failure occurred in the parachute system, and the main chute failed to deploy correctly. Soyuz 1 smashed into the ground at high speed and burst into flames, killing Komarov. Some of his remains were recovered and interred with full honours in the Kremlin Wall. He was survived by his wife, the former Valentina Yakovlevna Kiseleva, and two children: Yevgeny, born 21 July 1951, and Irina, born 10 December 1958.

Leonov, Alexei Arkhipovich   The first person to walk in space and a noted painter of space themes, Alexei Leonov grew up as one of nine children to coalminer Arkhip Alexeyevich and Yevdokia Minayevna Leonov. Born in the Tisoul district village of Listvianka, Kemerovo Region on 30 May 1934, he displayed an early talent for painting. Finishing school in 1953, he contemplated attending the Academy of Arts in Riga before being drafted into the military. He undertook basic pilot training at the Kremenchug Military Aviation School, then entered the Chuguyev HAFP School, graduating with an honours diploma in 1957. Leonov then served as a fighter pilot in various units. He would also make 115 parachute jumps, later becoming an instructor. Interviewed as a potential cosmonaut in October 1959, at which time he was serving as a senior MiG-21 pilot with No. 294 Regiment of the Air Army 24 in East Germany, he was selected on 7 March 1960. Awarded the title of Cosmonaut of the Air Force on 5 April 1961, he acted as an assistant CapCom for the flight of Yuri Gagarin, and in 1963 served as second backup pilot for Vostok 5. Assigned to the Voskhod 2 mission with Pavel Belyayev, which lifted off on 18 March 1965, Leonov performed the world's first EVA, tethered outside of Voskhod 2 for 12 minutes. He was subsequently assigned to the ill-fated Zond programme, training for a possible lunar flight with Oleg Makarov. Graduating from the Zhukovsky Air Force Engineering Academy in 1968, he was reassigned to the Salyut 1 programme, and was backup commander for Soyuz 10 and commander of Soyuz 11, before his crew was totally replaced for medical reasons. He also trained to command a Soyuz flight to a Salyut station in 1972, but these missions were cancelled. A second flight came in July 1975 as commander of Soyuz 19 together with Valery Kubasov, docking with an Apollo spacecraft on the ASTP mission. In 1982 he became deputy director of TsPK. Retiring from the military as a major-general in March 1992 he went into private business, as president of the Alfa Capital Fund trust group, until his retirement in 1998. He and wife Svetlana had two children: Viktoriya, born 21 April 1961 (died 1996), and Oksana, born 1967.

Nelyubov, Grigori Grigoryevich   Born in the Crimean (Ukrainian) city of Porfiryevk on 31 March 1934 to parents Grigori and Darya, Nelyubov completed his schooling in 1954. From then until February 1957 he was a student at the Stalin Naval Aviation School. He became a senior pilot in the 639th Regiment of the 49th Fighter Division of the Soviet Air Force's Black Sea Naval Fleet, later transferring to the 966th Regiment of the 127th Fighter Division. As a military pilot, third class, he was interviewed for the cosmonaut team, beginning his post-selection training on 7 March 1960. Seven months later, Nelyubov joined the advanced Vostok training group of six cosmonauts, and was a backup for Gagarin and Titov on their missions. He was awarded the title of Cosmonaut of the Air Force on 25 January 1961. Later that year, promoted to captain and in training for Vostok 3, he suffered medical problems after a centrifuge test and was temporarily removed from flight status. On 27 March 1963, together with cosmonauts Filatyev and Anikeyev, he was involved in a drunken altercation with military guards at Chkalovskaya station near the training centre, which led to all three being summarily dismissed from the cosmonaut team. He departed on 4 May and served as a pilot with the 224th Fighter Regiment of the 303rd Division, 1st Far East Air Force Army. No one would believe that he had been a cosmonaut, and he fell into a deep depression, drinking heavily. On 18 February 1966, wandering intoxicated by railway tracks at Ippolitovka station near Vladivostok, he was hit and killed by a passing train in an apparent suicide. He was buried in a simple grave in Zaporozhe, near Kiev, but his body was later relocated to a cemetery in Kremovo, near Vladivostok. After 1986, when his identity was revealed, a new headstone was erected, recognizing his place in spaceflight history. He is survived by his wife Zinaida Ivanova.

Nikolayev, Andrian Grigoryevich   Of Chuvash descent, Andrian Nikolayev was born in the village of Shorshely in the Mariinski Posad district on 5 September 1929, the son of collective farmers Grigori Nikolayevich and Anna Alekseyevna Nikolayev. With his schooling completed in June 1944 he wanted to follow his brother Ivan into a forestry career, attending an industry training school until 1947. Conscripted into military service he was able to join the Air Force, training as a gunner and radio operator at the Kirovabad HAFP School, later entering the Chernigov HAFP School to become instead a fighter pilot. A subsequent transfer took him to the advanced pilots' school at Frunze, where he graduated in December 1954. Over the next five years he flew with the 401st Air Regiment in the Moscow military area, serving for a time as a test pilot. As a cosmonaut, he was soon selected as one of the six men to receive advanced Vostok training, being awarded the title of Cosmonaut of the Air Force on 25 January 1961. He then served as backup pilot to Gherman Titov on Vostok 2. On 11 August 1962 Nikolayev became the third Soviet cosmonaut on a tandem mission with Pavel Popovich, launched a day later aboard Vostok 4, and set a separate endurance record, orbiting the Earth 64 times in 96 hours. He was also the first person to make a live television broadcast from space. On 3 November 1963 he married Valentina Tereshkova, and they would have a daughter, Elena Andrionova. The couple were divorced in 1982. He would graduate from the Zhukovsky Air Force Engineering Academy in 1968. After training as backup pilot on Soyuz 2 to Valery Bykovsky and to Vladimir Shatalov on Soyuz 8, Nikolayev returned to space on 1 June 1970 as commander of Soyuz 9, together with Vitali Sevastyanov. Retiring from active cosmonaut duties on 26 January 1982, Nikolayev became 1st Deputy of the TsPK training centre, and on 6 August 1992 transferred to the Air Force reserve as a major-general. On 3 July 2004 he suffered a massive heart attack in Cheboksary, Chuvasia, and died in hospital. After a farewell ceremony in Cheboksary he was buried in a chapel in his native village of Shorshely.

Popovich, Pavel Romanovich The future cosmonaut was born in the Ukrainian village of Uzin near Kiev, on 5 October 1930, the son of Roman Porfiryevich (a furnace stoker) and Feodosia Kasyanovna Popovich. Growing up close to an Air Force base, which fired an early interest, he completed his war-interrupted schooling in 1945 and entered a carpentry trade school in Belaya Tserkov. Two years later he enrolled at the Magnitogorsk Industrial Polytechnic, and in his spare time took lessons at a local flying club. In 1951 he graduated from polytechnic school, only to be drafted into the Soviet army. He used his flying background to transfer to the Myasnikov Air Force Flight School in Kacha, remaining there until 1954 before flying fighters in the Far East, Siberia, Karelia and Moscow regions. He was the first member of the cosmonaut group to report for training, on 7 March 1960, and was one of six cosmonauts to receive advanced Vostok training. He was awarded the title of Cosmonaut of the Air Force on 25 January 1961, and would attend the Zhukovsky Air Force Engineering Academy from 1961 to 1968. His first assignment was as pilot of Vostok 4, launched 12 August 1962 in a tandem mission with Vostok 3. In 1969, after spending three years training for the cancelled lunar-landing programme, he joined the highly secret Almaz group of cosmonauts. However, his second spaceflight would not come until 3 July 1974 as commander of the Soyuz 14 spacecraft, flying a 15-day mission with Yuri Artyukhin to the Salyut 3 space station. Resigning as an active cosmonaut on 26 January 1982, he became Deputy Chief of the TsPK training centre, in charge of scientific research support and testing. His first marriage was to renowned test pilot and engineer Marina Vasilyeva, and they would have two daughters: Natalya, born 30 July 1956, and Oksana, born 8 October 1968. Promoted to major-general in 1976, he wed Alevtina Oshegova four years later, and in 1987 became director of a firm involved in satellite remote-sensing techniques for agricultural sites and forests, land reclamation projects and bodies of water.

Rafikov, Mars Zakirovich Born in Dzhalal-Abad, Kirghizia on 29 September 1933 to Tartar parents Zakir (died 1943) and Marziya Rafikov, the future cosmonaut trainee completed his secondary schooling in 1948 and undertook preliminary flying instruction, graduating in 1951. After further instruction he attended Air Force School No. 151 in Syzran (now Samara), Kuybyshev Region, graduating as a lieutenant in the Soviet Air Force on 29 December 1954. He subsequently served in the 15th Fighter Aviation Division of the Soviet Anti-Aircraft Defence Force, in the same squadron as two of his future colleagues, Valentin Filatyev and Valentin Varlamov, and was promoted to senior lieutenant on 30 April 1957. Earlier he had met and married Ludmila Voizechovskaya and they would have a son, Igor, born in 1956. The marriage was a short one, and he would wed for a second time to Olga Borisovskaya while still based with the 15th Aviation Division. They would have two daughters, Zhanna and Elmira. Together with Varlamov and Anatoli Kartashov, he commenced his cosmonaut training on 28 April 1960. Further promoted to captain on 30 June, he was awarded the title Cosmonaut of the Air Force on 16 December 1961. Rafikov soon attracted attention for all the wrong reasons, when rumoured domestic violence combined with news of a pending divorce, drinking and overt womanizing. On 12 March 1962 he and Ivan Anikeyev were caught against orders and in uniform drinking in the Moskva Hotel. Anikeyev would survive with a reprimand, but Rafikov had crossed the line once too often and was summarily dismissed from the cosmonaut team on 24 March. Rafikov would rise in rank to major (22 April 1970) and serve as a fighter pilot in various locations including Germany. Qualifying as a military pilot, first class, he would later take part in ejection seat tests in Prikarppatye and Zakavkazye. Rafikov would serve in Afghanistan in 1980 before resigning from the Air Force on 7 January 1982 to work for a housing collective, and later training pilots in the DOSAAF civil aviation programme. Married for a third time to the former Raichan Abouchayeva, he died of natural causes aged 66 on 23 July 2000, in Alma Ata, Kazakhstan.

Shonin, Georgi Stepanovich   Georgi Shonin was born on 3 August 1935 in the Ukrainian village of Rovenki, and grew up in Balta, near Odessa, under the Nazi occupation. His father Stepan was a casualty of the war, leaving Georgi's mother Sofya a widow. Completing 7 years of secondary schooling in 1952, he then undertook two years of flying instruction at a specialized Air Force school in Odessa, after which he attended naval aviation schools in Leningrad and Yeisk, graduating in 1957. He then served with the naval arm of the Soviet Air Force in the Baltic and Arctic Far North, serving in the same unit as Yuri Gagarin. Both were selected for cosmonaut training, reporting on 7 March 1960. He was awarded the title Cosmonaut of the Air Force on 5 April 1961. His first assignment was backup pilot to Pavel Popovich for Vostok 4 in 1962, until he failed a centrifuge tolerance test. Later cleared, he was assigned to the 15-day Voskhod 3 mission, cancelled in 1966, and graduated from the Zhukovsky Air Force Engineering Academy in 1968. He then served as backup commander to Boris Volynov on Soyuz 5 before making his only spaceflight as commander of Soyuz 6, one of three Soviet manned spacecraft in orbit at the same time. In 1970 he was assigned as the first commander of the new Salyut 1 station, but was replaced by Vladimir Shatalov for disciplinary reasons. After serving as a CapCom on the ASTP joint mission, Shonin trained for a flight in the TKS spacecraft, and was briefly assigned as commander of the first three-man TKS in 1979 before failing a physical examination. He resigned from the cosmonaut team in April 1979 and transferred to the Air Force reserve in 1990, having reached the age limit with the rank of lieutenant-general. His next role was as the director of the Central Scientific Research Institute of the Soviet Ministry of Defence, located in Shchelkovo, outside

Moscow. He died of heart failure aged 61 on 7 April 1997 and is survived by his first wife, the former Lidiya Fedorovna Shumilova and children Andrei, born 22 May 1961, and Olga, born 23 May 1970 (their older daughter Nina, born 9 September 1955, died in 1990), as well as his second wife Lyudmila Valentinovna Atrashenko and their son Antoni, born 4 February 1979. He is also survived by his third wife, Galina Arkadyevna.

Titov, Gherman Stepanovich   Gherman Titov was born on 11 September 1935 in the rural village of Verkhneye Zhilino, southern Siberian Altai region, to Stepan and Alexandra Mikhailovna Titov. In July 1953, after graduating from Nalobikhinsk secondary school, he entered the army as an aviation cadet at the 9th Military Air School in Kustanai, Kazakhstan and was enrolled at the Stalingrad (now Volgograd) HAFP School two years later, graduating with a 1st Class Honours degree and qualification as a military pilot in 1957. He served as a fighter pilot in regiments of the 26th and 41st Air Divisions, Leningrad Military District for three years. On 28 October 1959, now a senior lieutenant, he was transferred to the 103rd regiment. Selected for cosmonaut training on 7 March 1960, he was one of the advanced group training for a Vostok mission. On 25 January 1961 he was awarded the title Cosmonaut of the Air Force. After serving as backup to Yuri Gagarin, Titov (now a major) flew on Vostok 2, launched 6 August 1961, to this day the youngest person to fly into space, and the first to suffer from space adaptation sickness (SAS). He attended the Zhukovsky Air Force Engineering Academy from September 1961 to 1968, graduating with honours. During this time he attended the Chkalov test pilot school and qualified as a test pilot, third class. Unfortunately, circumstances meant he would never fly into space again, and he left the cosmonaut detachment in 1970 after commanding the Spiral programme, after which he took on full-time tuition at the Armed Forces General Staff Military Academy. In 1972 he joined the Office for Space Facilities with the Soviet Ministry of Defence, where he held a number of senior positions within the military side of the space programme before retiring in 1992. Three years later he was elected to the State Duma. Aged 65, General-Colonel Titov died of cardiac arrest in the sauna of his home on 21 September 2000 and was buried in Moscow's Novodevichy cemetery with full military honours. He is survived by his wife, the former Tamara Vasilyevna Cherkas, and daughters Tatyana (born 23 September 1963) and Galina (born 14 August 1965). Their first child Igor died in infancy in 1960.

Varlamov, Valentin Stepanovich   An outstanding and popular candidate, Valentin Varlamov might have been considered for the first Vostok mission but for an accident that caused him to leave the cosmonaut team. Born to Stepan and Klavdia Varlamov in the Penza Region of Russia on 15 August 1934, he attended secondary school until 1952 and then, following his conscription into the military, undertook flying lessons. On 29 November 1955 he graduated from a two-year course at the Stalingrad HAFP School with the rank of lieutenant. Later promoted to senior lieutenant on 24 December 1957, he would fly in the same squadron as two future cosmonaut colleagues, Mars Rafikov and Valentin Filatyev. Together with Rafikov and

Anatoli Kartashov, he commenced his cosmonaut training on 28 April 1960 as a "military pilot, third class," and gained his Air Force captain's promotion on 30 June. He had also been selected as a member of the group of six men undergoing advanced Vostok training for an early flight. Just under a month later, on 24 July 1960, he rashly dived into a shallow lake while picnicking with other cosmonauts and badly dislocated a vertebra in his neck. Placed in traction, he was hospitalized for a month before reporting back for training, but a medical commission carried out an examination and he was ruled ineligible for spaceflight. His place in the so-called "Sochi Six" was taken by Valery Bykovsky. Eventually, on 6 March 1961, he resigned from the cosmonaut team, but continued to work at the TsPK training centre as deputy commander of the command point for controlling spacecraft. He became a senior instructor in astronavigation and was promoted to the rank of major on 18 July 1964. He would finally leave TsPK on 6 March 1966. Varlamov died on 2 November 1980, aged 46, following a cerebral haemorrhage. He was survived by his wife, Nina Fedorovna, and daughter Elena, born 28 February 1964.

Volynov, Boris Valentinovich   Born into a Jewish family in Irkutsk, the capital of East Siberia on 18 December 1934, Boris Volynov was three years old when his father Valentin died, leaving Yengeniya Izrailyevna a widow. She moved to Prokopievsk in the Kemerovo region where Boris finished secondary school in 1952 and then enrolled for Air Force schooling in Pavlodar, Kazakhstan. Graduating in 1952, he was sent to the Stalingrad Military Aviation School for advanced training, alongside another future cosmonaut, Gherman Titov. From January 1956 Volynov spent four years as a MiG-17 fighter pilot with the 133rd Air Division of the Soviet Air Defence Forces, based at Yaroslavl in northern Central Russia. In 1958, promoted to senior lieutenant, he married his childhood sweetheart, Tamara Fedorovna Savina. He began his cosmonaut training on 7 March 1960, and on 5 April 1961 was awarded the title Cosmonaut of the Air Force. In August 1962 he served as second backup on both Vostok 3 and Vostok 4, and then as backup pilot to Pavel Bykovsky on Vostok 5. More backup duties followed, for the first Voskhod mission. He would have commanded a Voskhod mission in 1966, but it was cancelled. He then backed up Soyuz 3 before making his first spaceflight as commander of the three-man Soyuz 5 mission, launched on 15 January 1969, during which his two fellow crewmembers transferred by EVA to Soyuz 4. He suffered severe facial injuries after a near-fatal solo re-entry, but recovered. He was in charge of the Almaz cosmonaut group from 1969. After directing the training of a July 1970 group of cosmonauts, Volynov served as back-up commander for Soyuz 14 in 1974. He flew as commander of Soyuz 21 with Vitali Zholotov, the spacecraft was launched on 6 July 1976 to dock with Salyut 5. Later he worked as a Salyut flight controller. From November 1982 until his retirement from the cosmonaut team and active military service as a full colonel on 17 March1990, he was given overall command of the cosmonaut team. He and his wife Tamara have two children: a son Andrei, born in 1958, and daughter Tatiana, born in 1965.

Zaikin, Dmitri Alexeyevich   Dmitri Zaikin became the last member of his cosmonaut group denied the chance to fly into space. Born on 29 April 1932 in Yekaterinova, near

Salsk in the Rostovskya Region, he began school in 1940, but the war caused many moves for the family. His father Alexei was killed fighting the Nazis at Stalingrad in 1942. His mother Zinaida would later remarry. After completing his seventh-grade studies he attended the Chernigov HAFP School at Armavir, learning to fly in dual-seat Yak-18 aircraft, and graduating on 18 August 1951. He was then 1 of 50 graduates selected to attend the Frunze HAFP School, graduating along with Andrian Nikolayev and given the rank of lieutenant on 29 December 1954. Then, over the next five years, he served as a fighter pilot in the Soviet Air Force in western Belo-Russia, promoted to senior lieutenant on 19 June 1957. He was first interviewed in September 1959, and was inducted into the first cosmonaut group on 25 March 1960. As he continued his spaceflight training, he was further promoted on 9 May that year to the rank of captain, and was awarded the title of Cosmonaut of the Air Force on 16 December 1961. Prior to the flight of Voskhod 2, Belyayev's backup pilot Viktor Gorbatko fell ill and was replaced by Zaikin. He then trained as a possible Voskhod commander until mid-1966, when that programme was cancelled. In line to one day become a Soyuz commander, he graduated from the Zhukovsky Air Force Engineering Academy in 1968, but was forced to stand down from cosmonaut duties in April that year after a medical commission detected an ulcer following routine tests. He remained at the TsPK training centre, and with the rank of colonel still assists in training cosmonauts for technological experiments. He was not identified as a cosmonaut until 1977 when he was named in a book by Georgi Shonin. Further details became known in an officially sanctioned *Izvestia* article in 1986. Married to the former Tatyana Vladimirovna Sukhoryabova, they have two sons: Andrei Dmitrivich, born 5 September 1962, and Dennis Dmitrivich, born 13 March 1971. He still lives in Star City.

Appendix B

Final cosmonaut candidates

These were the 29 candidates who appeared before the Credential Committee, which was empowered with the final selection of the first Soviet cosmonauts.

Anikeyev, Ivan N.	Selected	Leonov, Alexei A.	Selected
Belyayev, Pavel I.	Selected	Lisitz, Leonid Z.	Not selected
Bessmertny, Nikolai I.	Not selected	Nelyubov, Grigori G.	Selected
Bochkov, Boris I.	Not selected	Nikolayev, Andrian G.	Selected
Bondarenko, Valentin V.	Selected	Popovich, Pavel R.	Selected
Bravin, Georgi A.	Not selected	Rafikov, Mars Z.	Selected
Bykovsky, Valery F.	Selected	Shonin, Georgi S.	Selected
Filatyev, Valentin I.	Selected	Sviridov, Valentin P.	Not selected
Gagarin, Yuri A.	Selected	Timokhin, Ivan M.	Not selected
Gorbatko, Viktor V.	Selected	Titov Gherman S.	Selected
Inozemtsev, Grigory K.	Not selected	Varlamov, Valentin S.	Selected
Karpov, Valentin A.	Not selected	Volynov, Boris V.	Selected
Kartashov, Anatoly Y.	Selected	Yefremenko, Mikhail A.	Not Selected
Khrunov, Yevgeny V.	Selected	Zaikin, Dmitri A.	Selected
Komarov, Vladimir M.	Selected	—	—

Appendix C

The first cosmonaut team (TsPK-1)

Name of cosmonaut (alphabetical)	Rank	Date of birth	Age at selection	Date of official selection	Left cosmonaut detachment	Date of death
Anikeyev, Ivan	Snr. Lt.	12.02.1933	27	07.03.1960	17.04.1963	20.08.1992
Belyayev, Pavel	Major	26.06.1925	34	28.04.1960	10.01.1970	10.01.1970
Bondarenko, Valentin	Snr. Lt.	16.02.1937	23	28.04.1960	23.03.1961	23.03.1961
Bykovsky, Valery	Snr. Lt.	02.08.1934	25	07.03.1960	26.01.1982	—
Filatyev, Valentin	Snr. Lt.	21.01.1930	30	25.03.1960	17.04.1963	15.09.1990
Gagarin, Yuri	Snr. Lt.	09.03.1934	25	07.03.1960	27.03.1968	27.03.1968
Gorbatko, Viktor	Snr. Lt.	03.12.1934	25	07.03.1960	28.08.1982	—
Kartashov, Anatoly	Snr. Lt.	25.08.1932	27	28.04.1960	07.04.1961	11.12.2005
Khrunov, Yevgeny	Snr. Lt.	10.09.1933	26	09.03.1960	25.12.1980	19.05.2000
Komarov, Vladimir	Captain	16.03.1927	32	07.03.1960	24.04.1967	24.04.1967
Leonov, Alexei	Lt.	30.05.1934	25	07.03.1960	26.01.1982	—
Nelyubov, Grigori	Snr. Lt.	31.03.1934[a]	25	07.03.1960	04.05.1963	18.02.1966
Nikolayev, Andrian	Snr. Lt.	05.09.1929	30	07.03.1960	26.01.1982	03.07.2004
Popovich, Pavel	Captain	05.10.1930	29	07.03.1960	26.01.1982	—
Rafikov, Mars	Snr. Lt.	29.09.1933	26	28.04.1960	24.03.1962	23.07.2000

(continued)

Name of cosmonaut (alphabetical)	Rank	Date of birth	Age at selection	Date of official selection	Left cosmonaut detachment	Date of death
Shonin, Georgi	Snr. Lt.	03.08.1935	24	07.03.1960	28.04.1979	07.04.1997
Titov, Gherman	Snr. Lt.	11.09.1935	24	07.03.1960	17.06.1970	20.09.2000
Varlamov, Valentin	Snr. Lt.	15.08.1934	25	28.04.1960	06.03.1961	02.10.1980
Volynov, Boris	Snr. Lt.	18.12.1934	25	07.03.1960	30.05.1990	—
Zaikin, Dmitri	Snr. Lt.	29.04.1932	27	25.03.1960	25.10.1969	—

[a] On Nelyubov's grave in Kremovo, near Vladivostok, the date of birth is incorrectly given as 8 April 1934.

Appendix D

Guide to flight and programme assignments

PAVEL BELYAYEV

Flight	Mission assignment	Result
Vostok	Did some general Vostok training, but not considered for a specific mission	No Vostok mission
Voskhod 2	*Mission commander*	*Flew mission*
Almaz	Commander of first Almaz/Salyut space station	Replaced due to ill health

VALERY BYKOVSKY

Flight	Mission assignment	Result
Vostok 2	Replaced Grigori Nelyubov as 2nd backup pilot to Gherman Titov	Mission flown by Titov
Vostok 3	Backup pilot to Andrian Nikolayev	Mission flown by Nikolayev
Vostok 5	*Prime pilot on joint mission with Vostok 6*	*Flew mission*
Soyuz 2	Prime crewmember for proposed 3-man link-up flight with Soyuz 1	Mission cancelled following problems with Soyuz 1
Soyuz 7K-L1 (Flight 2)	Commander of planned second circumlunar spaceflight with Nikolai Rukavishnikov	Programme cancelled after Apollo 8 orbited Moon
Soyuz 22	Commander of 7-day mission with flight engineer Vladimir Aksyonov	Flew mission
Soyuz 31	Commander of Intercosmos flight with East German research pilot Sigmund Jähn to Salyut 6 space station	Flew mission. Crew returned from Salyut 6 on Soyuz 29
Soyuz 37	Backup commander for Intercosmos flight to Salyut 6 space station	Mission flown by prime crew

VALENTIN FILATYEV

Flight	Mission assignment	Result
Vostok	Did some general Vostok training but not considered for a specific mission	No Vostok mission

YURI GAGARIN

Flight	Mission assignment	Result
Vostok (1)	Prime pilot: first human spaceflight	Flew mission
Soyuz 1	Backup pilot to Vladimir Komarov	Did not fly: Komarov killed in landing mishap
Soyuz 3	Prime pilot	Removed from flight status following loss of Soyuz 1

VIKTOR GORBATKO

Flight	Mission assignment	Result
Vostok	Did some general Vostok training, but not considered for a specific mission	No Vostok mission
Voskhod 2	Commander of backup crew. Did some general Voskhod command training	Failed medical, replaced by Zaikin
Soyuz 2	Member of 3-man backup crew	Mission cancelled following problems with Soyuz 1
Soyuz 4	Member of original prime 3-man crew	Mission deferred after loss of Soyuz 1; crew changed
Soyuz 5	Member of 3-man backup crew	Mission flown by prime crew
Soyuz 7	*Member of 3-man prime crew*	*Flew mission*
Soyuz 7K-OK (Kontakt A)	Prime crewmember for final test flight of Kontakt lunar rendezvous/docking system	Mission cancelled
Almaz	Commenced mission training	Mission cancelled
Soyuz 21	Member of 2nd backup crew	Mission flown by prime crew
Soyuz 23	Member of backup crew	Mission flown by prime crew
Soyuz 24	*Commander of prime 2-man crew on link-up mission with Salyut 5*	*Flew mission*
Soyuz 31	Backup commander for Intercosmos flight to Salyut 6 space station	Mission flown by prime crew
Soyuz 37	*Commander of prime Intercosmos crew carrying Vietnamese research pilot Pham Tuan to Salyut 6 space station*	*Flew mission*

YEVGENY KHRUNOV

Flight	Mission assignment	Result
Vostok	Did some general Vostok training, but was not considered for a specific mission	Mission cancelled
Voskhod 2	Backup EVA crewmember to Alexei Leonov	Mission flown by prime crew
Voskhod 6	Commander of prime 2-man crew	Mission cancelled
Soyuz 2	Prime crewmember on 3-man flight	Mission cancelled following problems with Soyuz 1
Soyuz 5	*Prime crewmember on mission involving EVA to Soyuz 4 spacecraft*	*Flew mission; landed aboard Soyuz 4*
Soyuz 9 (Kontakt)	Proposed crewmember on flight to test the Kontakt lunar rendezvous/docking system	Mission cancelled
Almaz	Trained for possible commander assignment	Never assigned
Soyuz 38	Backup commander for Intercosmos mission to Salyut 6 space station	Mission flown by prime crew

VLADIMIR KOMAROV

Flight	Mission assignment	Result
Vostok 4	Backup pilot to Pavel Popovich	Mission flown by Popovich
Vostok 7	Named as mission commander after Vostok 5/6 tandem flight	Mission cancelled
Voskhod (1)	*Mission commander*	*Flew mission*
Soyuz 1	*Prime pilot*	*Flew mission: killed in landing mishap*

ALEXEI LEONOV

Flight	Mission assignment	Result
Vostok 5	2nd backup pilot	Mission flown by Bykovsky
Voskhod 2	*Prime crewmember with Pavel Belyayev*	*Flew mission; conducted first ever EVA*
Soyuz 7K-L1 (Flight 1)	Commander of planned first circumlunar spaceflight with Oleg Makarov	Programme cancelled after Apollo 8 orbited Moon
Soyuz 10	Backup commander of 3-man flight to occupy Salyut 1 space station	Mission flown by prime crew
Soyuz 11	Original prime crew, replaced due to suspected incapacity of Valery Kubasov	Flown by replacement crew; all 3 died in re-entry mishap
Soyuz 12 (DOS-1)	Commander of proposed mission to Salyut 1 space station	Mission cancelled following loss of Soyuz 1 crew
Soyuz 12 (DOS-2)	Commander of proposed mission to Salyut DOS-2 (Kosmos 557) space station	Mission cancelled due to loss of Salyut DOS-2 space station
Soyuz 12 (DOS-3)	Commander of proposed mission to Salyut 4 DOS-3 space station	Mission cancelled: DOS-3 space station failed to orbit
Soyuz 19 (ASTP)	*Commander of crew to link in orbit with Apollo manned spacecraft*	*Mission successfully completed*

GRIGORI NELYUBOV

Flight	Mission assignment	Result
Vostok (1)	2nd backup pilot to Yuri Gagarin (after Titov)	Mission flown by Gagarin
Vostok 2	2nd backup pilot to Gherman Titov (after Nikolayev)	Replaced as backup crewman by Bykovsky due to medical problem. Mission flown by Titov
Vostok 7	Backup commander to Vladimir Komarov	Mission cancelled, later revised with new crew, but again cancelled

ANDRIAN NIKOLAYEV

Flight	Mission assignment	Result
Vostok 2	Backup pilot to Gherman Titov	Mission flown by Titov
Vostok 3	*Prime pilot on joint mission with Vostok 4*	*Flew mission*
Soyuz 2	Commander of backup 3-man crew	Mission cancelled following problems with Soyuz 1
Soyuz 3	Original crewmember with mission commander Georgi Beregovoi	Reassigned to allow Beregovoi solo test flight
Soyuz 4	Commander of original prime crew	Mission deferred after loss of Soyuz 1; crew changed
Soyuz 6	Commander of 2nd backup crew	Mission flown by prime crew
Soyuz 7	Commander of 2nd backup crew	Mission flown by prime crew
Soyuz 8	Commander of backup crew	Mission flown by prime crew
Soyuz 9	*Commander of 2-man crew with Vitali Sevastyanov*	*Flew mission*

PAVEL POPOVICH

Flight	Mission assignment	Result
Vostok 4	*Prime pilot on joint mission with Vostok 3*	*Flew mission*
Soyuz 7K-V1 (Zvezda)	Commander of first planned Soyuz V1 combat spacecraft with Gennadi Kolesnikov	Programme cancelled
Soyuz 7K-L1 (Flight 3)	Commander of planned third circumlunar spaceflight with Vitali Sevastyanov	Programme cancelled after Apollo 8 orbited Moon
Soyuz 12 (Almaz)	Commander of first flight to Almaz (Salyut 2) space station with Yuri Artyukhin	Flight cancelled after loss of space station
Soyuz 14	*Commander of 2-man crew to Salyut 3 with Yuri Artyukhin*	*Flew mission*

GEORGI SHONIN

Flight	Mission assignment	Result
Voskhod 3	Prime crew (with commander Boris Volynov)	Mission cancelled
Soyuz 4	Backup commander to Vladimir Shatalov	Mission flown by Shatalov
Soyuz 6	Commander crew with Valery Kubasov	Flew mission
Soyuz 11 (DOS-1)	Original commander 3-man prime crew flight to Salyut 1 space station	Replaced by Shatalov due to illness
Soyuz 12 (Kontakt)	Commander of proposed 3-man crew to test Kontakt lunar rendezvous/docking system	Mission cancelled

GHERMAN TITOV

Flight	Mission assignment	Result
Vostok (1)	Backup pilot to Yuri Gagarin	Mission flown by Gagarin
Vostok 2	Prime pilot	Flew mission

BORIS VOLYNOV

Flight	Mission assignment	Result
Vostok 3	2nd backup pilot	Mission flown by Nikolayev
Vostok 4	2nd backup pilot	Mission flown by Popovich
Vostok 5	Backup pilot to Valery Bykovsky	Mission flown by Bykovsky
Vostok 7	Commander of proposed high-altitude flight	Mission cancelled
Voskhod (1)	Backup commander of three-man crew	Mission flown by prime crew
Voskhod 3	Commander of proposed 3-man crew	Mission cancelled
Soyuz 3	Commander, 2nd backup crew	Solo mission flown by Beregovoi
Soyuz 5	*Commander of 3-man crew involved in 1st EVA crew transfer to Soyuz 4*	*Flew mission; landed alone in Soyuz 5*
Soyuz 12 (Almaz)	Commander, proposed backup crew, for mission to Almaz (Salyut 2) space station	Mission cancelled
Soyuz 14	Commander, backup crew	Mission flown by prime crew
Soyuz 16A	Commander, proposed crew	Mission cancelled after Soyuz 15 docking failure
Soyuz 21	*Commander of 2-man crew with Vitali Zholobov to Salyut 3 space station*	*Flew mission*
Soyuz T-15A	Commander of proposed expedition to Salyut 7 space station	Mission cancelled

DMITRI ZAIKIN

Flight	Mission assignment	Result
Voskhod 2	Assigned as backup commander to Pavel Belyayev following illness of Gorbatko. Did follow-up Voskhod training	Mission flown by prime crew
Soyuz	In line to command a Soyuz mission	Medically disqualified

Appendix E

Parachute jumps completed by Valentin Filatyev, 1960–1963

An example of the parachute jumps completed over a three-year period by one member of the first cosmonaut team. (Reproduced with permission from "Russia's Cosmonauts: Inside the Yuri Gagarin Training Center" by Rex D. Hall, David J. Shayler and Bert Vis). Transcribed from Filatyev's original parachute record held at Star City, Moscow, in August 2004.

Date	Aircraft	First jump: altitude (m)	Delay (s)	Second jump: altitude (m)	Delay (s)
1960					
18 Apr	Antonov AN-2	800	None	800	None
19 Apr	Antonov AN-2	800	None	800	None
21 Apr	Antonov AN-2	800	None	—	—
22 Apr	Antonov AN-2	800	None	800	None
23 Apr	Antonov AN-2	800	None	800	None
25 Apr	Antonov AN-2	800	None	1,000	5
26 Apr	Antonov AN-2	1,100	10	—	—
03 May	Lisunov LI-2	1,300	15	1,300	15
04 May	Lisunov LI-2	1,300	15	–	–
07 May	Lisunov LI-2	1,300	15	1,300	15
08 May	Lisunov LI-2	1,600	20	1,600	20
09 May	Lisunov LI-2	1,600	20	1,600	20
10 May	Lisunov LI-2	1,600	20	1,600	20
11 May	Lisunov LI-2	1,800	25	2,100	25
13 May	Mil MI-4	2,100	30	2,600	40
14 May	Ilyushin IL-14	3,000	50	2,100	30
05 Aug	Mil MI-4	1,100	10	1,300	15
06 Aug	Ilyushin IL-14	1,800	25	3,000	50
11 Aug	Ilyushin IL-14	1,000	None	—	—
02 Sep	Ilyushin IL-14	2,100	None	—	—

(continued)

Date	Aircraft	First jump: altitude (m)	Delay (s)	Second jump: altitude (m)	Delay (s)
1961					
09 Mar	Ilyushin IL-14	1,000	5	—	—
16 Mar	Ilyushin IL-14	1,000	5	1,000	5
28 Sep	Antonov AN-2	1,300	15	1,300	15
29 Sep	Antonov AN-2	1,600	20	1,600	None
30 Sep	Mil MI-4	1,000	5	—	—
02 Oct	Antonov AN-2	1,300	15	1,300	15
04 Oct	Mil MI-4	1,000	None	—	—
05 Oct	Mil MI-4	1,600	20	—	—
1962					
30 Aug	Ilyushin IL-14	1,000	5	1,100	10
31 Aug	Ilyushin IL-14	1,100	10	1,100	10
01 Sep	Ilyushin IL-14	1,100	10	1,600	10
02 Sep	Ilyushin IL-14	1,600	20	1,600	20
03 Sep	Ilyushin IL-14	2,100	30	2,100	30
04 Sep	Ilyushin IL-14	1,600	20	1,600	20
05 Sep	Ilyushin IL-14	3,000	50	3,700	70
06 Sep	Ilyushin IL-14	3,700	70	—	—
1963					
12 Feb	Antonov AN-2	1,300	15	1,300	15
13 Feb	Antonov AN-2	1,600	20	1,600	20
14 Feb	Antonov AN-2	2,100	30	2,100	30
15 Feb	Antonov AN-2	2,100	30	—	—

Appendix F

Spaceflights by Group 1 cosmonauts

Cosmonaut	Mission call sign	Spacecraft	Other crewmembers	Launch date	Landing date	Orbits	EVA duration (hh:mm)	Mission duration (dd:hh:mm:ss)
Gagarin	Kedr	Vostok	—	12.4.1961	12.4.1961	1	—	00:01:48:00
Titov	Oryel	Vostok 2	—	6.8.1961	7.8.1961	17	—	00:01:18:00
Nikolayev	Sokol	Vostok 3	—	11.8.1962	15.8.1962	64	—	03:22:22:00
Popovich	Berkut	Vostok 4	—	12.8.1962	15.8.1962	48	—	02:22:56:43
Bykovsky	Yastreb	Vostok 5	—	14.6.1963	19.6.1963	81	—	04:23:06:00
Komarov	Rubin	Voskhod	Feoktistov Yegorov	12.10.1964	13.10.1964	16	—	01:00:17:03
Belyayev Leonov	Almaz	Voskhod 2	—	8.3.1965	19.3.1965	18	(Leonov) 00.12	01:02:02:17 01:02:02:17
Komarov	Rubin	Soyuz 1	—	23.4.1967	24.4.1967	18	—	01:02:47:32
Volynov Khrunov	Baikal	Soyuz 5	Yeliseyev	15.1.1969	18.1.1969	49	(Khrunov) 00:37	03:00:54:15 01:23:45:50
Shonin	Antey	Soyuz 6	Kubasov	11.10.1969	16.10.1969	80	—	04:22:42:47
Gorbatko	Buran	Soyuz 7	Filipchenko Volkov	12.10.1969	17.10.1969	80	—	04:22:40:23
Nikolayev	Sokol	Soyuz 9	Sevastyanov	1.6.1970	19.6.1970	286	—	17:16:58:55
Popovich	Berkut	Soyuz 14	Artyukhin	3.7.1974	19.7.1974	—	—	15:17:30:28

Appendix F: Spaceflights by Group 1 cosmonauts 381

Leonov	Soyuz	Soyuz 19 (ASTP)	Kubasov	15.7.1975	21.7.1975	96	—	05:22:30:51
Volynov	Baikal	Soyuz 21	Zholobov	6.7.1976	24.8.1976	—	—	49:06:23:32
Bykovsky	Yastreb	Soyuz 22	Aksyonov	15.9.1976	23.9.1976	127	—	07:21:52:17
Gorbatko	Terek	Soyuz 24	Glazkov	7.2.1977	25.2.1977	—	—	17:17:25:58
Bykovsky	Yastreb	Soyuz 31	Jähn	26.8. 1978	3.9.1978	—	—	07:20:49:04
Gorbatko	Terek	Soyuz 37	Tuan	23.7.1980	31.7.1980	—	—	07:20:42:00

Appendix G

Cumulative time in space

Cosmonaut	Cosmonaut No.	No. of flights	Total time in space (dd:hh:mm:ss)
Yuri Gagarin	1	1	00:01:48:00
Gherman Titov	2	1	01:01:18:00
Andrian Nikolayev	3	2	21:15:20:55
Pavel Popovich	4	2	18:16:27:11
Valery Bykovsky	5	3	20:17:47:21
Vladimir Komarov	7	2	02:03:04:55
Pavel Belyayev	10	1	01:02:02:17
Alexei Leonov	11	2	07:00:33:08
Boris Volynov	14	2	52:07:17:47
Yevgeny Khrunov	16	1	01:23:45:50
Georgi Shonin	17	1	04:22:42:47
Viktor Gorbatko	21	3	30:12:48:21

Appendix H

Highest honour: Hero of the Soviet Union

Awarded to members of the first cosmonaut team

Recipient	Awarded (1)	Awarded (2)
Gagarin, Y.A.	14.04.1961	—
Titov, G.S.	09.08.1961	—
Nikolayev, A.G.	18.08.1962	03.07.1970
Popovich, P.R.	18.08.1962	20.07.1974
Bykovsky, V.F.	22.06.1963	28.09.1976
Komarov, V.M.	19.10.1964	24.04.1967 (posthumous)
Belyayev, P.I.	23.03.1965	—
Leonov, A.A.	23.03.1965	22.07.1975
Volynov, B.V.	22.01.1969	01.09.1976
Khrunov, Y.V.	22.01.1969	—
Shonin, G.S.	22.10.1969	—
Gorbatko, V.V.	22.10.1969	05.03.1977

Index

(numbers in italics indicate photographs)

Printing: Mercedes-Druck, Berlin
Binding: Stein+Lehmann, Berlin